ARMY ANTS

Head-on view of major worker of *Eciton hamatum*.

# ARMY ANTS
*A Study in Social Organization*

T. C. Schneirla

*Late Curator*
*Department of Animal Behavior*
*The American Museum of Natural History, New York*

EDITED BY

Howard R. Topoff

*Department of Animal Behavior*
*The American Museum of Natural History, New York*

W. H. FREEMAN AND COMPANY
San Francisco

Fitz Memorial Library
Endicott College
Beverly, Massachusetts 01915

44139

QL
568
.F7
S3

Copyright © 1971 by W. H. Freeman and Company

No part of this book may be reproduced
by any mechanical, photographic, or electronic process,
or in the form of a phonographic recording,
nor may it be stored in a retrieval system, transmitted,
or otherwise copied for public or private use
without written permission from the publisher.

Library of Congress Catalog Card Number: 70-149408
International Standard Book Number: 0-7167-0933-3
Printed in the United States of America

1 2 3 4 5 6 7 8 9

# Contents

HOWARD R. TOPOFF
Foreword *xiii*

CARYL P. HASKINS
Introduction *xix*

1. Army Ants *1*
2. The Colony and Its Members *22*
3. The Bivouacs *44*
4. Raiding *69*
5. The Emigrations *101*
6. The Broods *123*
7. Functional Cycles and Nomadism *149*
8. The Queen *169*
9. Males and Young Queens *198*
10. Colony Division *218*
11. The New Colonies *245*
12. Individual and Colony *264*
13. *Aenictus:* Army Ant on a Small Scale *286*
14. The Doryline Colony as an Adaptive System *304*

Glossary *327*
References *333*
Index *345*

# Figures

1.1 Functional cycle of the army ant *Eciton hamatum*. 2

1.2 Types of individuals present in a colony of the driver ant *Dorylus (Anomma) wilverthi*. 4

1.3 Distribution of principal genera of doryline ants. 8

1.4 Locale of the statary bivouac of a colony of *Neivamyrmex nigrescens* studied in an Arizona canyon. 17

2.1 Representatives of the worker populations of species in four genera of doryline ants, all polymorphic except for the *Aenictus* species, which is virtually monomorphic. 25

2.2 Graph comparing the relative frequency distributions of pupal samples from colonies representing three species of doryline ants. 26

2.3 Polymorphic size-frequency distributions in *Eciton hamatum* and *Eciton burchelli*. 27

2.4 Test of species interactions based on odor. 30

2.5 Queen and worker types (with larvae) of the army ant *Eciton burchelli* in a laboratory nest. 33

2.6 The staphylinid beetle, *Smectonia gridelli* Patr. lives in colonies of the driver ant *Dorylus (Anomma) nigricans* of Ethiopia. 36

3.1 Nomadic bivouac cluster of a colony of *Eciton hamatum* in a cylinder approximately 40 cm in diameter. 46

3.2 Close-up view of a portion of the outer wall of a bivouac of *Eciton hamatum* during the nomadic phase. 48

3.3 Curtain- (or half-cylinder-) type bivouac formed by a nomadic colony of *Eciton hamatum* between the buttressed roots of a tree. 49

3.4 Curtain-type bivouac formed by a colony of *Eciton burchelli* in the nomadic phase, hanging from the base of a log to the ground. 50

3.5 Telephoto of a bulb-type bivouac formed by a colony of *Eciton burchelli* in the nomadic phase. 51

*Figures* vii

3.6 Bivouac construction using tarsal hooks. 56

3.7 Strands formed by workers of *Eciton hamatum* introduced to a laboratory nest, much as they begin a bivouac under natural conditions. 57

3.8 Records of intrabivouac temperatures taken near 10:00 A.M. daily through an activity cycle in the successive bivouacs of a colony ('49 H-34) of *Eciton hamatum.* 65

4.1 Trail following by workers of *Eciton hamatum.* 71

4.2 Raiders from a nomadic colony of *Eciton hamatum* have thickly covered a small bulb nest of wasps and, after having expelled the adults, are ransacking the brood cells, leaving in column with quantities of pupae. 74

4.3 Sketch of the radial foraging routes formed and used by a colony of *Eciton burchelli* in two successive statary phases and of emigration routes used in the intervening nomadic phase. 76

4.4 Comparison between patterns of column raiding and swarm raiding. 78

4.5 Diagrams of raiding systems of *Neivamyrmex nigrescens* and of *Aenictus laeviceps,* typical of the late nomadic phase. 81

4.6 Diagrams of raiding systems typical of *Eciton hamatum* in the nomadic phase and in the statary phase of a functional cycle. 82

4.7 Stages in the development of a swarm raid of *Eciton burchelli.* 84

4.8 Rate of development of a typical swarm raid of *Eciton burchelli* from dawn to noon. 90

4.9 Basal column of a raid of *Eciton burchelli* issuing from the bivouac around midmorning. 91

4.10 Submajor workers of *Eciton burchelli* carry in tandem a sizable piece of booty, the tail of a scorpion. 93

5.1 A sketch to show the case of a colony of *Eciton burchelli* in late afternoon, with the afternoon exodus forcibly diverted from the main raiding route of the day by a heavy return of booty-laden traffic. 104

5.2 Emigration columns of *Eciton burchelli* and *Aenictus laeviceps* carrying nearly mature larvae. 106

5.3 Larva-carrying column of workers in a nocturnal emigration of *Eciton burchelli.* 107

5.4 Section of column in a nocturnal emigration of *Eciton hamatum* passing along a vine. 108

5.5 Bivouac of a colony of *Eciton hamatum* during a nocturnal emigration just as the queen and her worker entourage are leaving the site. 109

5.6 Two views of nocturnal emigrations of *Eciton hamatum* at the time the queen passes with her entourage. 110

5.7 Section of a nocturnal emigration column of *Eciton burchelli,* widened at the passing of the queen. 111

5.8 Eastern part of Barro Colorado Island, Panama Canal Zone, mapped to show the itinerary of colony '46 H-B (*Eciton hamatum*) through a period of 112 days. 115

5.9 Sections of emigration routes in the driver ant *Dorylus (Anomma) wilverthi* on the surface become trenches through which the procession runs. 119

6.1 The bivouac of a colony of *Eciton hamatum* late in the nomadic phase with its wall opened by a stroke of the tweezers. 126

6.2 Series of brood samples from colonies of three doryline (group A) genera, representing the course of growth in major and minor larvae through the nomadic phase. 128

6.3 Representatives of polymorphic and monomorphic army ant broods. 129

6.4 Schema of brood developmental stages in *Eciton hamatum, Aenictus laeviceps,* and *Dorylus (Anomma) wilverthi,* representing the principle of developmental allometric convergence—differential growth in the brood decreasing time range at maturity over that at egg laying. 131

6.5 Polymorphic range of *Eciton hamatum.* 133

6.6 Developmental condition of different size groups of pupae. 135

6.7 Series of brood samples from colonies representing the army ant group A genera *Eciton, Neivamyrmex,* and *Aenictus,* indicating the general course of growth in larvae of major and minor size groups through the main part of the larval stage. 137

7.1 Schema of the functional cycles characteristic of three doryline genera (group A). 152

8.1 Queens of four doryline genera, drawn in contracted condition. 170

8.2 Functional queens of *Eciton burchelli* photographed in the contracted state and in the physogastric and egg-laying condition. 171

8.3 Colony queen of *Eciton hamatum,* marked on December 22, 1947, and returned to her colony on Barro Colorado Island, then rediscovered and studied further during the dry season of 1952. 175

8.4 A colony queen of *Eciton hamatum* with workers in a laboratory nest, much reduced in physogastry after having laid most of her current batch of eggs. 177

8.5 Gaster lengths of queens of *Eciton hamatum* and *Eciton burchelli,* as a crude indication of reproductive condition, measured at different times in the functional cycles of their colonies. 184

*Figures* ix

8.6 Schema to represent corresponding conditions in the behavior and function of a colony of *Eciton*, the coordinated development of its broods, and the queen's reproductive processes, the last based on Hagan's research. 186

8.7 Anatomy of reproductive apparatus of the queen of *Eciton hamatum*, with the right ovary and its ovarioles shown in part. 187

8.8 Close-up of gaster of physogastric queen of *Eciton hamatum* at the moment a new stream of eggs begins to emerge. 192

9.1 Specimens of mature larvae from the sexual brood of colony '48 B-XVII (*Eciton burchelli*). 204

9.2 Distribution of individuals in large samples of the sexual broods of three colonies of *Eciton hamatum* in the early, intermediate, and late stages of larval growth. 205

9.3 Larval growth of two brood types represented by series of samples from colonies of *Eciton hamatum* in the nomadic phase. 207

9.4 Emigration columns of *Eciton burchelli* carrying sexual larvae. 214

10.1 Schema of conditions in the bivouac of a colony of *Eciton hamatum* with a mature sexual brood just before overt colony division. 219

10.2 Bivouac of a colony of *Eciton burchelli* ('66 B-I), showing the sexual-brood pole where cocoons are being opened. 222

10.3 Queen cluster of the type formed about queens (including "sealed-off" young queens) in colony division. 223

10.4 Sketch of principal raiding systems and queen movements of a colony of *Eciton hamatum* on the day of actual division. 224

10.5 Curtain-type bivouac formed by colony '46 B-I (*Eciton burchelli*) which is in process of division, having just matured a sexual brood. 232

10.6 Sketch of statary bivouac site of colony '59 N-III (*Neivamyrmex nigrescens*), indicating the line of approach on July 15 and the paths later taken by two daughter colonies (III-X and III-Y) in the division carried out on and after August 11. 238

11.1 Schema indicating line of approach of colony '48 H-27 (*Eciton hamatum*) to statary bivouac site where it matured a sexual brood, then divided. 246

11.2 Alate males of *Eciton hamatum* in nocturnal emigration. 252

11.3 Alate males of *Eciton burchelli* (and other species), started at the base of a narrow incline in laboratory, readily mount on the path. 253

11.4 Mating pair (postflight, dealate male and young colony queen) of *Eciton hamatum*, preserved after having been coupled for 10 hours. 260

11.5  Callow queen and alate (preflight) male of *Eciton burchelli* in mating test.  261

12.1  Clustering behavior of *Eciton burchelli* during emigrations in the late, most excitable part of the nomadic phase of their colony.  274

12.2  Pattern of swarm raiding in *Eciton burchelli*.  277

12.3  Excited workers from a nomadic colony of *Eciton hamatum* move in a circular column in a laboratory nest.  282

12.4  Circular mill formed under natural conditions by a few thousand workers of *Labidus praedator* after being cut off from their raid by rain.  283

13.1  Monomorphic brood of a colony of *Aenictus laeviceps* as the colony is about to enter the statary phase.  287

13.2  A platter-type bivouac of the surface-adapted *Aenictus laeviceps*, formed under dry leaves and a rock.  290

13.3  Workers of *Aenictus gracilis* posed to match a field sketch of an attack on a worker of *Polyrachis bihamata*, a tree ant that is a frequent victim of raids by *Aenictus*.  293

13.4  Section of a raiding trail of *Aenictus gracilis* expanded over a leaf bridge on which some of the workers pause to drink at damp spots, others carry booty, and a group struggles with an especially large piece.  294

13.5  Queens of *Aenictus laeviceps*.  300

14.1  Schema of factors and relationships underlying the behavior pattern of army ants, based on *Eciton hamatum*.  305

14.2  Twenty-four-hour activity schedules typical of nomadic colonies in species representing two genera of New World army ants, for comparison with the schedule of *Aenictus laeviceps* of southeastern Asia.  311

14.3  Area bordering a canyon in southeastern Arizona in which a colony of *Neivamyrmex nigrescens* overwintered; sketched to show the main routes of nocturnal raids during the spring resurgence of this colony until May 13 when it became nomadic.  317

14.4  Schema of modifications and specializations postulated for the rise of existing genera of army ants described for groups A and B.  320

# Tables

2.1 Approximate population ranges of colonies. 23

3.1 General bivouac situation and phase of cycle in two species of *Eciton*. 48

4.1 Raiding traffic between 10:30 A.M. and noon in three colonies of *Eciton hamatum*, through intervals of 6 days or more in each activity phase. 88

4.2 Statary phase raids in colonies of three genera. 96

6.1 Approximated "average" populations of workers and of brood in normal colonies of doryline ants. 126

6.2 Contrasts in larval maturation among the dorylines. 144

7.1 Cycles completed by colony '46 H-B (*Eciton hamatum*). 151

7.2 Phases and cycles completed by colony '46 B-I (*Eciton burchelli*). 153

7.3 Number of cases observed for respective durations of nomadic and statary phases in functional cycles of three group A genera and of nesting stops between emigrations in a colony of *Dorylus* (*Anomma*) *wilverthi*. 155

7.4 Frequency of raids through the statary phase in a colony of *Eciton burchelli* —summary for four phases. 161

7.5 Inter-emigration stops recorded for 27 colonies of *Dorylus* (*Anomma*) *wilverthi*. 164

# Color Plates
(Follow page 138)

---

I. Curtain- (or half-cylinder-) type bivouac formed by a nomadic colony of *Eciton hamatum* between the buttressed roots of a tree.

II. Test of species interactions based on odor.

III. Emigration columns of *Eciton burchelli* and *Aenictus laeviceps* carrying nearly mature larvae.

IV. Polymorphic range of *Eciton hamatum*.

V. Alate males of *Eciton burchelli* (and other species), started at the base of a narrow incline in laboratory tests, readily mount on the path.

VI. Callow queen and alate (preflight) male of *Eeiton burchelli* in mating test.

VII. Monomorphic brood of a colony of *Aenictus laeviceps* as the colony is about to enter the statary phase.

VIII. A platter-type bivouac of the surface-adapted *Aenictus laeviceps,* formed under dry leaves and a rock with exposed clusters on its downhill side.

# Foreword

What could possibly prompt a psychologist to abandon the comfortable confines of a laboratory in New York City, travel to the tropical rain forests of Panama, climb into field clothes, and roam through the wilderness at night looking for colonies of army ants? Could it have been the excitement of studying a species of animal about which little was known? Could it have been that his interest in the complex ecology of the tropics made it all the more challenging for unraveling the relationships between a species and its physical and biotic environment? Or might it simply have been to clarify whether or not army ants really do devour everything in their path, including people!

In 1932, when T. C. Schneirla made his first trip to Barro Colorado Island in Panama to study the social biology of army ants, it was not primarily for any of the above reasons. Instead, he went to the tropics to find out whether the application of a scientific methodology that was used primarily by laboratory researchers could be used in the field to clarify some of the problems concerning the social behavior of army ants. The results of Schneirla's analyses, in generating new theories and hypotheses about the *causation* of the cyclic behavior patterns of army ants, stand as a tribute to the effectiveness of his application of these methods of scientific inquiry and to his contribution in placing psychological studies on a truly comparative basis.

During his years as a graduate student at the University of Michigan, Schneirla began a series of studies on learning and orientation in formicine ants. In the course of these laboratory and field studies, he led the way in taking the most detailed written records of his observations—an art that was facilitated by his proficient use of shorthand. His notes were always replete with codes representing different behavioral patterns and interactions and with diagrams recording the positions of animals and the distances between them.

Later, when Schneirla was pursuing comparative studies on learning in insects and mammals, he began to reexamine the old literature on the behavior of social insects. From reading an account of the dorylines in W. M.

Wheeler's 1910 book on ants, it became clear to him that this group of ants had proven to be particularly difficult to study even though their colonies were massive and their movements therefore well documented. To be sure, many field biologists and naturalists had previously described the long marches of army ants, and the published literature on their habits dates at least as far back as 1802. This early notoriety of the dorylines, and especially of their massive raids, is understandable since their huge colonies must have crossed paths with many scientists, explorers, and especially missionaries working in tropical regions of the world. It was the published accounts of these early observers that most perplexed Schneirla because their conclusions, that the migratory habits of army ants were caused by the colony's lack of food or by fluctuations in atmospheric conditions, seemed to be based on insufficient and very subjective evidence.

By the 1930s field research had already increased in popularity, and many scientists were bringing their research animals into the laboratory for experimentation. Schneirla, on the other hand, was pioneering in perfecting the application of the methods of laboratory research to field studies. He believed that there was no difference in rationale between observation and experimentation and therefore that both were valid methods of scientific investigation. In his field studies, therefore, Schneirla constantly applied the same rigorous methods he developed in studying ants in the laboratory, including the codification of behavioral items, the quantification of observational recordings, and the use of a multitude of diagrams and photographs.

Schneirla's analysis of the causation of army ant behavior concentrated on the interrelationships among all members of the colony—brood, workers, and queen—in order to determine just what produced the colony's typical complex social integration, rather than on deductive conceptualizations such as instinct, which influenced much of the behavioral research at the time. Using his keen observational ability and supplementing it with numerous field and laboratory tests, Schneirla set out to explicate the biological processes underlying the social organization of army ant colonies. By the end of his first field trip, Schneirla was able to theorize that the periodic migratory behavior of army ants was not sporadically triggered by a need for new sources of food for the colony, nor was it induced by sudden changes in the weather. Instead, he found that colonies of army ants exhibited regular fluctuations in activity that were regulated by changes in the stimulative interactions between the colonies' large developing broods and the adult workers.

Additional evidence to support this theory of army ant behavior was gathered during subsequent field trips to Panama in 1933, 1936, 1938, 1946, 1948, 1952, 1955, and 1967, to Mexico in 1944 and 1945, and to Trinidad in 1950. During these trips, Schneirla concentrated on comparing two surface-adapted species of the genus *Eciton,* although he also made numerous observations on *Labidus, Cheliomyrmex, Neivamyrmex,* and *Nomamyrmex.*

Later he broadened his comparative studies by including a detailed study of the genus *Neivamyrmex* in southern United States and conducted field trips each year from 1956 through 1959. In 1961, Schneirla traveled to the Philippines and Thailand to study the Old World genus *Aenictus.*

Central to Schneirla's study of social integration in the dorylines was his analysis of the role of the different kinds of stimulative processes responsible for the cohesive unity of all members of the colony. He broadened Wheeler's concept of trophallaxis significantly by interpreting it to include not only relationships resulting from the actual exchange of food (or of other chemical stimuli) but also to include the mutual exchange of *all* stimuli, including tactual, that are equivalent in their sensory and physiological effects of maintaining the social bond among the queen, the workers, and the brood. He then went on to show how the exchange of stimuli, which he called "reciprocal stimulation," was involved in processes of socialization in animals representing many other levels of phylogenetic evolution. His comparative analysis of the genesis of the social bond in cats and ants is now a classic in the literature.

Some recent workers have felt uncomfortable with Schneirla's expanded definition of trophallaxis to include numerous chemical and tactual communicative interactions because they feel that the new meaning is too broad to have value as an analytical concept. Schneirla certainly appreciated the fact that many complex chemical and tactual interactions were subsumed under his definition of reciprocal stimulation, but he continued to stress the fact that the evolution and development of social behavior could best be understood by examining the many different responses by all members of the colony to a variety of chemical and tactual stimuli, including those involved in food exchange, trail following, and "alarm" behavior. He maintained that those who restrict the definition of trophallaxis to the "mutual transfer of food" are nevertheless obligated to include all of these other kinds of reciprocal stimulative interactions in their analysis of social behavior.

It is unfortunate, therefore, that some investigators[1] feel that the concept of reciprocal stimulation (or even trophallaxis in its restricted sense) should be discarded as an explanation for the evolution of social behavior and be replaced with W. D. Hamilton's theory of haplodiploid altruism.[2] The essence of Hamilton's theory is that because male social Hymenoptera

---

[1] See, for example, E. O. Wilson, The superorganism concept and beyond. In *L'Effet de groupe chez les animaux.* Colloques Internationaux du centre national de la recherche scientifique. Organisé par M. R. Chauvin, M. Ch. Noirot, et P.-P. Grassé. Editions du centre national de la recherche scientifique, Paris (1968). Pp. 27–39.

[2] W. D. Hamilton, The genetical evolution of social behavior, I, II. *J. Theoret. Biol.* 7: 1–52 (1964).

are haploid, sister worker ants developing from eggs laid from a single-mated queen share three-quarters of their genes whereas the queen and workers share only one-half. Therefore, it is of great selective advantage for a worker ant to care for another worker, both in the larval and adult stages. This theory poses many more problems than it solves[3]; briefly, most of the interpretations of Hamilton's theory reflect a basic misunderstanding that is, unfortunately, still prevalent in the literature on behavior: It confuses the adaptive value of a behavior pattern with its causation. The implication is that Hamilton's theory and reciprocal stimulation are mutually exclusive concepts, but of course they are not. Even if the unique genetic mechanism associated with social Hymenoptera does make it highly adaptive for social behavior, we cannot accept the adaptive value of the genetic mechanism as a replacement for an analytical approach. As Schneirla repeatedly emphasized, it is a long and complex route from genes to behavior, and we cannot substitute adaptation for analysis.

Throughout this book, Schneirla's approach is based on examining structural, physiological, behavioral, and environmental factors and their interactions in producing the alternating behavioral cycles characteristic of army ants. The results of his work, arrived at through the use of the inductive method in organizing research and through his insistence on asking questions designed to analyze social behavior in terms of its underlying developmental processes, makes this book invaluable not only for entomologists but for all students of comparative animal behavior.

It is always difficult to complete another person's work, and the scope of this book, synthesizing almost forty years of research on the dorylines, compounded the task even more. At the time of Schneirla's death in 1968, the manuscript was essentially completed, with the exception of the last three chapters. These required editing in order to match the first eleven chapters stylistically. Chapter 14 posed a special problem, to which I will return later.

With the exception of the last figure, all of the illustrative material was already assembled, but captions had to be written for many of the photographs. In most cases these captions were synthesized from material already in the body of the text.

The most difficult task was the compilation of the bibliography. Throughout the book, Schneirla had placed empty parentheses in the text to denote bibliographic citations and had written the initials of the cited authors in the left-hand margins. Usually the author's name was followed by a date, and unless the author had written more than one paper in that year, the references were easily identified. Where the author had written more than

---

[3] I will present a detailed analysis of the usefulness of Hamilton's theory for understanding the evolution and ontogeny of social behavior in a separate paper.

one article in the cited year, the appropriate citation could usually be recognized and inserted in the text.

In the last three chapters, many parentheses were associated with an author but not with a date. In most cases, I was able to determine the appropriate citation, but in a few cases, the parentheses were not associated with an author. These were excluded from the text.

Although Schneirla decided to write this book in a semipopular style, he did not want to omit the more technical scientific information. On the advice of a consultant, he placed much of the material that was mainly of interest to specialists in an appendix following each chapter. By the time Schneirla had completed the appendix to Chapter 11, however, he was already concerned that the appendices might be disruptive to the reader. As a result, he was seriously thinking of incorporating this material into the body of the text, and he never did write appendices to the last three chapters. As a compromise, I have placed the appendices as footnotes on the appropriate pages of the text. Schneirla's glossary of terms appears at the end of the book.

The last point of clarification I wish to make concerns the system of colony designations. During his studies of army ants in the field, Schneirla often changed his system of numbering colonies. I have retained his original colony designation with no effort to standardize them because the reader will find that these are the colony designations always quoted in the literature.

Chapter 14, which includes a discussion of the evolution of the dorylines, was of particular concern to Schneirla, and this is the only chapter that he was planning to revise substantially. Although he clearly shows that all existing army ants have fundamentally similar nomadic and predatory behavior patterns, the question of the evolution of their patterns is, at present, still a very difficult one to answer. Schneirla concluded that the hypothesis of a monophyletic origin for the dorylines was most consistent with his functional and behavioral evidence. On the other hand, he was thoroughly familiar with the limitations of drawing conclusions concerning phylogeny by homologizing behavior. He was continuously consulting with leading students of ant taxonomy, but the issue was not clarified since the morphological evidence can be interpreted to indicate a monophyletic, diphyletic, or even triphyletic ancestry for the dorylines.

It was clear to Schneirla that additional evidence had to be obtained, especially from studies on the behavior of the legionary ponerines and their taxonomic relationships to the dorylines. Since his principal objective in this book was the comparison of colony function and behavior in those species of dorylines that have been sufficiently studied, he concluded that this was not the place to pursue in detail so complex a problem. His last letters and notes indicate that he did not consider the problem of origins a major point for this book and that he was preparing to offer only brief discussions of both the monophyletic and polyphyletic views in Chapter 14. Additional evidence that Schneirla planned to present these two views is gleaned from the last figure in the book. This figure, which is cited in the text but which is the only figure that he did not complete, presents a scheme of the evolution of the dorylines. Searching through Schneirla's notes, I found many cladogram-type diagrams that presented alternative phylogenetic schemes. These diagrams could be divided into two basic types: those based on a monophyletic hypothesis and those based on a polyphyletic hypothesis. Therefore, in preparing Figure 14.4, I combined the most significant ideas present in Schneirla's sketches. The only change that I have made in the wording was to substitute "ancestral forms" for "ancestral species" in the bottom block of the diagram. I made this substitution in order to keep open the question as to whether the dorylines evolved from a single species, a group of closely related species, or several relatively unrelated species.

At the time of Schneirla's death, I had just received my doctorate, and frankly, I was petrified at the thought of the difficulties involved in seeing the manuscript through to publication. My task was greatly simplified, however, by the very generous assistance of many colleagues and friends. I want especially to thank Dr. Ethel Tobach for patiently transcribing many of Schneirla's shorthand notes and Dr. Tobach and Dr. Lester Aronson for assisting me throughout all phases of the work on the manuscript. I wish also to thank John Miller for helping with the stylistic editing. The compilation of the bibliography was facilitated by the able assistance of Michael Boshes, Julie Joslyn, and Leslie Bernstein. Kathy Lawson and Patricia Richards helped in proofreading the manuscript and in preparing the index.

*March 1971*

Howard R. Topoff

# Introduction

For those who in years past have had the inestimable privilege of accompanying Dr. Theodore C. Schneirla on a nocturnal observing visit to the great, silent raiding rivers of one of the larger surface-foraging army ants, the experience must remain engraved forever. The opportunity was not only that of exposure at first hand to one of the most remarkable spectacles in all the insect world: the dynamic, flowing stream of the hundreds of thousands of behaviorally integrated individuals composing a single colony of one of the most intensely communally structured of all the social insects, and perhaps the only group in which evolution to huge colony populations has been combined with the apparently contradictory perfection of an exclusively predatory habit and carnivorous diet. Yet greater was the experience of viewing that spectacular phenomenon through the eyes of a student of army ant life who achieved a far deeper and more comprehensive understanding of its basic nature and its dynamics than had any predecessor. It is the sum of such observations and experiences in field and laboratory, extended over more than thirty-five years, that constitutes this book.

The army ants of the subfamily Dorylinae have been known to entomologists for almost two centuries since the formal description of the worker of *Eciton hamatum* as *Formica hamatum* by Fabricius in 1781 and his erection of the genus *Dorylus* to accommodate *Dorylus helvolus* in 1793. They were a major puzzle to taxonomists even earlier, as testified by the assignment of the extraordinary male of *D. helvolus* to the genus *Vespa* by Carl von Linné in 1764. Charles Darwin carefully studied the series of worker-soldier castes in *Anomma* from specimens furnished him by Frederick Smith, and for Darwin they posed one of several major challenges to the theory of natural selection, as his discussion of them in *On the Origin of Species* makes clear. By the first half of the nineteenth century the conspicuous raiding patterns so typical of the larger surface-foraging members of the subfamily were already attracting wide attention in South America, as witnessed by A. W. Lund's *Lettre sur les habitudes de quelques fourmis du Bresil,* written in 1831 to M. Audoin, the disciple and successor of Réaumur, and in the Old World by T. S. Savage's paper *On the Habits of the "Drivers" or Visiting Ants of West Africa* (1847). Through the second half of the nineteenth century and the first decades of the twentieth century these remarkable insects were brought under increasingly sophisticated scrutiny by the great myrmecologists of the period. Taxonomy was better established, discovery of their extraordinary dicthadiigyne queens was made in several species, and for several more species males were identified. Increasingly detailed descriptions appeared of the raiding and statary phases in several surface-living species.

Yet when Dr. Schneirla began his penetrating studies of the army ants in 1932, as his own studies later made amply clear, there was no real understanding of the behavior or of the life cycles of the dorylines at the level of either the individual or the community. Indeed, no investigator had ever brought to the task the zeal, the close observation, the long and continued and broadly based study that Dr. Schneirla was to give the subject. And no one had brought to the subject the same resources of preparation, viewpoint, and experience. A comparative psychologist and animal behaviorist in the broadest sense, as his classic text *Principles of Animal Psychology,* written with Norman R. F. Maier and first published in 1935, bears eloquent witness, Schneirla had also become a pioneer in comparative behavioral studies of higher ants well before he undertook the long program with the dorylines. His penetrating studies of learning behavior in ants of the genus *Formica,* using particularly subtle and highly adapted maze designs, were initiated in the laboratory of John F. Shepard in 1925 and were already well known when Schneirla began his work with *Eciton* on Barro Colorado Island. He was the man ideally prepared to deepen—and indeed to rectify—our understanding of the basic behavioral dynamics of this extraordinary group of social insects, whose evolutionary derivation, as he points out, is still so little understood.

The full and detailed account of those thirty-five years of searching work in the field and the laboratory is presented in this volume, which to the immeasurable loss of us all, appears posthumously. It is to be read in detail and repeatedly, for many of its most significant facts and insights must be ferreted out where they originated: in the particular observations and the general reflections of a great student of animal societies. This book is a tribute to his life's work and his memory.

*August 1970*

Caryl P. Haskins
Carnegie Institution of Washington

ARMY ANTS

# 1

# Army Ants

The army ants, members of the subfamily Dorylinae,[1] are the one major group among all ants in which much the same ways of life are found throughout. These are: (1) massive predatory raids against other insects in particular, (2) great nomadic movements of the colonies based upon the ability to form and readily abandon temporary nests, and (3) the ability of all colonies to change more or less regularly from a high level of activity and function to a low level, and the reverse.

The purpose of this book is to examine the raids, the emigrations, and the cyclic operations of these ants as an important scientific problem and to work out a theory for these aspects of doryline life, all of which depend upon a type of activity pattern featuring cyclic colony behavior that is really unique to these ants.

Figure 1.1 illustrates the principal features of one of the most specialized doryline adaptive systems, pointing out the roles of the brood and the colony queen in the colony behavioral cycles. Although we find significant differences in the cyclic patterns of the major doryline genera, we also find important similarities in their everyday colony behavior.

Before beginning our discussion of present-day investigations of the army ants, let us review some highlights of the way they first came to attention.

> Wherever they pass all the rest of the animal world is thrown into a state of alarm. They stream along the ground and climb to the summits of all the lower trees, seaching every leaf to its apex, and whenever . . . booty is plentiful, they concentrate . . . all their forces upon it, the dense phalanx of shining and quickly-moving bodies, as it spreads over the surface, looking like a flood of dark-red liquid. They soon penetrate every part of the confused heap, and then, gathering together again in

---

[1] From the Greek *dory,* meaning "spear." The name expresses well the effect of the dorylines on the animals they raid, whether or not the stings are weak and the numbers enormous, as is usual with the driver ants, or the stings potent with fewer attackers, as in most of the others.

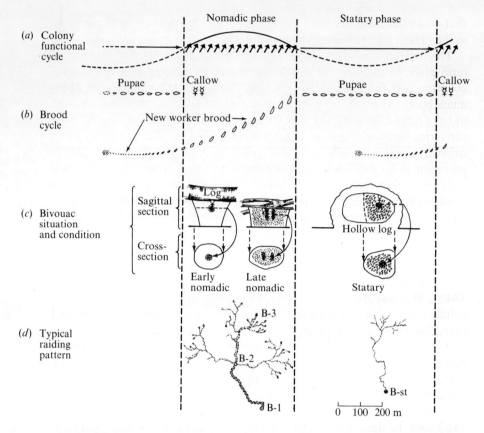

FIGURE 1.1
Functional cycle of the army ant *Eciton hamatum*. (*a*) Phases in the cycle indicated by a sine curve; arrows: large daily raids and nightly emigrations. (*b*) Concurrent development of successive (coordinated) all-worker broods, from eggs (*at left*) to larvae, to pupae, and finally to callow workers (☿ ☿, *at right*). (*c*) Types of bivouac in each functional phase, indicating typical placement of brood in each. (*d*) Patterns of raiding systems typical of the functional phases; B-1, B-2, and B-3: successive bivouac sites; B-st: statary bivouac site.

marching order, onward they move. All soft-bodied and inactive insects fall an easy prey to them and . . . they tear their victims in pieces for facility of carriage. . . . The margins of the phalanx spread out at times like a cloud of skirmishers from the flanks of an army. (Bates, 1863, pp. 362–363.)

So reported the Victorian field naturalist Henry Walter Bates on the maneuvers of army ants he watched in the Amazonian forest. Military metaphors abound in the literature on these ants and seem appropriate for events that prompted William Morton Wheeler (1910) to speak of these predators as the "Huns and Tartars of the insect world."

The designation "army ants," although an analogy, fits well these creatures given to carrying out attacks en masse, to moving about in orderly columns, and to changing camp at intervals—with tactics that bring them much booty. This analogy may be very ancient, so likely were these insects to have forced their attacks into the life of early man in the tropics. African terms signifying "drivers" (Siafu) and "visiting ants" and corresponding tropical-American names (e.g., Soldados, Arrieros, and Tepeguas) are so prevalent as to indicate a human interest in these creatures dating from man's earliest cultures. The army or visiting ants must always have been considered a menace by people shocked from sleep and routed from their dwellings by the intrusions of the ant masses (Savage, 1847; Loveridge, 1922). Others, making the best of the situation, accepted the invaders at least as an aid to housecleaning, vacating their homes until the swarm had flushed out roaches, centipedes, and other vermin from walls, roof, and elsewhere in the premises and had passed on to their neighbors (Savage, 1847). Man has also used these ants for other purposes peculiar to his local culture: as food, as a device for suturing wounds,[2] and even as a means of executing enemies.

At first, the living subjects and the scientists studying the specimens were worlds apart. It took students of insect taxonomy seventy years to surmise that these creatures probably were ants and almost a century to be definite about it. William S. Creighton (1950), an American ant taxonomist, noted that one and the same army ant—after having entered the literature in 1802 with the name *Formica coeca* (Latreille, 1802, in Wheeler, 1910)—had been assigned to no less than seven different genera and fifteen different species within just one century. Even then its list was not complete as another specialist (Borgmeier, 1955) found sufficient reasons to change the name to *Labidus coecus*. Even the most general comparison of the male, queen, and workers of the same doryline species (Fig. 1.2) indicates how difficult is the task of classifying these ants.

It was not until 1849 that the driver ant male was finally described definitely as an ant. This came about when T. S. Savage (1849), a medical missionary in East Africa, sent back a detailed account of his observations of the local visiting ants, ranging about in vast armies, driving everything

---

[2] In certain interior regions of tropical America, a practice of long standing has been to hold over the two edges of a deep cut the tong-jawed major worker of an *Eciton* species. Squeezing the ant's body between thumb and forefinger causes powerful muscles to contract so that the needle-sharp jaws are seated so deeply in the flesh as to pull the open sides of the wound firmly together. Once the body is pinched off, the ant suture stays firmly in place—held by muscular contracture and by the opposed hook-jaws—until removed by force. For the removal, one can insert the closed tips of tweezers between the closed jaws, then release the blades so that in springing apart they yank out the implanted mandibles with the head.

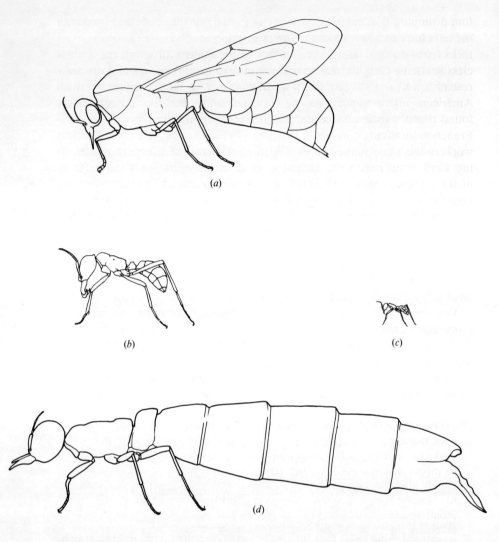

FIGURE 1.2.
Types of individuals present in a colony of the driver ant *Dorylus* (*Anomma*) *wilverthi*. (*a*) male, (*b*) major worker, (*c*) minor worker, and (*d*) queen.

before them, and appearing and disappearing seemingly with no fixed habitation.[3] Along with specimens of the workers, he sent about a dozen males

---

[3] Savage (1847) described the striking way in which these ants hang together in clusters by means of their hooked claws and slender legs; the severity of their bites, delivered by vast numbers; and the effectiveness of their attacks against every living thing. "They will soon kill the largest animal if confined. . . ." (P. 14.)

found running in column with them, suggesting that they must belong with the ants because they reentered the procession each time he tried to divert them from it. This was the insect that the innovator of plant and animal classifications, Carl von Linné of Sweden, examined in 1764 and, from its resemblance to wasps, named *Vespa helvola*. S. S. Haldeman (1849), an American insect taxonomist in Philadelphia, examined these males, found them "another condition" of the workers described in 1802 by the French naturalist Latreille (in Wheeler, 1910), and gave both males and workers the same name. Finally, after further stages of study and naming, Carlo Emery (1887), an Italian ant specialist, described all three forms of these driver ants—male, worker, and queen—under the name *Dorylus helvolus*.

Complications in naming were inevitable. From archaic times, the workers, queens, and males of doryline ants had specialized in very different ways, and they hardly seemed to belong together. In the many efforts in the twentieth century to identify them dependably, nearly every type of army ant acquired a long, and usually, confusing list of names, i.e., synonymy.[4]

Very gradually the interesting ways (or "habits," as they were called) of army ants in tropical regions came to world attention. The Darwinian era greatly stimulated interest in the natural history of animals, and detailed reports from field observations began to appear. Thomas Belt (1874), an English naturalist, from his studies in Nicaraguan forests, described the surface forays of the swarm-raiding species *Labidus praedator* as follows:

> One of the smaller species used occasionally to visit our house, swarm over the floors and walls, searching every cranny, and driving out the cockroaches and spiders, many of which were caught, pulled or bitten to pieces and carried off. . . . I saw many large armies of this, or a closely allied species, in the forest. My attention was generally first called to them by the twittering of some small birds, belonging to several different species, that followed the ants in the wood. On approaching to ascertain the cause of the disturbance, a dense body of ants, three or four yards wide, and so numerous as to blacken the ground, would be seen moving rapidly in one direction, examining every cranny, and underneath every fallen leaf. On the flanks, and in advance of the main body, smaller columns would be pushed out. (Pp. 17–18.)

This description, supplementing that of Bates (1863) from Brazil, covered the general aspects of predatory behavior in representatives of the same species operating in widely separated areas.

---

[4] Note, as examples, the taxonomic records of the tropical American species *Eciton hamatum* and *E. burchelli* as summarized by Borgmeier (1955).

François Sumichrast (1868), a French naturalist, observing these ants in Mexico, noted that their movements in columns might be at times "expeditions of pillage, sometimes changes of domicile, veritable migrations." Belt (1874) was the first to distinguish—for a different species, *Eciton hamatum*— the general aspects of emigration.

> I have sometimes come across the migratory columns. They may easily be known by all the common workers moving in one direction, many of them carrying the larvae and pupae carefully in their jaws. . . . Such a column is of enormous length, and contains many thousands, if not millions of individuals. I have sometimes followed them up for two or three hundred yards without getting to the end. (Pp. 24–25.)

Belt (1874) also who gave the first clear description of the bivouac or temporary nest of an American army ant (probably *Eciton burchelli*):

> They make their temporary habitation in hollow trees, and sometimes underneath large fallen trunks that offer suitable hollows. A nest that I came across in the latter situation was open at one side. The ants were clustered together in a dense mass, like a great swarm of bees, hanging from the roof, but reaching to the ground below. Their innumerable long legs looked like brown threads binding together the mass, which must have been at least a cubic yard in bulk, and contained hundreds of thousands of individuals. . . . (P. 25.)

Sumichrast (1868) described the nesting behavior of *Labidus praedator* as follows:

> During about three months, a colony of soldados had been domiciled under a little bridge formed by some rough trunks of trees bound together by a heap of vegetable mould. . . . [Then, one morning, the colony was gone.] Its inhabitants did not return until about four months later. (P. 40.)

Of another species, *Eciton burchelli,* he wrote, "The nests are found in cool, shady places in great woods or among rocks and are tunneled more often at the foot of, and among the roots of, old trees." (Pp. 40–41.) Actually, the bivouacs of these two dorylines differ greatly, both in situation and in their common forms (Chapter 3).

At about the same time, Wilhelm Müller (1886), a German naturalist visiting in southern Brazil, studied a colony of *Eciton burchelli* that held its bivouac for nearly three weeks in a hollow tree despite his efforts to drive it away with smoke. His was the first continuous observation on any army ant colony. The significance of these observations for the problem of doryline emigration, however, was overlooked for more than fifty years (Chapter 7).

From the last decades of the nineteenth century an increasing emphasis upon the relationship of classified specimens to the living material has played an important part in the study of ants. This change came about gradually, especially through the work of the myrmecologists Gustav Mayr (1886) in Germany, Carlo Emery (1895a, 1895b) in Italy, Auguste Forel (1891, 1893) in Switzerland, and William Morton Wheeler (1900, 1910) in the United States. The studies of these men introduced the modern era in which problems of taxonomy and function are studied together in the laboratory and field. One important turning point was Emery's analysis (1895a) of the "doryline problem"; another was wheeler's study (1928) of social organizations in insects. Accordingly, I will be discussing army ant behavior and its biological basis in this book from the premise that to study an animal adequately, problems of structure, function, and behavior must all be considered together.

More than 200 known species of army ants distributed through the tropics and subtropics of the world (Fig. 1.3) make up the Dorylinae (Wheeler, 1910), one of the eight or so main subfamilies of ants. This group with all of its subdivisions is distinguished from the other ants by: (1) structural and functional characteristics of its (generally) polymorphic workers, of its unique wingless queens of great reproductive capacity, and of its winged, wasplike males (Fig. 1.2); as well as by (2) distinctive predatory and nomadic behavior and related functions. Structural features of particular importance to our discussion, used by taxonomists to distinguish the dorylines from other subfamilies of ants, are given in the key at the end of this chapter. Interested readers will find keys for the species of army ants with generic classifications in works by Wheeler (1922), M. R. Smith (1942), Creighton (1950), Borgmeier (1955), and Wilson (1964).

I have not included in my keys the tribes or superordinate groups of genera, although they are mentioned in the following paragraphs. Many students believe that tribal distinctions have value as indicators of broad relationships within the subfamily, but for the purpose of studying function and behavior, I think that these distinctions based on structure alone are too loose and debatable to be valid. In the following chapters, therefore, genus and species names will suffice.

The names taxonomists assign to animal groups should be taken not merely as convenient tags for types of animals but as means of marking out relationships of similarity and difference among them. As a good example, Thomas Borgmeier (1955), a Brazilian insect biologist, made a significant advance for doryline study when, in reviewing the New World representatives of this subfamily, he reclassified as distinct genera several groups, formerly listed as subgenera, under the name *Eciton*. He separated these genera on the basis of substantial and consistent structural distinc-

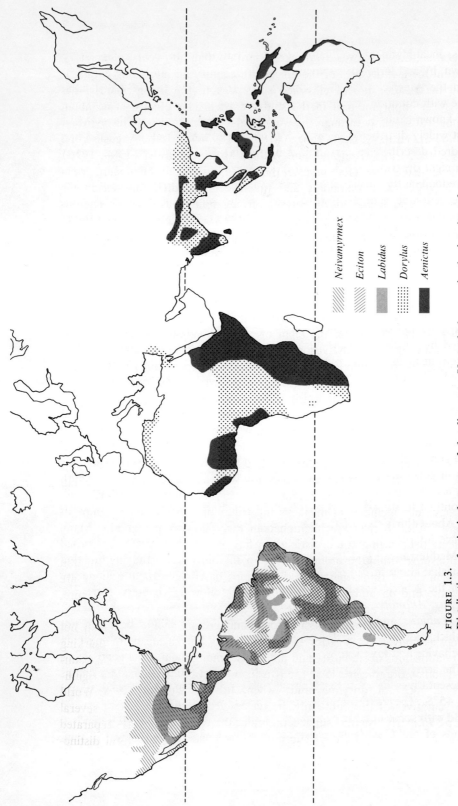

FIGURE 1.3.
Distribution of principal genera of doryline ants. Unmarked zones in the interior of a habitat may be either vacant or uncollected.

tions, meaningful not only for structure but, as our investigations have shown, by and large for behavior as well.

In the American dorylines, under two tribes, five genera are named, three with common, widely distributed species. In the tribe Ecitonini, the best-known genus is *Eciton,* with twelve described species; the largest and most widely distributed genus is *Neivamyrmex,* which has more than one hundred described species. The other genera in this group are *Labidus,* with eight described species, and *Nomamyrmex,* with just two. In *Eciton,* a predominantly surface-living genus, all species are tropical. The other three ecitonine genera, however, tend to be subterranean, with climatic adaptations to tropical, subtropical, and even subtemperate environments. *Cheliomyrmex,* the one genus in the relatively unknown tribe Cheliomyrmicini, is tropical and distinctly subterranean.

The problem of classification, unfortunately, has not been as well resolved for the tribe Dorylini of the Old World, which contains twenty five species of driver ants usually allocated after Emery (1895a) to seven subgenera under the single generic name *Dorylus.* The best known of these groups are the two subgenera: *Dorylus,* strongly subterranean and represented by species in both Africa and tropical Asia; and *Anomma,* also subterranean in nesting but more surface active than *Dorylus* and represented only in Africa. These two subgenera contain the species best studied to date in behavior and function (Cohic, 1948; Raignier and Van Boven, 1955).

The other Old World tribe, Aenictini, contains the single genus *Aenictus.* In this group are more than fifty recognized species, whose taxonomy has been greatly clarified by Wilson's revision (1964). The *Aenictus* are smaller ants in generally smaller colonies than most other dorylines, from which they are distinguished, especially by their essentially monomorphic workers in contrast to the polymorphic worker populations of the others (Chapter 2). About thirty-five of the named species of *Aenictus,* ranging from the surface adapted to the subsurface adapted, are distributed in Asia from the Mediterranean and India to the Philippines, New Guinea, and Eastern Australia; about fifteen species, all subterranean, are listed for Africa.

Of the generic names mentioned, the ones used most in this book are *Dorylus, Anomma,* and *Aenictus* of the Old World, and *Eciton, Neivamyrmex, Labidus,* and *Nomamyrmex* of the New World. These are the groups in which species have been studied sufficiently to permit a useful discussion of behavior and function.

The army ants, as Figure 1.3 shows, have a worldwide distribution, with representatives in each hemisphere roughly between the parallels 45°N. and 45°S. They are found in most tropical and subtropical parts of the world with some notable exceptions; in the New World: Chile, the volcanic islands of the Caribbean, mountain heights, and certain extensive desert and swampy areas; and, in the Old World: and Sahara, Afghanistan, and desert areas, as well as certain mountainous and swampy regions. In the New World, their habitats range from the surface-arboreal zone of *Eciton burchelli* to the deep subterranean zone of *Neivamyrmex pauxillus,* and in the Old World from the surface zone of *Aenictus laeviceps* to the deep subterranean zone of *Rhogmus fimbriatus.* The wide range of many dorylines above sea level is illustrated by *Dorylus (Anomma) nigricans,* which is active at an altitude of nearly 3000 meters (10,000 feet) in the mountains of Sudanese Africa, and by colonies of *Neivamyrmex nigrescens* at well over 1500 meters (5000 feet) in the mountains of the western United States (Smith, 1942; Cole, 1953; Schneirla, 1958).

The doryline genera of widest distribution are *Neivamyrmex* in the New World and *Aenictus* in the Old World. In each of these genera, species differ widely in their ranges. *N. pilosus,* for example, is common throughout an extensive range from Arizona and Texas to northern Argentina and Bolivia whereas *N. pauxillus,* more subterranean, is restricted mainly to Texas, Arizona, and northern Mexico. *Ae. ceylonicus* is a fairly common ant throughout a wide range from northern India into the Philippines, New Guinea, and Queensland in Australia; *Ae. wroughtoni,* in contrast, has been found only in India in an area to the south of Bombay. In the Americas, the genera *Eciton* and *Neivamyrmex* are widely sympatric (i.e., overlapping) in their ranges from east to west in the tropics, but the latter extends into the temperate zones as well; in the Old World, *Anomma* is restricted to tropical forests in Africa whereas the other and more hypogaeic *Dorylus* range more widely in Africa and are found in Asia as well. Species of *Aenictus,* somewhat restricted in Africa, have a wider range in Asia than representatives of *Dorylus.* Interesting to note also is the wide area separating the African and Asian representatives of *Aenictus,* with the distinctly surface-adapted species in this genus confined to Asia.

In expanding their zones, the dorylines undoubtedly were aided greatly by the nomadic and predatory behavior of their colonies in respect both to mobility and to competition with other insects. Each species evolved adaptations to a distinctive niche, i.e., a range of optimal conditions of temperature, humidity, and (especially) food supply under which its colonies could maintain themselves. The niches of most doryline species, as indicated by their nesting, are subterranean (hypogaeic). Exceptions are some of the species in *Eciton* and in the *Aenictus* of Asia, which are surface adapted (epigaeic), also indicated by their nesting.

Differences in distribution are naturally related closely to ecological properties. Thus, in the Americas, species of the dominantly surface-adapted genus *Eciton* are tropical and mainly forest dwellers whereas species of the subterranean-adapted genus *Neivamyrmex,* overlapping with *Eciton* in the forests, also extend into savannahs and cultivated areas (von

Ihering, 1912). Species of *Neivamyrmex* range considerably farther to the north and to the south of the tropics than *Eciton,* pushing even into the temperate zone where some of them survive annual winter conditions. Species of *Labidus,* also all hypogaeic and appreciably nocturnal, extend well to the north and to the south of the tropics but with a shorter range on both ends than *Neivamyrmex.*

All existing dorylines are efficient in their environmental adaptations (Chapter 3) and in the social organization of their colonies (Chapter 13). Without doubt, they have been long established as a dominant type of ant. As a comparison, Edward O. Wilson, biologist at Harvard University, and his doctoral student Robert Taylor (1964) described from a find by the British anthropologist, Louis Seymour Leakey, in the lower Miocene deposits of Kenya, a fossil aggregation of workers, brood, and nest remnants of a formicine tree ant (*Oecophylla leakeyi*). At the date of this remnant, reliably established as at least 30 million years ago, these ants had worker polymorphism (Chapter 2), complex brood-worker relationships (Chapter 6), tree nests, and other characteristics very similar to those of their modern counterparts. Comparably specialized social patterns may have by then also evolved in the dorylines although as yet no fossil records of these ants have been found.

Let us briefly consider the origin of the army ants. In his theory of the evolution of social insects, Wheeler (1928) stresses the role of the female as the key to early group functions. In existing tiphiid wasps, solitary insects which he considered to be similar to the hypothetical formicid ancestor, the fertilized female stings a captured insect, then lays an egg on it and leaves it there. In this earliest presocial stage, the young develop alone, subsisting on the booty through the larval stage. Early ancestors of ants, Wheeler concluded, evolved their grouping properties on this basis, through descendent lines in which the female was more and more attracted to her offspring and in further stages became equipped to remain longer and longer with the young. Finally, she remained with them throughout their development, with a little group forming at their emergence—a primordial colony.

It is generally agreed that ants, as completely socialized insects, arose from an archaic stock resembling a modern solitary wasp. Emery (1895a), on taxonomic grounds, favored a mutillidlike type as ancestor and dated the origin to early Cretaceous times, i.e., roughly 130 million years ago. Wheeler (1928), on ecological and other grounds, favored an earlier date. This might have been as early as the Triassic Period, he thought, as thermophilic (heat-loving) insects could have flourished in the hot, dry uplands then prevalent and might have thrived on a diet of the roachlike insects then on the increase (Carpenter, 1930).

Until a few years ago all of this was speculative as no really archaic fossil ants had turned up. But recently the discussion has taken a more realistic turn with the finding of two amber-embedded fossil specimens in the Raritan Bay formation of New Jersey. These two workers, probably from the same colony, combine wasplike characteristics with such other features as a well-formed petiolar node, which is definitely characteristic of ants (Wilson et al., 1967). At the time these creatures lived, which can be dated from the geological formation as about 100 million years ago, i.e., before the mid-Cretaceous period, they probably formed small colonies of interdependent, hence socialized, individuals.

Although all existing ants live in colonies, a minority of them are considered very primitive, suggesting what early forms were like. From these, one gathers that mechanisms basic to group unity and function must have evolved slowly at first. The nature of early deficiencies is suggested by observations made by the American biologists Caryl and Edna Haskins (1950), on species of the bull ant *Myrmecia* and other archaic ponerines that have survived in Australia. In these ants, they found interindividual communication poor, with other indications that a low-grade colony integration still prevails. These conditions may reflect the state of colonies in the earliest ants: small and very loosely organized.

Through the great advantage of group action in getting food and in maintaining a collective shelter, ancestral ants presumably advanced steadily in evolving social functions and behavior. From insects of this type, possibly very early in their evolution, may have sprung the predatory stock ancestral to army ants. How this may have occurred, we consider in Chapter 14, after having first reviewed evidence on the functions and behavior of modern army ants.

Our main object in this book is to consider and compare evidence on the patterns of function and behavior typical of modern army ants. That such evidence was not come by easily is suggested by the following statement with which Wheeler ended the doryline chapter of his classic book, *Ants— Their Structure, Development, and Behavior* (1910).

> In conclusion, attention may be called to certain problems that are suggested by our present meager knowledge of the dorylines. Besides the investigation of the species with a view to obtaining all the sexes and thus clearing up the taxonomy, we are in great need of a fuller insight into the domestic economy of these singular insects. As yet no one has been able to observe the methods of rearing the broods and the mating of the sexual forms. . . . Nor has it been possible to plot the territory covered by the annual migrations of any of the species, to determine the time spent in the bivouac or in the presumably more permanent breeding nests, or the precise relations which these nomadic ants bear to their myrmecophiles. . . . Another problem of more theoretical interest is presented by the dichthadiigynes, which are so unlike typical female

ants. . . . Too few female Dorylinae are known at the present time to enable us to decide this question, which must be left to future students. (Pp. 265–266.)

It was this chapter of Wheeler's, in particular, that led me some years later to select the army ants as an important area in "instinctive behavior" that needed investigating. In the preceding 150 years, much had been learned about the taxonomy of army ants and a little about their biology, but almost nothing about their behavior. For example, I found that their colony movements were then thought of as sporadic actions, forced when the ants had used up the booty available in the environs of their current nest (Heape, 1931). No hypothesis could have missed the truth more widely.

To test this hypothesis of food scarcity in particular in its bearing on the nomadism of army ants and to study their group organization in general, I began field research at Barro Colorado Island, Panama Canal Zone, in June of 1932. From the first day, the investigation resolved itself into studies of the raids and the emigrations of *Eciton hamatum* and *E. burchelli,* two common species in the American tropics, fully adapted to surface conditions and well suited to a comparative approach.

My findings on these two *Eciton* species (Schneirla, 1938, 1945, 1957a) provided a foundation for comparisons with representatives of the genus *Neivamyrmex* in the New World (Schneirla, 1958) and the genus *Aenictus* in the Old (Schneirla and Reyes, 1966). These and further studies set the basis for comparing colony functions and behavior of army ants all over the world in terms of an idea to be tested in this book: that colony behavior patterns in these ants differ fundamentally according to the degree of species adaptation to conditions on the surface (group A) or beneath the surface (group B).

Using this approach (Schneirla, 1965), I have developed each chapter in my analysis of army ant behavior and its biological basis according to three main stages: (1) a consideration of the behavior and functions of known surface-adapted species, usually starting with *Eciton;* (2) a study of these problems in comparing results for the different genera of doryline ants; and (3) a comparison of adaptive functions in surface-adapted army ants with those of subsurface-adapted types.

For an overall view, Chapters 2 through 8 are concerned with outstanding features common to the colony adaptive patterns of all army ants. Characteristics vital to normal colony functions are population size and pattern (Chapter 2); the temporary nest or bivouac as the base and operating center of the colony (Chapter 3); the species pattern of raiding and its variations according to colony condition (Chapter 4); emigrations and their relation to conditions outside and within the colony (Chapter 5);

the unique doryline broods and their role in the colony functional pattern (Chapter 6); cyclic variations in colony behavior and their dependence upon colony condition, genus, and species (Chapter 7); and the reproductive and other functions of the single colony queen (Chapter 8). (See Fig. 1.1 for the organization of the functional cycle of the highly specialized *Eciton hamatum.*)

In these eight key aspects of colony function, important similarities as well as important differences are discernible among the chief genera of army ants. These similarities and differences, I think, can be studied to best advantage by contrasting the mainly surface-related genera, group A: *Eciton, Neivamyrmex,* and *Aenictus,* with the mainly subsurface-related genera, group B: *Dorylus* and possibly also *Labidus* and others.

In Chapters 9, 10, and 11, I offer evidence on the production of males and young queens and important events leading therefrom to the formation of new colonies, using a different approach to problems of army ant adaptations from that of earlier chapters. Chapter 12 discusses some outstanding aspects of colony organization, and Chapter 13 tests the overall theory of the book by comparing a doryline genus considered generalized with others considered highly specialized. Chapter 14 brings the entire study to a head in a comparison of colony adaptive systems in the different army ants, which leads to outlining a theory of their evolution.

The Dorylinae, although the one ant subfamily most uniform in the functional patterns of colonies in all its genera, at the same time presents an array of differences. Fundamental to all doryline colony functions, nevertheless, is the life history of the workers: their main stages from life to death. The worker begins life as an egg and develops in a brood containing many tens of thousands of similar individuals (Chapter 6). After an embryonic stage, these pass through a larval stage in which they are fed by workers with booty collected in frequent raids. In many dorylines, colony emigrations then also occur. Next, the brood passes through its pupal (nonfeeding) stage, with the colony in a stable bivouac. When the pupae mature, this brood emerges as a great population of new workers (or callows); then the colony enters a new condition of activity. The young of sexual broods develop in similar stages (Chapter 9).

Because doryline broods excite their colonies at a low or a high level according to their stage of development, we find the brood-stimulative effect a major factor in the colony adaptive patterns of all army ants. The energizing effect of the brood on the colony is the central concept in the theory of doryline behavior presented in this book.

The general method of these investigations has been one of systematic field and laboratory observations supplemented by controlled tests. There is no basic difference in the scientific rationales of observation and experiment as both of these in principle are disciplines of investigation subject to appropriate controls (Schneirla, 1950). In his observations, a scientist investigates the phenomenon under study by direct perceptual operations; in experimentation, he investigates it by appropriately planning situations, apparatus, and methods in order to increase his degree of control over independent and dependent variables. Observation becomes systematic and reliable when it is planned and carried out in ways consistent with the methods of experimentation.

A scientist gains insight into his phenomenon as his experience with it increases, provided that his theory—his logical interpretation of the phenomenon—grows and changes in step with and in keeping with the evidence. He checks on the most significant questions raised by his observations by reperceiving aspects considered crucial, by comparing under different conditions the observations found significant for each main question, by taking notes and other records in a reliable way, and by supplementing his observations whenever possible with field and laboratory tests.

A versatile program of tests is indispensable to research of this kind. One example of field tests is the rotation of a trail-bearing leaf at a "Y" junction in a raiding system to determine how returning booty carriers orient themselves (Chapter 4); another is the removal of a queen from her colony for a specific interval to test the role of her odor in colony behavior (Chapter 8). Examples of laboratory tests are the use of the circular column situation to investigate trail using (Schneirla, 1944c) and the use of the two-disk technique to test queen odor (Schneirla et al., 1966). As an example of control procedures, in the latter case the effect of the queen's odor is tested by presenting to workers both the surface on which the queen has rested and a facsimile surface on which the queen has not rested. Tests are often crucial for interpreting evidence from field observations; also they often open the way to specialized experiments.[5]

The best introduction to studies of colony function and behavior is by the short survey method whereby one observes such functions as raiding, emigration, and bivouac operations in standardized ways through intervals up to a few days in each of several colonies operating in a given area. By these procedures one may compare in detail colonies of the same species operating under equivalent or under different conditions. To this end, a series of daily studies can be scheduled according to the problem, the species, and the conditions of each colony under investigation. Studies

---

[5] In systematic field work, observations lead to tests, and tests to experiments. Thus, finding that the removal of a colony queen of *Eciton* caused the workers, after some hours, to backtrack into old trails, led to a procedure for gaining information about previous movements of newly discovered colonies (Schneirla, 1949) and to behavioral contrasts of colonies with all-worker and with sexual broods (Schneirla and Brown, 1952). Hypotheses for queen-pheromone experiments (Chapter 8) result.

of the same colonies over longer periods of time by continuous survey methods open the way for intercolony comparisons and for comparisons of the same colony under different conditions. In these operations, we devise appropriate procedures of following and mapping the operations of colonies (Fig. 5.8). Combining these two approaches in my Panama project of 1946 (Schneirla, 1949), I studied one colony of *Eciton hamatum* and one of *E. burchelli* each for more than four months with other colonies of each species also on record for shorter times.

My field notes in this investigation were taken as a rule in shorthand to get a running account of events without looking away from the object of study. To aid comparisons, notes were condensed and organized daily. An example of a project in which all of these procedures were used opportunely was the analysis of swarm raiding (Chapters 4 and 13). Study of this problem followed a schedule for observing the phenomenon in each of several colonies under different conditions (e.g., time of day; nomadic or statary phase) and for observing large sections of the phenomenon (e.g., the swarm as a whole) and significant details (e.g., changes at the front border) in the proper order and at times that were found to be significant (Schneirla, 1940).

The continuous study of a colony usually requires closely following all of its activities (e.g., raids and emigration) with physical continuity from day to day, as in tracing out emigrations. When many colonies of *Eciton* were studied over considerable periods on Barro Colorado Island and physical continuity in tracing the movements of each colony was impossible, I marked the colony queens permanently (Chapter 8). To study the details of such colony activities as raiding and emigration, it is essential that procedures be established for measuring distances by pacing, for marking routes (as with string), and for recording notes of actions on contour maps. For the study of actions around bivouac sites, especially when details of spatial relations are crucial (e.g., Chapter 10) or when a colony may emigrate under surface cover, I used the cordon procedure (Fig. 1.4), which involves a regular patrolling of the colony site by means of narrow inspection lanes (often plotted concentrically) around it.

In view of the importance of the brood for colony function, we took samples of developing generations regularly from each colony under study. In relatively small army ants (e.g., *Aenictus*) these can be removed from the emigration with a suction apparatus; in large army ants (e.g., *Eciton*) they have to be taken directly from a bivouac or by special means from the emigration. Getting brood samples from statary colonies presents special problems (Chapter 6).

For observations and tests in the laboratory, workers and brood of surface-nesting species may be taken from the bivouac with a trowel and long tweezers; or, in subsurface-nesting species, from the emigration with

FIGURE 1.4.
Locale of the statary bivouac of a colony of *Neivamyrmex nigrescens* studied in an Arizona canyon. A pile of rocks at rear center marks the original entrance of the colony to its subterranean nesting site (preempted from a colony of fungus-growing ants). In the foreground is one part of the cordon, an inspection lane used for studying and mapping the activities of the colony. String at left rear marks one of the principal raiding routes.

suction apparatus. The sample is placed directly in a plastic sack and in the laboratory is introduced to a nest or a large glass-topped arena for study. Ants and brood may be transferred readily by rotating the sack a few times. The ants thereby form a ball in which they are then rolled quickly into the receptacle (Schneirla et al., 1966). One useful set-up is a glass-covered arena (with a damp pad on the floor and a red glass over one corner), connected by a plastic tube with a glass-topped cell in which food is supplied, or with a test set-up, whichever is in order. By working out improved controls of diet and sanitation in particular, we progressed toward housing entire army ant colonies in the laboratory for a variety of experimental projects.[6]

The importance of carefully devised procedures and techniques for research with these ants cannot be exaggerated. No strategy that one can work out for them can be quite equal to the elusive complexities of their adaptive system. Within reasonable limits, however, even with these creatures, one can control behavior by taking pains to understand it.

To the principal theme of this book, doryline behavior and its basis, problems of social organization in insects are central. Among the social insects many types of behavioral adaptations have evolved with differences as great as those between the plastic behavior of *Formica* species (Schneirla, 1929) and the stereotyped though complex mass patterns of the army ants. The doryline adaptive pattern is notable not just for its distinctive specializations but also for radically subordinating the individual as a social unit to the functions of the group (Chapter 12). In studying these problems, we are concerned with questions as wide-ranging, for example, as the relation between predation and nomadism (Chapters 4, 5, and 7), and relation between reproductive processes and cyclic functions (Chapters 6 and 7), and the organization of behavior in collective operations (Chapters 4, 5, and 13). These problems all concern the nature of insect social behavior.

As we investigate the army ant phenomenon in terms of these problems, we are also studying questions of instinctive behavior, i.e., questions concerning the nature of species-typical systems of behavior approached from the standpoint of their developmental and evolutionary origins. I contended that the word "instinct" denotes only a set of problems concerning the ancestry of any species behavior pattern and how this pattern arises and varies in the animals that display it—*not* a proved

---

[6] In 1938, at Barro Colorado Island, I prepared a set-up to investigate conditions of larval development. Of three colonies of *Eciton hamatum* captured with larval broods at roughly the same early stage, housed in mesh-wire nests, and fed equivalently, two received water containing calcium gluconate at different molarities and the third received distilled water as a control. The experiment failed chiefly because it showed that when colonies receive insufficient food, they consume their broods.

innate organization of behavior, which is really the problem under study.

Each of the following chapters deals with a major type of problem concerning the army ant adaptive system. The general aim is to present relevant evidence objectively while keeping the principal question clearly in view. Theory in this work has an essentially inductive basis with the deductive method used to devise hypotheses, to apply postulates, or to extrapolate conclusions. Theory in this context is a pattern of interpretations and of postulates designed to account for the phenomenon under investigation in terms of available evidence. The aim here is to clarify the nature and biological basis of army ant behavior.

## Taxonomic Keys for the Workers

A. Workers of the main subfamilies of ants
   (Creighton, 1950)

1. Gaster with a distinct constriction between the first and second segments or, if this constriction is faint, the mandibles are linear and the petiole is produced into a conical dorsal spine ................................................. 2
   Gaster without any constriction between the first and second segments ........................................ 3
2. Antennal scape short and very stout, even at the base; the scape flattened throughout or with a greatly enlarged tip which bears a prominent lateral furrow for the reception of the funiculus .............. Cerapachyinae
   Antennal scape not as above; usually long and slender, but if short and enlarged at the tip, at least the basal third of the scape is slender ........................................................ Ponerinae
3. Abdominal pedicel consisting of two segments .................... 4
   Abdominal pedicel consisting of one segment ..................... 6
4. Frontal carinae narrow and not expanded laterally so that the antennal insertions are fully exposed when the head is viewed from above ........................... 5
   Frontal carinae expanded laterally so that they partially or wholly cover the antennal insertions when the head is viewed from above ............................................................ Myrmicinae
5. Eyes very large, suboval or reniform, and consisting of several hundred fine ommatidia .................. Pseudomyrminae
   Eyes vestigial or absent; if present consisting of a single ocelluslike structure ............................. Dorylinae

6. Distinctly circular orifice below posterior tip of gaster
   usually surrounded by fringe of tiny hairs ............... Formicinae
   Orifice below posterior tip of gaster is slitlike; the
   hairs, when present, not forming an encircling
   fringe ........................................... Dolichoderinae

B. Workers of the genera of New World dorylines
   (Borgmeier, 1955)

1. Antennae 11- or 12-segmented; workers polymorphic;
   eyes nearly always present ..................................... 2
2. Petiole 1-segmented; mandibles long, slender, and
   falcate; eyes vestigial; tarsal claws with median
   tooth; hypogaeic ........................ Genus *Cheliomyrmex* Mayr
   Petiole 2-segmented ............................................ 3
3. Second segment of antennal funiculus at least twice as
   long as the first; major workers with large heads
   and (except in *Eciton rapax*) tong-shaped
   mandibles; workers 8 to 13 mm long,
   according to species ........................ Genus *Eciton* Latreille
   Second segment of funiculus less than twice as long
   as the first; major workers not as described
   above .......................................................... 4
4. Tarsal claws toothed .......................................... 5
   Tarsal claws not toothed; largest workers, according to
   species, 5 to 7 mm long .............. Genus *Neivamyrmex* Borgmeier
5. Scapes and funiculae of antennae strongly thickened
   (about one-third as wide as long); stings weak;
   hypogaeic ....................... Genus *Nomamyrmex* Borgmeier
   Scapes and funiculae thin and short; stings well
   developed; hypogaeic ..................... Genus *Labidus* Jurine

C. Workers of genera and subgenera of Old World dorylines
   (Wheeler, 1922)

1. Pedicel of two segments; workers monomorphic;
   without eyes ........................... Genus *Aenictus* Shuckard
   Pedicel of two segments;
   workers polymorphic; without
   eyes ........................ Genus *Dorylus* Fabricius ........ 2
2. Antennae 12-jointed in the soldier and in the large
   and medium size worker (Indo-
   Malayan) ........................ Subgenus *Dichthadia* Gerstaecker
   Antennae 9-jointed (Ethiopian, North Africa,
   Indo-Malayan, New Guinea) ............ Subgenus *Alaopone* Emery
   Antennae 11- or 10-jointed ..................................... 3

3. Pygidium with a semicircular impression, the margins of which are sharp; antennae 11-jointed ................................................. 4
   The impressed area of the pygidium without distinct margins ............................................................ 5
4. Antennae short and thick; all the joints of the funiculus, except the last, much wider than long (Ethiopian) ............................. Subgenus *Dorylus*
   Antennae elongate; at least some of the joints of the funiculus longer than wide (Ethiopian ... Subgenus *Anomma* Shuckard
5. Subapical tooth of mandibles simple; antennae 11-jointed; worker major 13 mm long (same distribution as the genus) .... .... Subgenus *Typhlopone* Westwood
   Subapical tooth of mandibles double or truncate; worker major 8 mm long (Ethiopian) ......Subgenus *Rhogmus* Shuckard

# 2

# The Colony and Its Members

Like all ants, dorylines are social. They live in colonies that are unitary groups made up of interdependent individuals. Although wasps and bees retain many solitary forms, the ants are group-specialized, varying more widely among themselves in pattern and complexity of group function than any other social insect. At one extreme, *Formica* and many other formicines combine complex forms of colony organization with degrees of individual independence in action so great as to enable workers to forage alone at long distances from the nest. At the other, the dorylines, although also high in colony integration, are very low in individual independence of function.

Colony organization is so strong in the army ants that group action and interdependence stand out in everything they do. Clearly, in their evolution, factors aiding collective action had the highest priority in natural selection. No members of the colony except the males have any existence apart from the colony.

Each doryline colony has a single queen, a large brood of developing individuals, and a great population of workers. For short times, broods of males and young queens are also present. There is also a diversified number of beetles and other insects, the dorylophiles—literally, "doryline lovers"—variously associated with the ants. Ant birds, flies, lizards, and other animals, attracted consistently or occasionally to the forays of the swarm raiders, are in a sense marginal associates of the colony.

The firm unity of each army ant colony is shown by its separation from other colonies of the same species, never mixing its members with others no matter how frequent or complex their meetings in raids or emigrations may be. Colonies of the same species present similar odors: In *Aenictus* they smell much alike to humans; in *Eciton* they seem to differ more. The colony odor of *E. hamatum* smells "meaty" to some observers, to others like potato blossoms; *E. burchelli* smells musky and somewhat fetid. Even so, the workers normally distinguish their own colony odor readily from those of other colonies of their own and of other species. This is shown in laboratory tests, in which workers readily join their nestmates but flee from, stand off from, or attack members of other colonies. Odor is important as

TABLE 2.1.
*Approximate population ranges of colonies*

| Species | Estimated range of worker populations | Reference |
|---|---|---|
| *Dorylus (Anomma) wilverthi* | 10,000,000 to 20,000,000 | Raignier and Van Boven, 1955 |
| *Eciton burchelli* | 500,000 to 2,000,000 | Schneirla, 1957b; Rettenmeyer, 1963 |
| *Eciton hamatum* | 150,000 to 500,000 | Rettenmeyer, 1963 |
| *Neivamyrmex nigrescens* | 80,000 to 140,000 | Schneirla, 1958 |
| *Aenictus laeviceps* | 60,000 to 110,000 | Schneirla and Reyes, 1966 |

the ants react similarly to pieces of blotting paper on which workers of the respective groups have been active. (A test of reactions to odor is illustrated in Fig. 4.1.)

Normal conditions for a unified, organized colony are those in which first of all the colony queen is regularly present. Observations and tests show that in species of *Eciton, Neivamyrmex,* and *Aenictus* the key factor for colony unity is contributed by the functional queen. The colony remains indissoluble and apart from others of its species as long as the queen is present, but not after she has been removed. Tests show that the essential element is her odor (Chapter 8).

The populations of doryline colonies, mainly their worker personnel, are among the largest in all ants and nearly the largest in social insects, excepting some tropical termites (Emerson, 1939a; Allee et al., 1949). General estimates can be exaggerated in either direction according to the observer's perceptual bias. My own early figures (Schneirla, 1934) were much too low as became clear later when captured sections estimated as given fractions of a colony under study were actually counted.[1] Reliable information on populations is still difficult to obtain as each of the available census methods has its own shortcomings (Rettenmeyer, 1963). Table 2.1 offers some reasonable approximations of the ranges of colony populations in five species of the four doryline genera represented in Figure 2.1. The figures in Table 2.1 show that *Anomma* and *Aenictus* are

---

[1] An estimated fraction of the original colony is spread out in a large flat pan filled to even depth with alcohol, then is divided into two equal parts by a thin partition. By repeating this procedure, eight equivalent parts are obtained. Parts selected randomly for counting and sizing can be compared for reliability. Methods of field assay by marking and recapture, developed for vertebrates and already used for certain insects, are promising (Brian, 1965); also sections of an emigration column can be captured in a long box, counted, and by proportion used to estimate colony populations (Raignier and Van Boven, 1955).

the genera with the largest and the smallest colony populations, respectively, in all investigated army ants.

Doryline colony populations are the largest in any of the subfamilies of ants. Colonies in the ponerines, for example, range from a few individuals to a few hundred or at the most one or two thousand (Brian, 1965), and in *Formica* and other formicine species (Pickles, 1937; Talbot, 1948) from a few hundred or thousand to 300,000. Colony populations in *Dorylus,* which include those of *Anomma,* may be the largest in all social insects.[2]

The dorylines have two types of females; the queen or reproductive female and the workers, the neuter females. From one genus to another, both workers and queens are remarkably different, particularly in size, as Figures 2.1, 2.2, and 8.1 show. The worker populations, except in *Aenictus,* are polymorphic, being a graded series differing both in overall size and form (e.g., structural proportions) from the large to the small extreme. By contrast, workers of *Aenictus* are quasimonomorphic, i.e., closely similar both in size and in form.

The English biologist Julian Huxley (1927, 1932) demonstrated that worker polymorphism in ants can be related to allometric growth—i.e., regular differential growth in body parts of the same individual—as studied in other animals,[3] suggesting also that the presence in a colony of workers graded in these respects can be most simply explained on the basis of variable nutrition (Chapter 6). E. O. Wilson (1953) confirmed allometric variation in detail for ants, finding it a primitive adaptive character in the dorylines but virtually absent in the ponerines. It is strongly developed in *Dorylus* (Raignier and Van Boven, 1955; Hollingsworth, 1960) and *Eciton,* much less marked in *Neivamyrmex,* and almost absent in *Aenictus.*

Figures 2.2 and 2.3 illustrate other important characteristics in the worker populations of different army ants concerning the distribution and frequency of worker size types from smallest to largest. These distributions are all unimodal (i.e., they have single peaks), and workers

---

[2] By contrast, legionary ants probably remotely related to the dorylines differ notably from the latter in certain aspects of their populations. Although as a rule colonies of legionary species in such ponerine genera as *Leptogenys* have more workers than those of closely related nonlegionary species (Wilson, 1958a), their numbers—ranging from a few hundred to at best two or three thousand—are far less than those of any doryline, even *Aenictus.* Also although some of them, like *Leptogenys,* have ergatoid females, others, e.g., *Simopelta* (Borgmeier, 1950; Gotwald and Brown, 1966) have ergatoids in some species but dichthadiigynes in others.

[3] In the allometric formula, $y = bx^k$, which Huxley (1927, 1932) derived from measurements of adult series of two ants (one of them *Dorylus (Anomma) nigricans*), $y$ is the size of the part compared (e.g., head width across eyes), $x$ is a general body dimension (e.g., thorax length), and $b$ and $k$ are constants (Wilson, 1953). Further investigation (Schneirla et al., 1968) suggests that the two constants have a different relevance for developmental stages from that in adult allometry (Chapter 6).

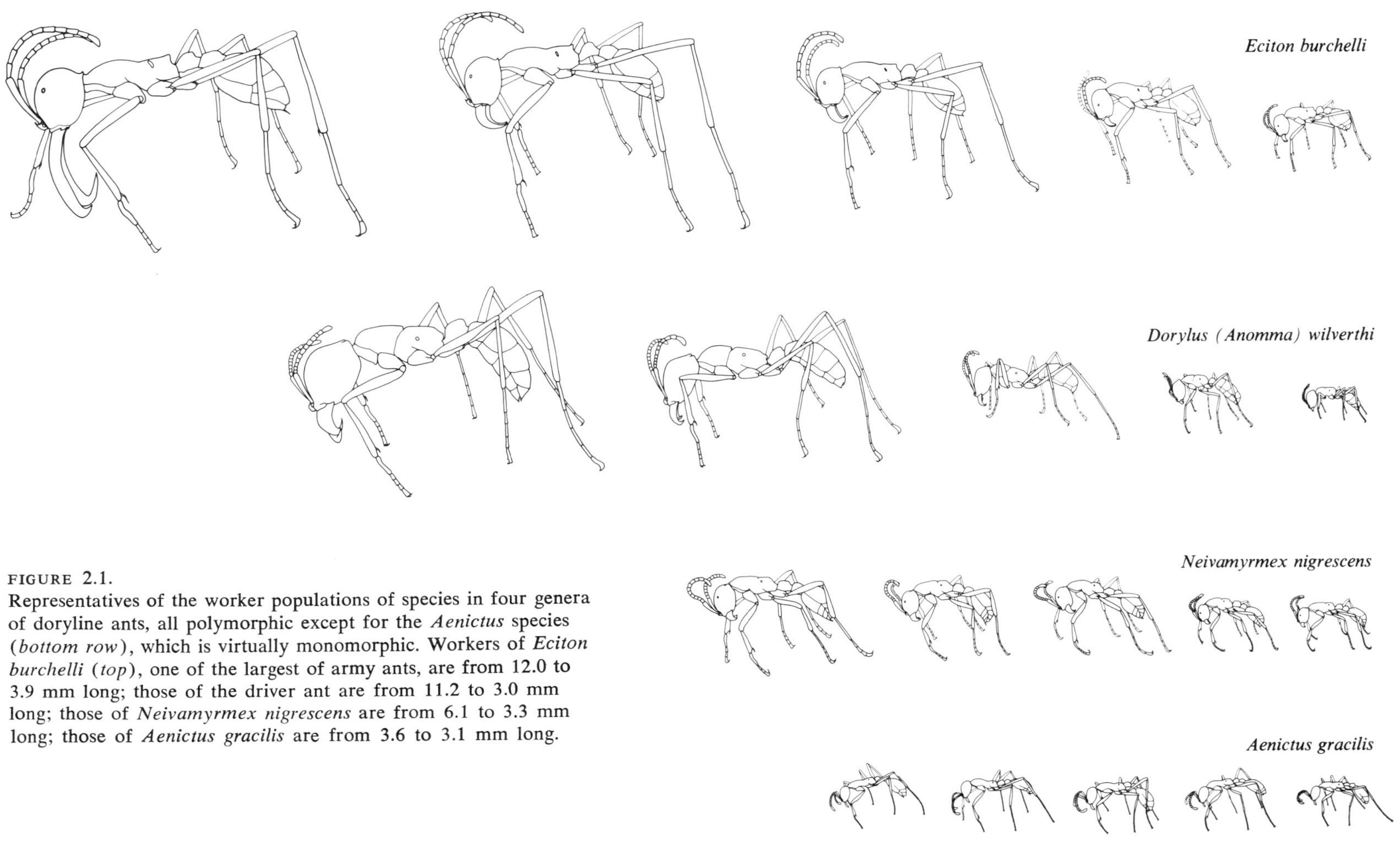

FIGURE 2.1.
Representatives of the worker populations of species in four genera of doryline ants, all polymorphic except for the *Aenictus* species (*bottom row*), which is virtually monomorphic. Workers of *Eciton burchelli* (*top*), one of the largest of army ants, are from 12.0 to 3.9 mm long; those of the driver ant are from 11.2 to 3.0 mm long; those of *Neivamyrmex nigrescens* are from 6.1 to 3.3 mm long; those of *Aenictus gracilis* are from 3.6 to 3.1 mm long.

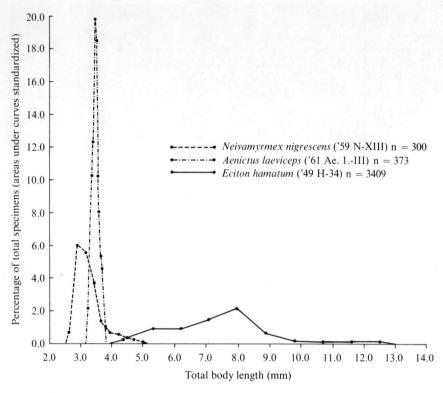

FIGURE 2.2.
Graph comparing the relative frequency distributions of pupal samples from colonies representing three species of doryline ants. The range of *Neivamyrmex nigrescens* is 2.4 mm; for *Aenictus laeviceps* it is 0.6 mm; and for *Eciton hamatum* it is 8.1 mm. (Topoff, in press.)

smaller than the average are the most numerous. The range of worker size types, greatest in *Eciton* and moderate in *Neivamyrmex,* is minimal in *Aenictus*. The significance of allometry and of individual size differences for colony function is suggested elsewhere in this book, primarily in Chapters 6 and 13.

Colony functions (i.e., such normal operations as emigration) in the dorylines, as in other social insects, are based on the properties of individuals. The workers, who carry out all tasks except reproduction, make up the bulk of the colony. As an important principle of doryline adaptation, the workers, although expendable as units, are indispensible as a class. Their behavior as individuals, in colonies of species throughout the subfamily, is much more uniform and stereotyped than in most other ants. They are easily recognized by their actions. All army ants run in a typically

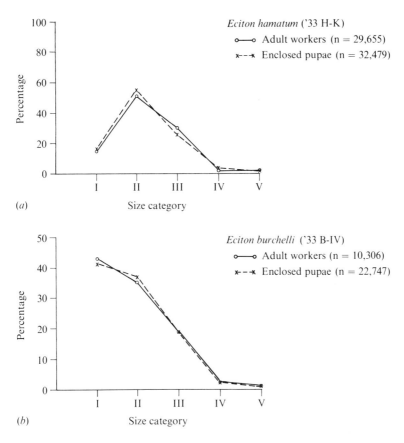

FIGURE 2.3.
Polymorphic size-frequency distributions in *Eciton hamatum* and *Eciton burchelli*. (*a*) Large samples of the populations of adult workers and an all-worker brood from the same colony of *Eciton hamatum,* compared as to distributions of size types from the minor workers (I) to the major workers (V). The distributions are not reliably different.
(*b*) Corresponding samples from a colony of *Eciton burchelli*.

regular gait but in a meandering path, usually with antennae bent downward and held close to the trail. As tests show, outside the bivouac they are odor-track bound, proving themselves one of the foremost insects in the making and following of scent trails. Army ant behavior in its various forms seems to depend especially upon the intimate association of contact and odor stimuli (Forel, 1899). In fact, the workers in their orientation depend mainly on contact odor and very little on light sensitivity. This contrast, which holds even among compared species of the same doryline genus, conforms with the general rule that among ants there is an inverse relation between size of eye and reliance on odor trials.[4] The workers in most of their behavior react to nearby objects (nestmates, brood, booty, or substrate) in terms of their odorous and tactile effects, as Figure 2.4 indicates for running in a column.[5]

The elongate, more or less cylindrical shape of the worker's body facilitates the characteristic doryline way of life which involves much maneuvering in narrow spaces by groups. This aspect of body form generally increases as the degree of species adaptation to subsurface conditions rises. Also important is the toughness of the body covering, or exoskeleton, which forms a heavier armor in surface-adapted than in subterranean species.

The army ant worker's antennae serve prominently in all of her operations. Upon entering fresh ground, a raiding worker vibrates her antennae rapidly and moves them from side to side, palpating objects and ground thoroughly. Such actions contrast with the slow play of her downward-curved antennae as she follows stretches of trail. Watching these ants run in a column, one is impressed to note how efficiently the proximal senses (contact, odor, and movement sensitivity) control behavior. So regularly and with such complexity does this sensory pattern dominate action that each individual seems to maintain her place with reference to others in the column as though she were enclosed within a contact-odor sheath (Fig. 2.4c). In raiding, which on the other hand is less stereotyped, we see versatile forms of contact-odor ranging from darting into insect burrows to subduing resistant prey or dragging heavy booty objects.

A major factor in army ant behavior is a brain that, although low in resources for association, serves well to bring stimuli into direct control of action (Schneirla, 1946; Vowles, 1958, 1961). Accordingly, army ants are particularly responsive to present, persistent stimuli as in carrying brood or booty for long distances or in holding positions in the bivouac structure; they show a minimum of the new, individual behavior so prominent in the formicine and other higher ants (Chapter 13). Hence the doryline worker's brain must be much smaller in relation to body size than that, for example, of *Formica* (Brun, 1959).

Although often spoken of as blind, all army ants are sensitive to light,

---

[4] It is notable that the dorylines, nearly the most dependent of all ants on odor-trails, have lost their eyes altogether (as in *Aenictus* and *Dorylus*) or have had them reduced to tiny ocelluslike structures.

[5] Forel (1899), having found a colony of *Neivamyrmex carolinense* in North Carolina, scattered some workers and brood about and was amazed to see how promptly the ants formed columns and carried off their young into "a cavity suited to their needs." "As if by word of command the workers follow and understand each other, and in very little time everything is safe." Organized movement of army ants in interlacing columns on a contact-odor basis is well illustrated by studies of circular columns (Schneirla, 1944c) and of swarm raiding (Chapter 12).

and many of them have eyes. New World species nearly all have two tiny degenerate compound eyes called "lateral ocelli," equipped with single lenses (Werringloer, 1932). Even the workers and queens of Old World species, which lack eyes, have a subdermal sensitivity to light (Chapter 3). Light sensitivity, although low in controlling the army ant's orientation, plays roles in arousing group movements that vary according to the species' adjustments to surface or to subsurface conditions.

Among the more prominent anatomical adaptations, the mandibles of doryline workers differ according to species and size type (Fig. 2.1). Those of most driver ants are well suited for cutting and shearing flesh, those of most New World forms for holding, squeezing, and macerating tissue. Readiness to bite and hold on typifies all army ants but differs according to species strength, speed of action, and other factors. The tongue is generally well equipped for rasping, scraping and, with the jaws, for malaxating booty. Its accessories, the labial and maxillary palps (Borgmeier, 1957) are mobile and richly supplied with gustatory and other sensory endings. These structures, along with reactivity to motion, to odors, and general characteristics of action, all influence species differences in the type of booty taken.

In the army ants, as in all insects, the thorax (i.e., alitrunk) and abdomen (i.e., gaster) are separated by a constriction. These main body sections, in workers and queens, are connected by one or two small nodelike segments, depending on the genus. For the workers, having two connecting segments in the pedicel instead of just one clearly promotes body flexibility in such actions as stinging (Pullen, 1963). This is an important adaptation in workers that must subdue strong, quickly moving prey aboveground before it escapes, and the surface-active species of *Aenictus* and *Eciton* possess both the double segment and a well-developed sting. But *Anomma* and *Dorylus,* which are more hypogaeic and attack prey as a rule in far greater numbers, have one connecting segment and weakly developed stings. Mobility of the worker's small abdomen contributes not only to stinging but also to such other adaptive actions as carrying larvae or booty or laying trails, in which flexibility of the body is advantageous.

The abdomens of doryline workers, nearly the smallest in all ants in relation to body size, expand and contract so slightly as to suggest that the workers feed little at any one time. By contrast, in the workers of many other ants, as in *Formica,* the size of the abdomen often changes according to whether the ant is satiated or unfed. The abdomen of the doryline queen differs radically from that of the workers by periodically undergoing extreme changes in size during reproduction (Chapter 8).

Workers of *Eciton* and other surface-active species have long and strong legs that enable them to follow trails at a rapid, steady pace while straddling larvae or booty held beneath their bodies. By contrast, work-

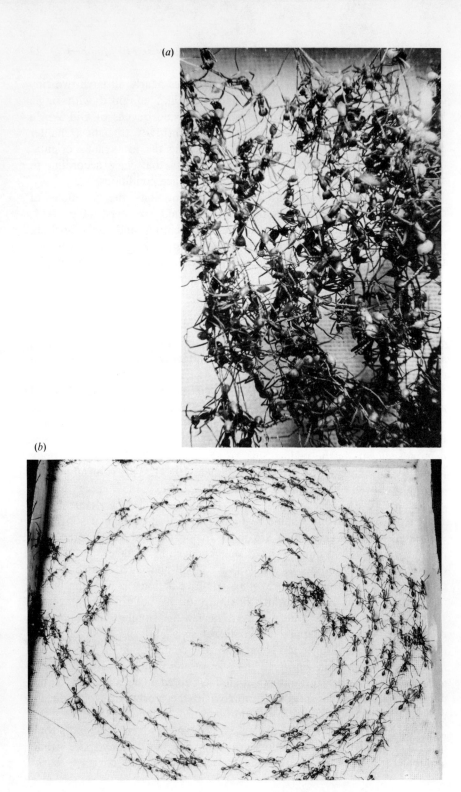

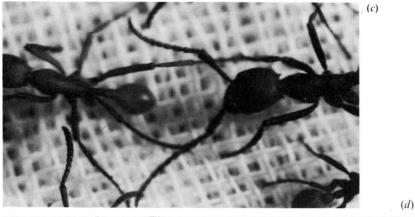

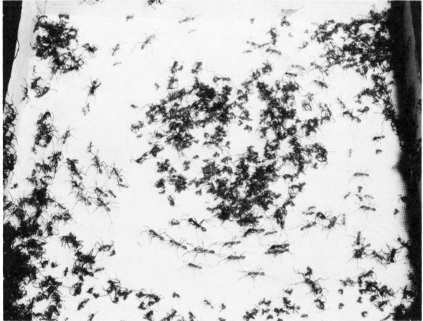

FIGURE 2.4.
Test of species interactions based on odor. (a) Workers of *Eciton burchelli* clustered in fabric hanging next to the front glass wall of a tall, narrow cell. The back partition, of cheesecloth, separates them from a similar cell in which hang workers of *Eciton hamatum*. For several days, these groups are exposed to each other's odor. (b) The workers in each cell are then released into a separate arena (through a plastic tube). Within an hour the workers of *Eciton hamatum* had formed this circular column. (c) Workers of *Eciton burchelli* then admitted to this arena, joined the circular mill of *Eciton hamatum*, running smoothly in it for about 20 minutes. (d) Then, as workers of the two species became disturbed (presumably by the other's odor), fighting began and the mill broke up. Preexposed workers of the two species are longer together, without fighting, than are unexposed control groups. (See Plate II, following page 138.)

ers of hypogaeic species have shorter legs which, with their slender bodies, aid movements along narrow galleries and turns within tight spaces. In species adapted to surface life, the worker's legs end in strong tarsal hooks (Fig. 3.6a), which grip the ground firmly in running, in pulling against strong and agile prey, and in other surface-essential actions. In *Eciton* long legs with stout tarsal hooks enable workers to construct their characteristic chains and clusters of interlocked bodies (Fig. 3.6). Also important in such actions, although not obvious, are sensory effects from stretching the legs and body which by reflex immobilize workers hooked in place and thereby enhance the clustering of bodies essential to nesting and many other activities of these ants.

Army ants also have physiological assets important for their normal behavior. Optimal (i.e., propitious) conditions of temperature and moisture, for example, underlie the speed and the persistence of activity that characterize army ant group operations at their peak. Under these conditions, these ants are speedy in running and in responding to other workers and to booty; they are persistent in overcoming prey, in carrying difficult pieces to the bivouac, and in forming and holding clusters. The properties of sensitivity, nervous conduction, action, and secretion thus promote rapid, strong actions ranging from running in a column or mass to attending the brood or queen. Workers of the surface-active dorylines are as a rule most "alert" and efficient in action at temperatures around 25°C and at humidities above 75% relative humidity. At much lower or high temperatures and in extreme dryness, actions are more variable. When atmospheric humidity falls to a point near 50% relative humidity, the workers die, beginning with the smallest (Schneirla et al., 1954). Raiding workers of *Eciton* tend to be less vigorous on dark, cool days than on bright, warm days, and in operations outside the nest they usually slow down through midday in what we call the "siesta reaction" (Chapter 4).

Colony activities and daily schedules vary in all army ants in ways clearly related to species-typical physiological properties. The optima of hypogaeic species, for example, seem to involve lower temperatures and higher humidities than those of epigaeic species. In all dorylines these properties vary impressively according to brood condition through entire colony populations, in ways important for general behavior (Chapter 7).

The army ant queen is a unique individual. She is much larger than the workers, and although like them she is wingless throughout life, she differs radically from them both in general appearance and in details of structure (Fig. 2.5). She is very different from them also in behavior and in being the sole female reproductive agent of her colony. Doryline queens are outstanding among insects in the great quantities of eggs they lay and in their regularity in producing them. Moreover, there are lifelong features

FIGURE 2.5.
Queen and worker types (with larvae) of the army ant *Eciton burchelli* in a laboratory nest.

of reproduction which young queens begin soon after they emerge from the pupa (Chapters 8 and 14).

Normally the queen is held in the midst of her colony from the time her life begins until she dies or is superseded by one of her own daughters. If she is lost, the colony disappears as an independent community, for it cannot replace her (Chapters 8 and 9). Her odor strongly attracts the workers to her (Schneirla et al., 1966) and serves as a mainstay of unity in her colony (Schneirla, 1949; Schneirla and Brown, 1950). The importance of the queen to her colony is discussed in greater detail in Chapter 8.

The males of army ants are much larger than the workers and are so different in appearance that at first sight they do not seem to belong in the same group (Fig. 1.2).[6] The strong resemblance all doryline males bear to wasps may be a clue to the ancestry of these ants. The males appear in the colonies under exceptional conditions, developing in sexual broods with the potential queens (Chapter 9), which they greatly outnumber. They are the only winged dorylines. Although they attract the workers

---

[6] The males of driver ants (Fig. 1.2), called "sausage flies" in certain parts of Africa, have the distinction of being the largest ants in the world.

strongly, the males take little part in the regular colony life except in feeding, clustering, and running in the emigrations. They seem almost to be interlopers as their stay in the colonies is relatively brief—just a few weeks until they are mature and able to fly—and their departure sudden. Their role in the colony, not well known as yet, is discussed in Chapters 9, 10, and 11.

The males have relatively small heads with correspondingly small brains. They have magnificent visual receptors, however, in three ocelli (simple eyes) on top of the head and two large compound eyes on the sides. There is no doubt that vision plays a role in the postflight behavior of surviving males although its importance may differ in epigaeic and hypogaeic species (Forbes et al., in prep.).

In most army ants, the male's sickle-shaped mandibles resemble those of the queen more than those of workers. His great humped thorax, packed with wing musculature, bears two pairs of wings, which are lacking in the females. In all doryline genera, the male's thorax and abdomen are linked by a single node—as against one or two in the females, according to the genus. The male's abdomen, roughly the size of the queen's in its most reduced condition, contains his testes and sperm-packed internal reproductive system, as well as, posteriorly, an impregnation apparatus that is one of the most distinctive accessories (Forbes and Do-Van-Quy, 1965). But relatively few of the males seem to survive long enough to use these organs (Chapter 11).

Most of the successive generations of young produced by an army ant colony are made up of workers. These all-worker broods are large, and in *Eciton, Neivamyrmex,* and *Aenictus* they appear at regular intervals as distinct new additions to the colony population (Chapter 6). In *Anomma* and others, however, they are less distinct as generations and appear at less regular intervals. Separation of the all-worker and the sexual broods is bound up with processes controlling sex and "caste" that normally prevail in the dorylines with great precision. Studies of normal colony functions (Chapter 6) clarify the conditions that determine all-worker broods and also provide a basis for understanding how the exceptional sexual broods arise (Chapter 9).

Any census of a doryline colony should include the highly diversified group of dorylophiles, the other insects that live with the ants. These include numerous species of beetles, flies, and others, all numerous in swarm raiding and in hypogaeic species. Best adjusted in the colony are those insects that live in a relatively mutualistic relationship with the ants, for example, beetles of staphylinid and histerid species (Wheeler, 1928; Kistner, 1958; Akre and Rettenmeyer, 1966). Many of these beetles feed with the workers, groom them and are groomed in turn, run in the emigrations, often appear in the raids, and live in the colony almost as members

(Fig. 2.6). This assortment of arthropods found with the army ants includes beetles and other insects that are merely "tolerated" and still others given to outright parasitism against the ants or their brood. Probably most of them poach on the booty. So well fixed are their affiliations with the dorylines that many of them are found only in the colonies of their respective host ant species (Seevers, 1959, 1965).

The principal subpopulations in a doryline colony are the queen and her guard group, the worker population, and the brood. In all army ants, normal changes in colony condition and function, repeated regularly or irregularly according to the species, involve all of these groups in progressively modified, reversible relationships. These changes—the phenomena of cyclic behavior and function (Chapter 7)—are unique to the army ants and are crucial to the normal life of their colonies. To function normally, an army ant colony requires its queen, successive generations of developing young, and of course a population of workers. The workers in any normally functioning army ant colony are strongly attracted both to their queen and to the brood although these affiliations vary somewhat according to changing conditions of both queen and brood during the cycle (Chapter 7). Attachments to the queen are always relatively strong as is indicated by the workers' invariable reacceptance of her after a test removal whereas when queens of foreign colonies are introduced, they are killed. But when a sexual brood is produced, odor affiliations of workers to the parent queen are weakened and may even be broken completely (Chapter 10).

Normally the subgroups are held together by strong linkages or "social bonds" (Schneirla and Rosenblatt, 1961)—dependent on properties of the queen, the brood, the worker population, and the bivouac—that keep the colony functioning well as a unit. The unifying effect of the queen's odor is vital; her reproductive properties also play a major role, as do stimulative effects of the brood. The worker population is the matrix through which the colony integrates and carries out its normal cyclic functions (Chapters 7 and 13).

Odor plays a crucial role in the colony life of the dorylines. Colonies of the same species normally never mix their members by virtue of the distinctive queen odor of each colony (Chapter 8). But workers that have been kept away from their colonies do so mix, which indicates that normal nestmate affiliations depend upon frequent odor reinforcements gained in the course of regular colony life. The potency of odor similarities is suggested by the test depicted in Figure 2.4, in which even workers of different species mix without conflict for a time, provided they have been preexposed to each other's odors.

Doryline broods attract the workers and especially in the larval stage strongly arouse and activate them through a variety of stimuli. These

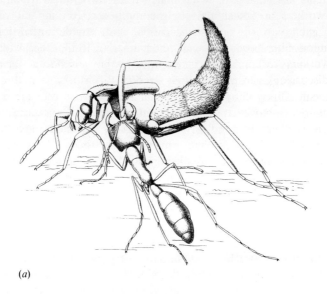

(a)

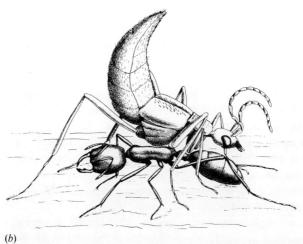

(b)

FIGURE 2.6.
The staphylinid beetle, *Smectonia gridelli* Patr. lives in colonies of the driver ant *Dorylus* (*Anomma*) *nigricans* of Ethiopia. It feeds on the ants' brood and frequently engages in reciprocal stimulative relations with the ants, as observed by the Italian myrmecologist Savero Patrizi (1948). In (*a*) an ant is the active agent; in (*b*) evidently the beetle.

excitatory effects are at a peak when the brood is in the larval stage, at which time workers are most actively grooming and feeding the young brood. This excitation of workers depends upon tactile and odorous effects and upon pheromones ("social hormones") from the brood, to which the workers respond vigorously in a variety of ways. Wheeler (1928), in discussing such occurrences, stressed the aspect of mutualism in them by means of his concept of "trophallaxis" (i.e., food exchange). It now seems clear, however, that the complex and changing relationships between workers and brood that we have described are better called "communicative" rather than simply "mutualistic." Furthermore, because these relationships of workers and brood entail a bilateral arousal basic to colony function, they are best termed "reciprocal stimulation."

In the communication system of a nomadic colony, there operate complex processes in which the members of one subgroup affect members of others by their behavior and secretions and in their turn are influenced by the others. A liaison among workers prevails in raiding, for example, even when they are widely distributed in space. Within a trail end group, the workers are integrated in advances or retreats, in capturing booty, and in returning it to caches or to the base. Within the foray as a whole, these communicative processes are sustained by frequent direct stimulative exchanges between advance foragers and those on the base trails. Through their comings and goings, the latter in their turn serve as links with the workers, brood, and queen in the bivouac. Communicative processes that support and steady the operations of raiding and other colony functions ultimately are grounded in ongoing worker relations with brood and queen. This is shown by the inevitable dissolution of a colony that has lost its queen; it is shown in normal colonies during the statary phase of the cycle, when a low level of stimulation from the brood is marked by a great reduction of such group actions as raiding.

In species of *Eciton, Neivamyrmex,* and *Aenictus* that we studied, and probably also in other army ants, there is a subgroup of workers we call the "queen's guard," usually clustered around the queen in the bivouac and prominent in her retinue in emigrations. These workers may be among the oldest in the colony, as is indicated by their color which in *E. hamatum* and *E. burchelli* is darker than that of other workers. When in a colony division (Chapter 10) a queen is superseded, the guard workers remain with her to the last. Presumably these are the workers normally most involved in grooming and feeding the queen. Their role in communicative processes is revealed in queen-removal tests (Chapter 8).

The conditions of ancestral predoryline ants may be approximated by *Leptogenys* spp. and some of the other existing ponerine legionnaires in which colony populations are small, workers are monomorphic, and two or more gynecoid (egg-laying worker) types are present. Significantly, the first two of these conditions are found in *Aenictus,* the genus that Emery (1920) considered archaic. Figures 2.1 and 2.2 contrast a worker series of *Aenictus* with those of other distinctly polymorphic dorylines. From the simpler basis of monomorphic workers in small colonies, polymorphic worker differences in colonies of increasing size may have evolved in conjunction with specialized patterns of colony function like those in existing *Eciton.*

Life in doryline colonies entails many worker tasks, ranging from operations of raiding and of maintaining bivouacs to those of feeding and transporting the brood. In the monomorphic *Aenictus,* with limited exceptions, these tasks probably are carried out equivalently by all of the workers at various times. As one exception, workers as they grow older may normally shift from one type of task to another, as when they enter the queen's guard. Relative differences in metabolic level may account for variations in worker function even in *Aenictus;* this is still untested, but in any case, worker populations of *Aenictus,* like those of other group A dorylines, change regularly from a busy daily schedule in each nomadic phase to a much reduced schedule in the succeeding statary phase. This difference is in fact more pronounced in *Aenictus* than in other army ants.

Most of the other dorylines, which are polymorphic (cf. Figs. 2.1 and 2.2), have worker populations in their colonies ranging smoothly in size and structure from the major types to the minor types. In this series, there are gradual variations in ratios of body length to head width and other local dimensions and in qualitative aspects of structure as well. The most prominent of these in *Eciton*—in all but one species [7]—is the contrast between the large major workers with their huge heads and long, ice-tong-shaped mandibles, and the intermediate and small workers with their relatively small heads and short mandibles.

To find how well polymorphic differences are maintained in a colony, it is of interest to examine how far the distribution of individuals in a single developing generation may resemble that of the adult population. Accordingly, we compared in this respect a large sample of adult workers with a large sample of nearly mature pupal all-worker brood from the same colony of *Eciton hamatum.* Specimens from each of these samples were sorted into five size groups representing approximately equal steps between the small and the large extremes. As Figure 2.3*a* shows, the adult and the brood distributions were closely similar; each was unimodal,

---

[7] The exception is *Eciton rapax,* the species with the largest workers and one of the most hypogaeic in the genus.

with the small intermediate size types most frequent, the large intermediates and the minor workers next, and the major workers the least frequent.

The data in Figure 2.3b, obtained by sorting equivalently two large samples from the same colony of *Eciton burchelli*, one of adult workers and the other of mature all-worker pupae, point to closely similar distributions in the adult in the brood populations of that colony. The frequencies of size types in this case are much like those in *E. hamatum*, except that in *E. burchelli* a larger part of the distribution lies on the side of the small individuals.

The close similarities of adult-worker and brood populations in these colonies, each judged normal for its species, indicate that usually highly stable conditions hold for the distribution of the worker size types (or "castes") in these army ants and lead to similar proportions of these types in brood and adult populations. The major types, for example, make up less than 2% of the whole in both the brood and adult populations; also, individuals of small size types are most numerous in both of these. Comparable results are to be expected for comparisons of adults and brood in colonies of other polymorphic dorylines.

What determines frequencies of worker types in a colony is a complex problem. One approach is to investigate how differences among the workers in size, structure, and physiology may affect the typical colony functions of the respective types.

In polymorphic army ants the intermediate workers seem to perform nearly all types of tasks. They are versatile trail layers, and because of their size and strength combined with agility, in raids, these workers are better adapted to rout out, attack, and carry off a wider variety of prey than are the largest and smallest workers. The small workers, on the other hand, are less involved in the rough operations of combat and portage; in attacking nests, however, they can slip past defenders into brood chambers and other recesses inaccessible to their larger sisters. In dealing with booty and other objects, the workers handle most frequently those objects that correspond to their respective sizes,[8] though often they work in groups that bundle along booty much too large for any one of them to manage alone (Fig. 4.10).

---

[8] F. E. Lutz (1929), then curator of entomology at the American Museum of Natural History, captured large and small workers from a column of the leaf-cutter ant, *Atta sexdens*, together with the leaf segments they were carrying from tree to nest, then weighed each ant and its burden. A close relationship between size of ant and size of burden was indicated by a correlation coefficient of +0.8 between the two series. He concluded that this result shows "that the small ants cannot carry excessively large burdens, and soon drop them when they are picked up . . . although large ants often carry burdens which are excessively small." In army ants, a comparable relationship may be predicted between the size of individual ants and the burdens they carry (Fig. 5.4).

The largest workers of most polymorphic dorylines (excepting the majors of *Eciton*) commonly take part in the raids in many ways, ranging from attacking prey to transporting booty. These workers are the ones that tear down the largest victims and the hardest pieces of booty that are impossible for smaller workers to manage. But the major workers of *Eciton* are automatically excluded from nearly all transport work since with their great double-fishhook jaws they cannot pick up, hold, or release objects. The submajors, in contrast, with their shorter, more workable mandibles carry and manipulate a variety of objects. Their size and strength are great assets to the colony in emigrations when these ants are transporting the nearly mature larvae of sexual broods (Fig. 9.4a), burdens so bulky that even the largest intermediate workers cannot handle them.

Although all workers deal with the brood in a variety of ways, it is their positions in the polymorphic series that mainly governs what they do. In the emigrations, one sees increasing signs of a correspondence between size of load and size of carrier, the more the larval brood grows. In the bivouac, the smallest workers generally handle and feed the small larvae of a young brood, with their larger colony mates involved more and more as the larvae become larger. Differences can arise, thereby, in feeding and in grooming that are related to the sizes of both larvae and of nurse workers. These relationships, resulting from polymorphic specializations in the brood (Chapter 6), may be largely responsible for the similar size-type distributions we have found between adult worker and all-worker brood populations.

There are many other differences in the colony functions of workers related to their size types. In forming bivouacs, for example, the largest workers—because their great size and strength, long legs, and large tarsal hooks enable them to hold fast against great structural tensions—serve as specialized foundation units. These tensions are too great for the smallest workers, which, on the other hand, contribute to structures already well begun by taking interstitial positions. Thus, in the cool of the night, they increase the insulating properties of the nest by resting in gaps left in the walls of a nearly constructed bivouac.

There are differences in levels of physiological action (i.e., metabolic rate) among workers of different size types in polymorphic populations, important both to function and to resistance against stress and hardship. Tests with *Neivamyrmex nigrescens* show that the smaller workers use more oxygen in proportion to total body weight than the larger workers (Topoff, in preparation); also, in tests of *Eciton,* with increasing atmospheric dryness, the smallest workers die first, the intermediates next, and the major workers last (Schneirla et al., 1954). Presumably related to these physiological differences are characteristic individual differences in reactions of attack, defense, or flight. In general, when under a similar condition of intense disturbance, the major workers of all army ants attack readily, the intermediates next, and the minors least readily, with their readiness to retreat in reverse order. Consequently, an important role of the majors and larger intermediates may be called one of heavy defense as they usually rush to the scene of a disturbance ready to bite and sting. This accounts for their gathering in rows bordering a column under conditions of high excitement and moving off when the disturbance subsides. Thus major workers of *Eciton,* although relatively few in the general population, are common in the queen's entourage in emigrations. Keyed high in this situation, they cluster over her at any disturbance, thereby becoming special defenders of the colony's reproductive assets.

Polymorphic differences within colony populations relate comparably to behavior and function but with distinctive differences among genera and species. In *Eciton,* for example, the majors and other worker types of *E. burchelli* are the quickest of all to attack, with those of *E. hamatum, E. mexicanum,* and *E. dulcius* ranking respectively in a decreasing order after them. In the raids, majors of *E. burchelli* and *Labidus praedator* attack sources of disturbance from which those of *E. dulcius* and of *Nomamyrmex* commonly flee.

Among American dorylines, *Eciton* is one of the most highly polymorphic; hence colonies of its species present great contrasts in function among their workers. *Neivamyrmex* has a lower degree of polymorphism, and correspondingly we find less marked variations in the behavior patterns of its worker size types. Colonies of *Aenictus,* with nearly monomorphic worker populations, show the greatest functional uniformity of any army ant studied. Corresponding differences exist among these genera in the patterns of raiding and other specialized functions (Chapters 4 and 5) as well as in the overall adaptive systems of their colonies (Chapters 12 and 14).

The polymorphic colony populations of *Anomma* and *Dorylus* exhibit smoothly graduated worker series comparable in general to those of *Eciton,* with the smaller size types most frequent and with relatively wide ranges between the majors and minors. Here, much as in *Eciton,* functional differences correspond to differences in size and structure.

Population size appears to be an important condition of species-typical adaptive patterns in army ants (Chapter 14). Significantly, colonies of the hypogaeic *Labidus praedator* and of the even more hypogaeic *Nomamyrmex esenbecki,* judging by the duration of their emigrations (Schneirla, 1957a), may be among the largest in American dorylines. There is no strict correspondence between population size and degree of subterranean adaptation, however, as the distinctly surface-adapted *Eciton burchelli* has much larger colony populations than some of the hypogaeic *Neivamyrmex* species.

Army ants as a group differ greatly from most ants and other social insects both in the magnitudes of their populations and in the factors controlling size and pattern of their populations. Not only do we find that their colony populations are larger on the whole and more precise and regular in the conditions controlling their magnitude and composition but that they *begin* differently. In most other ants, colonies are founded by individual queens in the claustral manner (Wheeler, 1933). Their worker populations begin gradually, then, as in most social insects, build up more rapidly in the course of time, next taper off during a period that varies with species and conditions, and finally decline to extinction (Bodenheimer, 1937; Brian, 1965). The army ants, in contrast, form new colonies through the division of parent colonies, as do many honeybees. New colonies begin with large numbers of mature workers centered on a functional queen but differ from swarming honeybees by taking along large brood populations as well. Thereby, they are, so to speak, off to a running start. Ordinarily, by virtue of the division process, their colonies never die except for catastrophies.

Colonies of army ants do not begin in sheltered nests as do most other ants but, notably in surface-active species, must cope at once with open environmental conditions in nomadic operations. Fortunately, each new queen is able to begin her reproductive functions promptly and fully (Chapters 8 and 9), with large increments assured for the worker population at regular intervals. In *Aenictus,* about 30,000 new workers are added every seven weeks, in *Eciton hamatum,* about 80,000 workers every five weeks. Predatory nomadism is a taxing life, and doryline colonies operate in pace with it.

The status of a colony of army ants, as in other social insects, varies in relation to many factors, including seasonal conditions, predators and food supply, and the queen's age and condition. But their adaptive pattern is a distinctive one as predation exposes them to heavy regular losses in personnel, and colony movements expose them more to drought and other natural hazards than do the more stable living situations of most social insects. If one finds a colony that appears small but seems to have the species-normal proportions of polymorphic size types, it is probably a new daughter colony. If, on the other hand, the discovered colony is small but has an exceptionally large proportion of major workers, it is likely to be an older one that has passed through hardships to which small individuals are most susceptible.

Colonies of polymorphic dorylines have closely similar proportions of worker size types in brood and in adult populations. The reason seems to be that functional and behavioral processes typical of these ants insure readjustment—and a restoration of the normal proportions of worker size types—even to serious deviations through accident or hardship (Chapters 6 and 14).

From the circumstances of army ant life, the realities of population control seem to be rather harsh. There must be a number of drains on the worker personnel of any colony, including especially the heavy losses that it must constantly incur through the great hazards of raiding. The average life expectancy of doryline workers, perhaps no more than a month or two, may be among the shortest in social insects. To counterbalance heavy population drains, however, the great numbers of new worker replacements added to the colony at frequent intervals through new broods must be adequate since most of the colonies in any area seem to keep going fairly well.

But every species has its typical range of population sizes within which colonies can operate; in every area we find large colonies that are clearly thriving, as well as the other extreme of very small colonies that may be near the hypothetical brink.[9] The latter experience, particularly, reminds us of the well-supported generalization that the doryline way of life persistently subordinates individual welfare to group welfare and individual survival to group and species survival.

---

[9] Our surveys indicate that a standard number of colonies is approximated in many natural areas year after year. A survey of Barro Colorado Island by a queen-marking method (Schneirla and Brown, 1950) suggests an average of nine colonies of *Eciton hamatum* per square mile and six colonies of *E. burchelli* per square mile.

# 3
# The Bivouacs

Colonies of most ants make their own nests, settling according to species in places ranging from niches in trees to cavities in the soil. Ant nests, whether fabricated, dug, or taken over from some other creatures, are usually expanded as the colony grows. The same place generally serves the colony as its shelter from an early stage of the colony's life to its end (Bodenheimer, 1937). But a chief difficulty with such permanent nests is that in time they become too small and may restrict the growth of the colony (Brian, 1965). It is otherwise with army ants.

Dorylines do not have permanent nests. Changing habitation is essential to their mode of life, as the term "bivouac" suggests, and has had priority in their evolution. The bivouac of an army ant colony, defined as "the colony living in its current nesting situation," is the state of the colony more than a particular place.

For a long time it was known that the "visiting ants . . . form temporary nests or habitations for themselves which they abandon from time to time (Belt, 1874). But these nests remained so elusive that Bates had to say (1863), despite long experience in the Amazonian forest: "I have traced an army sometimes for half a mile or more . . . but I never met with a hive." For the American naturalist Edward Norton (1868) also, working in Mexican forests, the task was difficult. One day, however, he "found under a fallen trunk a prodigious number of workers of *Eciton* . . . heaped and piled upon each other like the bees in a swarm." Belt (1874) too, successful in Nicaragua, stated that the *Eciton*

> . . . make their temporary habitations in hollow trees and sometimes under large fallen trunks. . . .
> 
> A nest that I came across . . . [under a large fallen trunk] . . . was open on one side. Ants clustered together in a great mass, like a swarm of bees, but reaching to the ground below. Their innumerable long legs looked like brown threads binding together the mass, which must have

been at least a cubic yard in bulk, and contained hundreds of thousands of individuals, although many columns were outside . . . I was surprised to see in this living nest tubular passages leading down to the center of the mass, kept open just as if it had been formed of inorganic material. Down these holes the ants who were bringing in booty passed with their prey.

This description fits well the bivouac of a colony of *Eciton* formed in the open under conditions of what we now call the nomadic phase (Fig. 3.1).

Sumichrast (1868) thought the same colony of *Eciton burchelli* might have temporary nests "distinct from those where they found the reproducing sexes and where is the place of the growth of the larvae and their metamorphosis." Müller (1886) studied a colony that held its place in a hollow tree for many days although he tried to expel the ants with smoke. Sumichrast (1868) observed the nest of a colony of *Labidus praedator* in Potrero, Mexico, where for about three months the ants were "domiciled under a little bridge formed by some rough trunks of trees bound together by a heap of vegetable mould." Von Ihering (1912), through his experiences with this same species, gave up the idea of short-term "Wanderneste" for that of (long-term) "Dauerneste," and others then applied the unclear concept of a stable brood nest to all army ants.[1]

Differing opinions, both as to the nature of the nests and how long they might be occupied, were based on fragmentary and conflicting evidence. There were reports of underground nests (von Ihering, 1894, 1912) as well as descriptions of nesting clusters on the surface (Norton, 1868); conjectures for nesting stops ranged from four to five days for *Eciton hamatum* (Belt, 1874) to fifteen days for *Anomma* (Vosseler, 1905). As we now know, there are important differences in the locations as well as in the mode of changing nests by army ants, depending upon: (1) the physiological condition of the colony, and (2) how well the species has adapted to surface or to subterranean conditions.

In studies on two species of *Eciton* (Schneirla, 1933, 1945), I found radical differences in the nests and in the behavior of the colonies at different times. At one interval a colony was found in the *nomadic* condition: highly active, with a brood in the larval stage, and usually forming

---

[1] The idea of long-term nests applies in more than one way to the dorylines. There are of course the statary bivouacs of group A species (Chapter 7), in which colonies stay about three weeks. In *Dorylus* (*Anomma*) *nigricans* and other group B ants, nesting stays of a few months may not be uncommon (Raignier and Van Boven, 1955). Colonies of *Neivamyrmex* species, and probably others also, occupy the same nests through long overwintering periods (Schneirla, 1963). In still other cases, different colonies of the same species may have occupied the same site at well-separated times (Gallardo, 1915).

FIGURE 3.1.
Nomadic bivouac cluster of a colony of *Eciton hamatum* in a cylinder approximately 40 cm in diameter. Vines and other vegetation had to be cleared away for the photo.

a new exposed surface cluster nightly in a new site. In the next interval the same colony would be in the *statary* condition[2]: low in activity, with

---

[2] The obsolescent English word "statary," which means "standing in place," best denotes the non-nomadic condition of doryline colonies when extrabivouac action drops sharply to a level below that of the nomadic phase (Schneirla, 1933). The proper word is not "sedentary" as the colony is not then inactive; nor is it "stationary" as the bivouac may shift a little. Generic differences in these respects are best covered by the word "statary."

a pupal brood, and staying in the same enclosed bivouac for many days. It was clear from these results that Wheeler (1925) had taken the first recorded queen of *E. hamatum* from the bivouac of a nomadic colony—an open mass "suspended from a looped liana and some twigs against the trunk of a young stilt palm"—whereas the stable nest Müller (1886) described for *E. burchelli* was the bivouac of a statary colony.

To study bivouacs, one must have a system for locating them. Finding bivouacs is a much simpler problem if the searcher knows the important behavioral clues to their location. Bates' difficulty (1863) may have resulted from his having hunted for the nests late in the afternoon when his main work of the day was done but when tracing the nest of a colony from raiding columns is most confusing (Schneirla, 1933). The principal clue, usually clearest for *Eciton* in the morning, is the direction taken by most of the booty-laden ants. Early in the raid, these ants go mainly toward the nest, but in the afternoon traffic conditions become complicated and confusing. Once the searcher knows that colonies of *Eciton* emigrate nocturnally, he has a new means of finding bivouacs. With experience, the range of clues can widen. One can even find bivouacs by catching downwind the distinctive species odor. *E. burchelli*, for example, has a musky, somewhat fetid odor, and its statary colonies may become so redolent as to be detectable some meters away.

As my research progressed, results increasingly supported the generalization that each colony of *Eciton* passes through alternate functional phases and that its bivouacs differ correspondingly (Schneirla, 1938, 1945). Table 3.1 sums up the evidence for several projects (Schneirla, 1949; Schneirla and Brown, 1950; Schneirla et al., 1954) with respect to the number of "exposed" and "sheltered" bivouacs found under different conditions.[3] In both *E. hamatum* and *E. burchelli* exposed bivouacs predominated in nomadic colonies. These are usually clusters open to view, e.g., under a raised log (Figs., 3.1, 3.2, 3.3, and 3.4). In contrast, sheltered bivouacs predominated in statary colonies. These are usually clusters shut off from the open, e.g., within a hollow log. In rainy weather virtually all of the nomadic bivouacs in *E. hamatum* and four-fifths of those in *E. burchelli* were exposed whereas three-fifths and four-fifths, respectively, of the statary bivouacs were sheltered; in dry weather, exposed nomadic bivouacs were fewer for both species. Clearly,

---

[3] The judgments made in Table 3.1 depend upon such criteria as the amount of cluster visible, the amount of physical shelter apparent, and the like. These results may understate the degree of exposure in nomadic clusters but may overstate this aspect in statary bivouacs. For example, a nomadic cluster in vegetation on a hilltop may be open to strong air currents whereas a statary cluster under a ravine bank, fully open to view (e.g., colony '46 H-D, *Eciton hamatum;* Schneirla, 1949) may be as well sheltered as though physically enclosed.

FIGURE 3.2.
Close-up view of a portion of the outer wall of a bivouac of *Eciton hamatum* during the nomadic phase. Note that most of the ants hang head-downward.

the type of bivouac and its location depend upon the colony's condition and its ecology, i.e., its relationship to its environment.

We consider these two species surface-adapted because most of their bivouacs are formed aboveground and most of their colony activities occur on or above the surface. *Eciton burchelli* is even somewhat arboreal

TABLE 3.1.
*General bivouac situation and phase of cycle in two species of* Eciton

| Weather | Species | Nomadic phase | | Statary phase | |
|---|---|---|---|---|---|
| | | Number of sites | % judged exposed | Number of sites | % judged sheltered |
| Rainy | E. hamatum | 129 | 97 | 27 | 59 |
| | E. burchelli | 50 | 82 | 21 | 86 |
| Dry | E. hamatum | 214 | 86 | 38 | 66 |
| | E. burchelli | 152 | 69 | 31 | 88 |

FIGURE 3.3.
Curtain- (or half-cylinder-) type bivouac formed by a nomadic colony of *Eciton hamatum* between the buttressed roots of a tree. (See Plate I, following page 138.)

in these respects, as at times its colonies raid and also form bivouacs as high as 30 m from the ground in the great trees of tropical forests (Fig. 3.5) whereas those of *E. hamatum* are seldom found more than one meter from the ground. This difference relates to aspects of colony behavior to be discussed later in this chapter.

Although *Eciton* as a genus seems the most surface-adapted of all dorylines, its species present a range from *E. burchelli* and *E. hamatum*, which are high in surface adaptation (Schneirla et al., 1954), to *E. mexicanum*, *E. vagans*, and *E. dulcius* (Bruch, 1923; Schneirla, 1947; Borgmeier, 1955; Rettenmeyer, 1963), which, although surface-active, form their nomadic bivouacs in darker, moister, better sheltered places than the first two. Correspondingly, the statary bivouacs of *E. burchelli* and *E. hamatum*, although sheltered, are usually settled aboveground whereas those of the other species mentioned are generally placed more deeply within or beneath logs or in erosion chambers and similar recesses underground.

*Aenictus*, like *Eciton*, runs a range from pronounced surface adaptation in the nesting of *Ae. laeviceps* and *Ae. gracilis*, in which most of the

FIGURE 3.4.
Curtain-type bivouac formed by a colony of *Eciton burchelli* in the nomadic phase, hanging from the base of a log to the ground.

nomadic bivouacs are formed on the surface of the ground, to a distinctly hypogaeic condition in *Ae. aratus* and others, which form nomadic bivouacs deep under logs or in recesses underground (Schneirla and Reyes, 1966). Most of the African representatives of *Aenictus* seem distinctly hypogaeic in nesting (Arnold, 1915).

Wheeler (1910) noted that some dorylines lead much more exposed lives than others. Thus, in contrast to the dominantly epigaeic *Eciton*, *Neivamyrmex* is hypogaeic. Throughout the range of this genus, both at its southern border in South America (Gallardo, 1915; Borgmeier, 1955) and its northern border in the United States (Smith, 1942; Schneirla, 1958), colonies of its species commonly bivouac in such places as mammal burrows, old stumps or roots, preempted ant nests under rocks, and comparably sheltered recesses. The ants may often extend natural cavities and form galleries by digging (Wheeler, 1900). The species differ widely in their relationships to the surface. In southeastern Arizona, for example, colonies of *N. nigrescens, N. opacithorax,* and *N. wheeleri* are active on the surface, in that order, which is also the order of their bivouacking, i.e., the first usually nearest the surface, the third deepest. Several other species sympatric with those may be even more hypogaeic, as their nests are unknown. In the case of *N. pauxillus*, whose minute, pale-colored workers are seen infrequently under deep rocks, the nests, males, and queens are thus far unreported. Comparably, in Argentina,

FIGURE 3.5.
Telephoto of a bulb-type bivouac formed by a colony of *Eciton burchelli* in the nomadic phase. This cluster, which approaches 90 cm in length, hangs about 20 m from the ground. (Photo courtesy of R. E. Logan.)

*N. spegazzinii* bivouacs much closer to the surface than the rare *N. orthonotus,* found (Eidmann, 1936) as far as 4 m below the surface in the fungus chambers of a nest of *Atta.*

Principles to bear in mind, in comparing species' nesting behavior, concern the ecological equivalence and the local availability of sites. Thus colonies of the same species may normally bivouac underground in one area but at times occupy well-sheltered sites above the surface in another area. In the southeastern United States, for example, colonies of *Neiva-*

*myrmex nigrescens* and *N. opacithorax* often bivouac above the surface within old pine stumps, but in Arizona canyons they nest underground almost without exception (Schneirla, 1958). Although colonies of the tropical *N. legionis* virtually always nest underground, colonies of this species at times cluster aboveground within hollow trees (Borgmeier, 1955). Comparisons of local differences in nesting bring out wide ecological differences among species within each doryline genus.

*Labidus* as a genus may be more hypogaeic in nesting than *Neivamyrmex*. Although columns of *L. praedator* are often seen on the surface, particularly on dark humid days,[4] their nests are seldom discovered. The bivouacs of *L. coecus* must be very deep and secluded. Although Wheeler (1910) found this a common ant in central Texas, its nests eluded him. A few nests have been discovered elsewhere, in open country as well as in forests, deep underneath logs, in and below stumps (Weber, 1941; Borgmeier, 1955),[5] within old termite nests and in the galleries below, and even in caves (Kempf, 1961).

There are two other tropical American genera that also seem very hypogaeic. Although *Nomamyrmex esenbecki* is frequently seen raiding in shady forests, its nests have not been found (Borgmeier, 1955). More elusive and perhaps even more hypogaeic is *Cheliomyrmex,* on which there is only one report of a temporary nest (Wheeler, 1921).

Although the driver ants are all subterranean in their nesting, comparable species differences appear. Colonies of *Dorylus (Anomma) wilverthi* commonly nest within 1 to 2 m of the surface with their clusters more concentrated than those of *D. (A.) nigricans,* which usually settle at depths between 2 and 4 m (Vosseler, 1905; Raignier and Van Boven, 1955). The other driver ants all nest relatively deep in the ground.[6]

---

[4] *Labidus praedator* is known as the "rain ant" in Brazil because its columns are often seen on the surface "shortly before the start of heavy rain" (von Ihering, 1894). Borgmeier (1955) offered evidence that the surface columns of this species appear regularly from one day to several days before a rain. My own experience is that both dim light and humid conditions attract this ant to the surface either before or after rains (Schneirla, 1947).

[5] Von Ihering (1912) studied for some time a colony of *Labidus coecus* settled beneath the hard shell of a deserted termite nest and in the earth below. The American biologist Neal Weber (1941) found a nest of this species in British Guiana in a rotted stump partly exposed to the sun. The ants with their brood occupied the basal half of the stump (Chapter 6). The bivouac extended below the ground level, where a young brood was taken as well as a gravid queen.

[6] The Belgian biologists Albert Raignier and J. Van Boven (1955), of Louvain University, investigators of the driver ants, compared in detail the nests of two species of *Dorylus (Anomma).* Bivouacs of *D. (A.) wilverthi* usually are formed under the roots of trees within 1.5 m of the surface, in an area about 2 m wide from which galleries run laterally and upward. The workers cluster centrally in a mass of "clearly perceptible temperature," often in a single cavity, with queen and brood below or in side galleries. Bivouacs of *D. (A.) nigricans* usually are deeper (about 2 m), and the ants are distributed variably over a broader area without the central mass of higher "physiological temperature." (Pp. 189ff.)

This review of nesting is related to the suggestion in Chapter 1 that two main trends dominated doryline evolution, one toward an epigaeic, surface-adapted existence, the other toward a hypogaeic, subsurface-adapted existence. The divergence may have centered on nesting.

In group A dorylines, a surface-adapted pattern appears strongly in *Eciton* and *Aenictus* though less so in *Neivamyrmex*. Most other genera tentatively included in group B are hypogaeic in nesting roughly in the following order: *Dorylus* (*Anomma*), *Labidus, D.* (*Dorylus*), and *Nomamyrmex*. Basic to these divergent trends in the vertical level of nesting adaptations may be species differences in optima for light, temperature, and humidity, in particular. The problem is complex and calls for further research.

Sensitivity to light, important for colony arousal in surface-adapted dorylines (Chapter 4), may be significant as an indicator of these differences. Relevant to this matter is the German biologist Werringloer's valuable histological investigation of visual resources in workers of various New World dorylines (1932). Workers of most of these ants, which ranged from epigaeic to deep hypogaeic species, possess a modified compound eye.[7] In *Eciton burchelli* this eye is the largest, with the largest visual nerves and best central connections of Werringloer's research series. In contrast, the eyes of *Labidus coecus* are tiny, have no pigment in their visual cells, and have only a threadlike visual nerve. At the extreme in this series, workers of *L. mars* lack eyes altogether and have only an indistinct mass of cells to mark an optic lobe in the brain. Thus, in these army ants, the visual equipment of workers is best developed in surface-adapted species but seems to be relatively more degenerate as the degree of subterranean adaptation increases.

The chief agents that keep workers of hypogaeic species below ground in the daytime may be intense light and atmospheric dryness[8] acting with air currents and other surface stimuli that make these ants agitated and tense when they come to the surface. Workers of *Dòrylus* and *Labidus* alike indicate disturbance in their group operations on the surface by clustering their bodies or by building structures of earthen

---

[7] This receptor, absent in workers and queens of *Dorylus* and *Aenictus* but present in nearly all New World dorylines, although embryologically a compound eye, is a small single-lens eye known as a "lateral ocellus" (Werringloer, 1932).

[8] Workers of *Neivamyrmex nigrescens,* normally nocturnal in their surface activities, often carry out daytime raids on dark or overcast days. In the southeastern United States they operate under surface cover even at night, probably in response to prevalent low atmospheric humidity. By contrast, in southeastern Arizona, where nocturnal humidity is usually high, these ants operate at night in the open, but usually under cover when the surface and air are dry.

We find workers and queens of *Aenictus laeviceps* and *Ae. gracilis* reactive to light, doubtless through photodermal receptors since these ants lack eyes. In their regular statary phase (Chapters 7 and 13) these ants rarely appear on the surface except after dusk.

pellets at the sides of the columns (Cohic, 1948). Significantly, the smallest workers, physiologically the most active of all, predominate in this activity. Surface wall-building arises both by day and by night in several (group B) species, thus seeming to be a response to air currents and other surface conditions as well as to light.

Subsurface nesting is not based simply on disturbing surface conditions that keep the ants down. All insects are very sensitive to temperature and humidity changes around them,[9] the subsurface dwellers most of all. The hypogaeic adaptations of a species may center on its having lower optima for temperature and higher optima for humidity than do surface-nesting species. Conditions governing the distinctive optima of *Eciton burchelli* hold open for its colonies a life aboveground; those of *E. dulcius* adapt its colonies to a life mainly below ground and also inhibit nesting and activity at the surface.

Clustering is a group response basic to the settlement of new bivouacs. Significantly, workers of the same colony of *Neivamyrmex nigrescens* withdraw more rapidly from bright light, are more responsive to chemical stimuli, and cluster more readily in dim light when their colony is statary than when it is nomadic (Topoff, in prep.). In diurnally active *Eciton*, the clustering of local groups that precedes bivouac formation begins toward nightfall (Schneirla, 1938). Formation of walls by clustering at the borders of columns, normally absent in surface-active *Aenictus,* is prominent during their rare movements aboveground in the statary phase (Schneirla and Reyes, 1966). These results support the idea that army ant workers cluster readily when weak external stimuli summate with body-stretching effects capable of inhibiting general activity through neural channels.[10]

Clustering is a characteristic reaction of all army ants (Chapter 2). Savage (1847) described this behavior as it occurred in a colony of *Dorylus (Anomma)* sp. that he drove from a cavity with smoke only to find the ants blanketing the base of a small tree two days later at the same spot.

> From the lower limbs (four feet high) were festoons or lines the size of a man's thumb, reaching to the plants and ground below, consisting entirely of these insects. Others were ascending and descending on them. . . . One of these festoons I saw in the act of formation; it was

---

[9] Insect ecologists generally consider that temperature is the predominant factor in the environmental adjustments of these animals, including nesting (e.g., Herter, 1924; Chapman, 1931; Allee et al., 1949; Bursell, 1964), and that temperature effects are closely related to those of humidity (e.g., Ludwig, 1945; Wigglesworth, 1965).

[10] The German zoologist Schips (1920) compared the immobilization of ants in the hanging strands of *Dorylus (Anomma) nigricans* with the state of "hypnotic stiffness" demonstrable in such other insects as the phasmid "stick insect."

a good way advanced when I first observed: ant after ant coming down from above, extending their long legs and opening wide their jaws, gradually lengthening out the living chain until it reached the broad leaves of a plant below. An ant finally fixed itself from below on the leaf, and served as an attachment. . . . In about two hours I visited the spot again, when the hanging lines of festoons were gone, and about half of the mass of ants also; some below the surface, others on the predatory excursions. . . . (P. 7.)

Clustering reactions appear most prominently in the formation of bivouacs by colonies of surface-living army ants. Nomadic bivouac clusters of epigaeic *Eciton* species typically hang from raised objects and are more or less round in cross-section with walls curved downward. When a colony settles against a flat vertical surface, as against a buttressed tree root, it usually clusters into a single mass hanging to the ground (Figs. 3.2, 3.3, and 3.4). If one breathes against it, the surface quivers, taking on, as the American naturalist William Beebe (1919) said, "the appearance of the fur of some terrible animal." Within the interior are intermeshed strands of workers, enclosing queen, brood, and workers in many chambers and galleries walled by living clusters. Through their interspaces move workers engaged in a great variety of chores.

We may observe bivouac construction to advantage when a colony of the highly specialized *Eciton hamatum* changes nesting sites (Schneirla, 1933). After a day-long raid, toward dusk hanging clusters begin to form, dangling from the under surface of some raised object (e.g., a log) near a booty cache, where a few ants have anchored themselves by their leg hooks. As others are attracted, newcomers run downward along the strand and extend it by fastening themselves by legs or mandibles to others near the end, usually linking legs by the tarsal claws (Fig. 3.6). In *Eciton* these claws are strong, sharp, and so curved that they catch readily upon others, most often upon those of the long rear legs. Tests with ants linked into a meshwork across a nest frame (Fig. 2.4) show that the placement of these hooks and the weight of ants pulling on other ants rather than a general behavioral disposition—e.g., "positive geotaxis," as Wheeler (1900) thought—explain why workers fastened into a bivouac wall usually hang head downward.

As darkness approaches, workers of *Eciton* begin to form strands hooked from overhanging surfaces. Once the first ants grip the surface and hang from it, their tarsal claws are set deeper as the weight of the strand grows. Then the pull of ants upon ants in the chain—under conditions of steady environmental stimulation at low intensity—can soon effect a reflexive immobilization as fastened parts of the chain become quiescent through body stretching. Strands gradually become ropes and ropes fuse into a heavy fabric (Fig. 3.7).

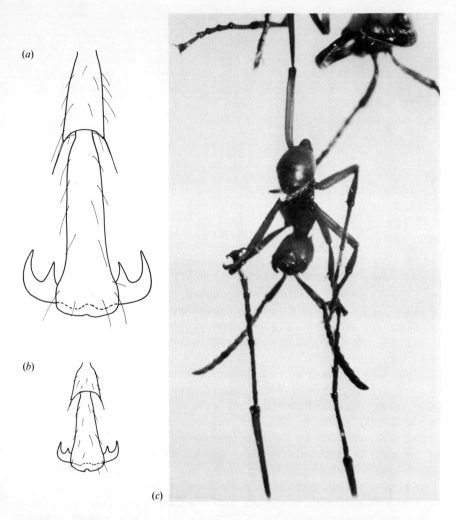

FIGURE 3.6.
Bivouac construction using tarsal hooks. Enlarged view of tarsal claws in a major (*a*) and minor (*b*) worker of *Eciton*. (*c*) Part of a hanging strand formed by workers of *Eciton hamatum* in an observation cell in the laboratory.

Forming a bivouac in *Eciton* is a slow process. Ant fringes often begin at several places, but adjacent clusters that form quickly usually are the most attractive. Although isolated strands may break of their own weight, adjoining ant ropes, as they lengthen, combine into strong fabrics. The separate strands of *E. burchelli* can become very long without breaking, but in *E. hamatum,* not as strong an ant, strings longer than 25 cm usually break before they reach the ground.

FIGURE 3.7.
Strands formed by workers of *Eciton hamatum* introduced to a laboratory nest, much as they begin a bivouac under natural conditions. Part of a colony transferred to New York in 1949 for study and for museum exhibition. (Photo courtesy of American Museum of Natural History, New York.)

When these *Eciton* form bivouacs, strands are pulled together as they grow downward so that the completed bivouac walls commonly taper inward from top to bottom—a rain-shedding adaptation. Clusters hanging from a broad ceiling grow best around an area where new clusters center upon the odor and mechanical effects of group action. Once a pillar has thus begun, it thickens as strands lengthen and combine around it. Figure 3.1 is an example of the cylindrical type of bivouac formed thereby, which may be considered the standard for *Eciton*. Under the ceiling, within about 36 cm of the ground, the cluster tends to be circular in cross-section, 25 to 40 cm in diameter, and to get narrow toward the bottom, its shape often modified by vines and other physical supports. This is the bivouac pattern common among colonies of *E. hamatum* early in the nomadic phase (Fig. 1.1).

In our observations (Schneirla et al., 1954) the cylindrical (or curtain) type was approximated by 209, or 63%, of 332 bivouacs of *Eciton hamatum,* and by 97, or 38%, of 199 bivouacs of *E. burchelli*. The bivouac differs in size and shape according to terrain (e.g., frequency of logs, rocks, vines, and brush) and forest cover and also to size and condition of the colony. Other kinds of bivouacs include the curtain type, most frequent in large colonies and in all colonies late in the nomadic phase. Curtain type bivouacs, found in 8% of the cases of *E. hamatum* and 8% of *E. burchelli,* arise when the space below the ceiling is open on just one side. Another type, the pouch (Fig. 3.5), forms through clustering at some distance from the ground without intervening support; then festoons and walls are drawn together by degrees to round out the bottom as a sack. Pouch types occurred in 28, or 17%, of 199 cases in *E. burchelli* but in only one case of 332 in *E. hamatum*. This difference arises because colonies of *E. burchelli* more often raid high in the vegetation and thus begin clusters well above the ground, and their workers, being larger, stronger, and more numerous than those of *E. hamatum,* can build this structure.

Bivouac engineering differs with time in the nomadic phase. In *Eciton* early in the phase the tough outer wall can sustain the colony even without such physical supports as vines. The interior strands then usually are fibrous, dividing off alveolar spaces that contain the queen and her worker group (centered above), the brood (centered below the queen), and the booty. At this time in the phase, the strength and sustaining properties of the outer wall of the bivouac can be demonstrated by slowly lifting a log under which a bivouac has formed, usually with the result that the wall comes away with the log as though it were an empty sleeve while the brood and most of the interior contents spill out from it. Toward the end of the nomadic phase, the colony clusters become larger, more variable in structure, and break apart more readily. Clustering ability thus varies with colony condition.

The bivouacs of surface-adapted species of *Eciton* are the most specialized in all army ants. The clusters of *Neivamyrmex nigrescens*—as seen, for example, in a cavity exposed as a rock is rolled away on a steep hillside—are made by smaller ants in smaller colonies and often resemble bivouacs of *Eciton* in miniature. The bivouacs of *Neivamyrmex*, however, from observations of their internal structure and brood distribution, appear simpler in pattern than those of *Eciton*. Bivouac structure gives useful clues to colony organization.

As a tentative comparison, with increasing subsurface adaptation in doryline species, abilities to organize the colony in well-centered nesting masses may lessen. Workers of *Neivamyrmex*, with their untoothed and simpler tarsal claws, are adapted to subterranean nesting but not to making the strong compact clusters typical of *Eciton* on the surface. Still other differences are suggested between epigaeic and hypogaeic species. As colonies settle deeper in the ground in the latter, their bivouac clusters may become more diffuse when integration of the colony is low (Chapters 2 and 13). Accordingly, Raignier and Van Boven (1955) found colonies of *Dorylus* (*Anomma*) *wilverthi* with a higher central ("physiological") temperature—indicating a more concentrated massing of workers and brood in their nests—than the more diffused masses of the deeper nesting *D.* (*A.*) *nigricans*. Comparably, the subsurface bivouacs of the relatively small colonies of *Neivamyrmex nigrescens*, formed near the surface, may be better centralized as a rule than the great assemblages formed by colonies of *Labidus* and *Nomamyrmex* deeper in the ground (Borgmeier, 1955; Rettenmeyer, 1963).

Contrasts in the degree of bivouac specialization are found among surface-adapted doryline genera. *Eciton* forms complex hanging bivouacs capable of much structural and behavioral differentiation internally. *Aenictus*, by contrast, commonly forms platter-type bivouacs in ground litter, not much more than a thick heap of ants massed around brood and queen, and simple in internal structure (Schneirla and Reyes, 1966), a type rare in *Eciton* (Schneirla et al., 1954). *Aenictus* holds its bivouacs far more variably and generally for shorter intervals, rebuilding them rapidly and with fewer complications than *Eciton*. We find evidence for comparable differences in colony organization in these two group A genera (Chapter 13).

Bivouac changing in *Eciton* is a devious process. Its first stage is the development of a great raid started at dawn; the second is a new major exodus after midday (Schneirla, 1933). In this second exodus the workers, through traffic developments (Chapters 4 and 5), finally settle on just one of the alternative raiding trails leading from the bivouac. Further events depend upon what traffic conditions are met at successive trail division points, any one of which may become the site of the new bivouac.

Clustering on a large scale begins at a center of attraction (e.g., a trail division cache) when light and other external stimulation fall to low intensity and when colony odors and tactile stimuli dominate. The dropping off of raiding in *Eciton* with darkness is aided by a decreasing responsiveness of the ants to booty stimuli as falling temperatures and intensity of light quiet them down. In army ants, responsiveness to colony odors increases as light intensity falls.[11] In this situation then the potency of stimuli from trails, from other workers, and from booty or brood held or carried increases until those effects dominate action. Columns then push on steadily until progress is blocked whereupon jostled workers either turn with the tide or join clusters of those already immobilized through being knocked about in the growing traffic jam.

Thus, the problem of the new bivouac is solved very gradually. As the light fades and hanging clusters begin to form here and there at trail junctions, how clustering progresses at these places depends upon local and general traffic conditions and upon stimulative, ecological, and physical conditions affecting the ants there. These factors affect local behavior so differently as to introduce a process of "selection" whereby the site of the bivouac is determined.

Usually in *Eciton,* several clusters thus begun are seen to break down successively as traffic presses onward. Favoring persistence of a cluster are such attractive local features as odors of workers gathered over booty and brood, vines and other objects canalizing the entrance to a niche, optimal temperature and moisture, and perhaps odors lingering from earlier stays of other colonies at the site. The selective process is also influenced by internal pressures forcing workers from overcrowded places too small to accommodate most of the colony. Ongoing columns may thereby gain headway and continue the exodus (Schneirla, 1938). When the carrying of booty and brood into a cluster is well underway, the collective odor accelerates the rush, and once the queen has entered, the issue generally is settled.

After a new bivouac of *Eciton* is formed, activity continues within its walls through the night. The queen settles with her group, the brood is gradually distributed, and feeding proceeds in centers of brood and workers here and there. Further shiftings occur as rearrangements of the brood progress. All of these events seem to be most complex late in a nomadic phase when new events arise involving both the brood (Chapter 6) and the queen (Chapter 8).

---

[11] In a laboratory test (Schneirla et al., 1966) two lots of workers from the same colony of *Eciton hamatum* were admitted to separate arenas, one dimly lit, the other brightly lit. Each arena contained two discs of blotting paper, one of them clean, the other a paper on which the colony queen had rested for ten minutes. The ants under dim light clustered on the queen-odorized disc within ninety seconds; those under bright light did not respond reliably to the discs within this time.

In comparison with the detailed processes described for *Eciton,* bivouac changing occurs in *Aenictus* as virtually one rapid operation (Chapter 13). Not only does raiding shift into emigration without the time-conditioned stages holding for *Eciton,* but in *Aenictus* the new cluster forms much more simply and quickly. To understand these differences, we must study the functional aspects of bivouacking.

In all army ants the bivouac has four main roles. It serves: (1) as a base and a center of operations for the colony, (2) as a shelter for the colony, (3) as an incubator for the brood, and (4) as a population reservoir. These functions are all highly specialized in *Eciton* and relatively simple in *Aenictus.*

The role of the bivouac as a colony-operating center is well illustrated in the nomadic phase when the settling of each new nest gives the colony a fresh vantage point for invading the environs and capturing booty. In *Eciton,* each nomadic bivouac serves the colony as its behavioral center during a day-long raid and the subsequent emigration to a new site. Of all social insects, surface-adapted army ants change their base of operations the most often and regularly. Group B dorylines, by contrast, hold their bivouacs as a rule over longer and more variable intervals (Chapters 5 and 7).

Through their bivouacs the army ants can adjust to difficult conditions in their general environment. The resources of a bivouac as a shelter are illustrated by the surface clusters of epigaeic *Eciton* species. Massed in their bivouac around brood and queen, workers of these ants, in particular the formidable majors, concentrate so heavy an attack of biting and stinging that they can readily drive off voracious coatimundis, anteaters, and other potential predators.[12]

The bivouac is also a physical shelter against climatic extremes, permitting the colony to obtain by active means an equable internal microclimate. The climatic conditions of the general environment must be met frontally each time the colony exposes its entire personnel in an emigration. But in epigaeic *Eciton* species this hazard is reduced greatly by restricting emigration and the change of bivouac to one night when atmospheric conditions in tropical forests are more uniform from place to place and closer to the optimum for army ant broods than at any other time. Between midmorning and midafternoon, atmospheric dryness is generally dangerously high for a larval brood. In their nocturnal emigrations army ants must move into a suitable location for the entire next day. Clustering places that are adequate at night—for example, under loose brush in a fallen tree clearing—may in the morning expose the colony to blazing sunlight and dry air.

---

[12] Significantly, the one *Eciton* in which major workers lack the great tong-shaped mandibles is *E. rapax,* perhaps the most hypogaeic species in the genus.

Three protective measures exist to protect the ants against these hazards. First, by settling their bivouac at night in a moist place rather than a dry one, the ants can insure their colony a humid, cool, and shady site the next day. In laboratory tests, doryline workers move readily with their brood from a hot and dry section of a nest into a cool and moist section, indicating thereby a keen sensitivity to atmospheric temperature and humidity. These responses to atmospheric conditions are typical of social insects in general (Uvarov, 1931; Allee et al., 1949; Bursell, 1964; Wigglesworth, 1965).

A second protective measure concerns the physical properties of the bivouac site. In surface-adapted *Eciton,* the raised places from which bivouacs hang normally serve as ceilings that block off wind, rain, and direct sunlight. The bivouac walls themselves provide physical protection by shedding rain as would a tight layer of oily heads. Significantly, it is the older and the larger workers that are more resistant than the others to desiccation and other extremes of the environment that predominate in the outer bivouac wall.

The third protective measure is the active responses of the ants to nonoptimal conditions, thereby readjusting the bivouac. Typical of surface-adapted army ants are responses to local disturbances in which all workers behave more or less alike. When a hot, bright sunfleck, for example, penetrates the forest canopy and strikes a section of bivouac wall, ants hanging there first begin to stir about, disengage themselves, and drift away. If the disturbance continues, more and more workers resettle away from the site of disturbance. When the disruption is intense and lasting, as when rain strikes one side of the bivouac, the colony may move a little or may even transport its brood by column to resettle in a new place close by.[13]

Most army ants appear to be especially sensitive to dryness in the air or soil, and laboratory studies demonstrate their great vulnerability to desiccation.[14] In the dry season colonies may lose heavily or even perish

---

[13] To study these reactions, I used a mirror to reflect sunlight upon the bivouac edge, measuring the time required for the affected part of the wall to break down prior to a readjustment of the general cluster. Other tests are discussed in Chapter 5 (see *Shift* in Glossary).

[14] In laboratory tests of large worker groups of *Eciton hamatum* (Schneirla et al., 1954), workers running in a continuous circular column began to falter when atmospheric moisture fell below 60% relative humidity. Then as the air was made drier (by adding a hygroscopic salt), they began to die. First to drop out were the minor workers, then the intermediates, and finally, when air in the experimental arena approached a low of 45% relative humidity, the major workers. Meanwhile, workers running in a control arena in which atmospheric humidity stayed near 100% relative humidity throughout continued their circling without important changes. The entire experiment, planned and carried out by my collaborators the zoologists Robert and Frances Brown, lasted about eight hours.

through overexposure to drought. In the rainy season, conversely, the great danger is being caught in a flood. Even so, if the losses are not too great, the colony may recover because its bivouac serves as a population reservoir.

As our observations (Schneirla et al., 1954) clearly show for surface-adapted species of *Eciton* and as can be surmised for subterranean species, the colony bivouac serves continuously as a brood incubator. In the nomadic phase, through worker responses to the larvae, the bivouac expands as the brood grows. Night after night the colony, in taking new sites, occupies more space. Enlarging the bivouacs naturally aids the housing and feeding of the brood, solving problems of physically restricted living space that often hamper colony growth in other social insects (Brian, 1965). To an appreciable extent, colonies are able to "select" and also to actively modify their bivouac environments.

Bivouac sites are not come upon by chance. Although the tropical forest may loosely be called a stable environment, most of the animals inhabiting it are likely to fare badly when forced to live outside their species niches (i.e., optimal conditions). Actually, any tropical forest is a mosaic of zonal patterns in which adjacent sites may differ greatly as living nooks for any animal. As a result, colonies of any species of ant (or other potential doryline booty) are likely to be distributed unevenly through the forest (Wilson, 1958c). In raiding, army ants usually penetrate areas rich in booty and so emigrate there. Very often they simply move into the nests of ants they raid, dispossessing the regular tenants. In this way they settle in places that are as adequate for them and their brood as for other insects already living there. In a real sense, therefore, army ants are guided to suitable living places by their prey.

Beyond the resources of army ant bivouacs already mentioned, others are gained through active controls exerted by the colony over its environment. Some of these controls resemble to an extent those known for a few of the social bees. When a hive of honeybees becomes overheated, the workers increase air circulation and the rate of evaporation by means of a wing-beating reaction called "fanning" and so cool the brood area (Grout, 1949); or when hive temperature falls low, they conserve heat in the brood area by clustering tightly over the comb (Dunham, 1931). Thus, they control their environment by reducing its extremes behaviorally, as do army ant colonies in the nomadic phase. Also, some of the social bees living in the wild have fairly stable conditions to begin with in the tree cavities or other physical shelters the colonies occupy. Similar advantages are enjoyed by colonies of surface-adapted army ants in their statary phase[15] and by colonies of hypogaeic species at all times.

---

[15] In a continuous five-day hygrothermographic record, temperatures and humidities in a hollow log at a spot just previously occupied for twenty days by a statary colony

In contrast, the nomadic bivouac masses of surface-adapted army ants are often exposed to the raw elements. Typically, in just one nomadic phase, a colony of *Eciton hamatum* occupies from sixteen to eighteen different sites, in most of which it forms open clusters. These bivouacs vary predictably with brood changes. Early in the phase, when the larvae are small, the young brood is held centrally in a mass. The bivouac then is small and compact, and the brood is well buffered by the outer wall. As the larvae grow, the workers distribute the brood more widely, with the result that subsequent bivouacs are larger and more variable structures with looser walls.[16] Also, as the nomadic phase continues, bivouacs are located increasingly in cooler and moister sites (Jackson, 1957). These changes in bivouac ecology may result from heightened physiological action in workers during the phase, paralleling increases in the mass, the heat-producing properties, and the stimulative potency of the brood. The workers, by modifying their behavior as the brood develops, make the bivouac a brood incubator with changing properties that meet the physiological conditions of their mass of young through its larval stage (Chapter 6).

Our studies of bivouac ecology show how well the colony controls the internal environment of its temporary nests. Measurements taken daily at regular intervals (Schneirla et al., 1954) showed that both nomadic and statary bivouacs of *Eciton hamatum* were almost always higher by 1°C to 2°C in temperature than the external environment and considerably less variable (Fig. 3.8). Our collaborator, the American biologist William Jackson (1957), found from continuous studies of these bivouacs that although their internal temperatures followed outer daily changes throughout the cycle, they maintained a higher level with reliably less fluctuation (one-third smaller standard deviations). In tropical forests, a daily range of about 7°C is common with dawn temperatures around 22°C rising to near 30°C during midday (Kenoyer, 1929). The intrabivouac range in nomadic *E. hamatum*, which in contrast seldom exceeds 3°C, thus involves an appreciable leveling off of temperature despite outer variations. Comparable intrabivouac controls may hold for humidity.

The bivouacs are made and changed actively by the ants; hence they have environment-buffering and brood-incubating functions based largely

---

of *Eciton hamatum* varied less daily than readings at a control spot just outside the log. In the daytime, temperature inside the log was generally lower and moisture higher then outside; through the evening and most of the night, the reverse usually held (Schneirla et al., 1954).

[16] Early in the phase, a clear thermal gradient prevails in the bivouacs of *Eciton hamatum*, which its peak central and usually in the brood area (Jackson, 1957). Late nomadic bivouacs of this species are porous with a much reduced thermal gradient, and midnomadic bivouacs are intermediate.

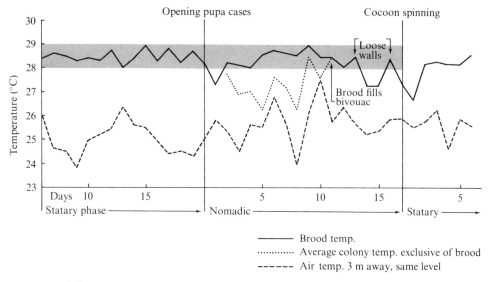

FIGURE 3.8.
Records of intrabivouac temperatures taken near 10:00 A.M. daily through an activity cycle in the successive bivouacs of a colony ('49 H-34) of *Eciton hamatum*.

on properties of the ants[17] and of their brood. To cite an example of environmental control, as the forest air cools late at night, ants hooked in the bivouac wall draw closer together while others move into spaces in the wall. Through these actions, the ants thicken and tighten their wall so that internal heat is conserved and cool air is shut out. As another example, after sunrise on each day in the nomadic phase, as more and more ants leave during the raiding, the bivouac wall becomes more porous. The result of this and of correlated changes is to speed the circulation of air through the bivouac, which both increases internal evaporation and cools the interior against rising temperatures outside. Bivouac sanitation is also improved thereby as gaseous wastes that eddy out in the discharged air are replaced by fresh air. A related feature of bivouac sanitation is the disposal of unused parts of booty and other solid wastes.[18] These the workers generally remove from the bivouac and drop nearby.

---

[17] As I found in a laboratory test (Fig. 2.4a), workers of *Eciton burchelli* and *E. hamatum*, in clustering and reclustering in a hanging fabric, turn their bodies according to directional differences in such conditions as temperature, air currents, and light.

[18] Usually there is little refuse around nomadic bivouacs of *Eciton hamatum* and of other army ants that take soft-bodied booty, which they largely consume. From

Such behavioral and physiological specializations controlling the colony microclimate seem most actively involved in the nomadic bivouacs of surface-adapted *Eciton*. In subsurface-nesting species, ecological stability is gained through movement of soil by the workers (e.g., influencing evaporation rate) and through insulation afforded by the layer of earth itself; also, the colony at times readjusts to its environment by means of "shifts" (Chapter 5). Underground nesting readjustments arising as responses to surrounding conditions (e.g., shifting upward during the day and downward in the evening) appear in the springtime resurgence of winter-dormant species (Schneirla, 1963). Reactions to the environment that depend upon colony condition can be observed in group A species in the early and the late days of each nomadic phase. Colonies of *Aenictus laeviceps,* for example, then cluster in sheltered sites on the surface or below although in the long intermediate part of this phase they usually form open clusters aboveground. By contrast, in the final days of the statary phase these ants begin to shift upward from subsurface bivouacs held during most of this phase, now exposing clusters at surface exits (Schneirla and Reyes, 1966). Comparably, the colonies of such hypogaeic species as *Labidus praedator* (Schneirla, 1947) and *Dorylus (Anomma) wilverthi* (Raignier and Van Boven, 1955) exhibit much surface activity just before emigrations, usually with much digging and heaping of earth around the exits.[19] In such behavior, both epigaeic and hypogaeic army ants actively modify their nests through worker responses depending upon colony condition (Chapter 7).

It is likely that in all dorylines, as in group A species, colony actions of changing bivouacs and settling at one site rather than another depend mainly upon effects from the brood. In *Eciton* and *Aenictus,* toward the end of the nomadic phase, colonies increasingly take sheltered sites. Our findings indicate that such changes in nesting behavior arise through distinctive effects of the larvae, then nearing maturity, upon the workers, altering their responses to environmental stimuli. The workers then, as our notes indicate, become increasingly "nervous," and, in such activities

---

statary bivouacs, one or more two-way columns lead to places nearby where the hard, unconsumed remnants of booty are abandoned with other refuse. *E. burchelli,* because it captures a wider range of booty, has much more extensive refuse dumps near its bivouacs, as does *Labidus praedator.* Colony refuse is much dispersed near the subterranean bivouacs of *E. vagans* and *E. dulcius* (Rettenmeyer, 1963).

[19] On rare occasions, colonies of *Labidus praedator* appear at the surface—under the end of a log, for example—from vertical tunnels around the exists of which a mound of earth soon accumulates from excavations below. Workers generally cluster at these apertures in the early morning and late afternoon. One such mound reached a diameter of 2 m before the colony disappeared. Similar occurrences are common in colonies of *Dorylus (Anomma) wilverthi* within the few days preceding an emigration (Raignier and Van Boven, 1955), evidently marking a time of maximal excitement in the colonies.

as linking themselves into strands in forming a new bivouac, they can be disturbed with increasing ease. The brisk air currents common in tropical forests toward evening usually prevent bivouacking in exposed places by agitating the ants and breaking up even the first stages of clustering. But when the ants start massing in a more sheltered spot in a burrow or in a hollow log away from gusts, disturbances are minimized and the operation proceeds to completion. The workers now seem changed physiologically and differently responsive to surrounding conditions since they cluster much more frequently in cool, damp places than in the intermediate part of the nomadic phase. This behavioral condition, crucial for bivouacking, may center on changes both in level of excitability and in sensory thresholds.

These different readjustments of a colony depending upon its condition may cast light on species differences. For example, our records (Schneirla et al., 1954) show that *Eciton burchelli* formed more sheltered bivouacs than *E. hamatum* in both rainy and dry weather and in both phases of the functional cycle (Table 3.1). The explanation may lie in a species difference in worker excitability.

The point may apply to species differences in frequencies of forming surface and underground bivouacs. Underground bivouacs, relatively infrequent in both *Eciton hamatum* and *E. burchelli,* seem to arise according to weather conditions. In times of rain, only one of 156 bivouacs of *E. hamatum* was underground whereas in "dry" weather 6% of 204 bivouacs in the nomadic phase and 21% of 38 in the statary phase were underground (Schneirla et al., 1954). This difference in bivouacking, paralleled by *E. burchelli,* suggests that under arduous surface conditions (i.e., in dry weather) these normally surface-living species approach hypogaeic species in their pattern of nesting.

The described behavioral difference, as our results for elevated sites suggest, may depend upon colony condition. Of 28 elevated sites recorded for nomadic colonies of *Eciton burchelli* in dry weather, 8 were formed in the first 6 days and 20 in the last 6 days of the phase; and of 16 recorded in rainy weather, 4 were formed in the first 8 days and 12 in the last 4 days of the phase. Arboreal bivouacs thus seem to occur most frequently in this species at times late in the nomadic phase when the colonies are most excitable. At such times these ants raid in the higher vegetation in greater numbers than at other times in the phase and are more likely to cluster and bivouac there near the day's end. By contrast, *E. hamatum,* except for large colonies, rarely raids very high aloft and is unlikely to cluster very far from the ground. The higher excitability and numbers of *E. burchelli* thus may promote a tolerance for bivouacking in more exposed places than are usual in *E. hamatum.*

Although the difference between *Eciton burchelli* and *E. hamatum* is appreciable, it is small compared with that between either of these species and such others in the genus as *E. vagans* and *E. mexicanum* (Schneirla, 1947; Rettenmeyer, 1963). It is also clear from the foregoing discussion that the species of *Eciton* resemble one another more in their ecology and their patterns of nesting than do the many species of the genus *Neivamyrmex*.

Army ants have evolved two distinctive ways of life that are expressed in their nesting: a surface-related, epigaeic pattern predominant in *Eciton,* and a subsurface-related, hypogaeic pattern exclusive in *Dorylus*. The generic contrast in these respects relates particularly to degrees of tolerance for surface conditions based on sensory equipment and excitation levels. Within each genus the species also differ in their characteristic levels of nesting. Contrasts within the genus are perhaps broadest of all in *Aenictus,* whose species range from the distinctly epigaeic to a far greater number of deeply hypogaeic forms.

Somewhat paradoxically, these two patterns of environmental adjustment arise regularly in the colony functions of all army ants studied. They are clearly marked in the nomadic and the statary conditions which alternately dominate all colonies of *Eciton, Neivamyrmex,* and *Aenictus;* they are also present, although less distinctly marked, in *Dorylus* and others. Such recurrent, alternating patterns of nesting in reality present no paradox, for we find them a regular feature of colony life in all dorylines, actually based upon reproductive functions.

The ability of army ants to abandon their nests and resettle at new sites can depend only upon a strong attachment to features of the colony itself, which simplifies the problem of breaking ties with any physical site the colony may occupy. In reality the bivouac, which we have defined as the colony in its nesting situation, is much more the body of ants itself than the particular site currently occupied. The several impressive functions of the bivouac are all attained through colony organization and appear each time the colony resettles its members, brood, and queen in a new bivouac and in each major collective action outside the home site. We can, therefore, examine to advantage the nature of colony organization by studying first the external colony operations of raiding and emigrating, then returning to investigate detailed occurrences in the colony center itself.

# 4

# Raiding

Adapted from archaic times to a life of organized group predation and carnivorous diet, all army ants take their booty in mass forays into areas around their temporary nests. Their pillaging expeditions or raids are the largest organized operations carried out regularly away from home by any animal except humans.

I define this army ant raiding behavior as large-scale predation in the pattern of all dorylines, characterized by a regular relationship to emigrations and cyclic colony function. There are other ants, related archaically to the dorylines and known as legionary ants (Wheeler, 1936), which behave similarly although in ways probably limited by the small size of their colonies.

These other ants, as Wheeler (1910) noted, carry out raids "suggestive of the predatory forays of the dorylines." As examples he offered the marching of *Leptogenys* (or *Lobopelta*) *chinensis* in narrow files (Bingham, 1903) against termite nests and (Wheeler, 1928) the forays carried out by ants of the cerapachyine genus *Phyracaces* against other ants whose brood is carried home as prey. Later, from a description (Arnold, 1914) of an emigration by the termite raider *Megaponera foetens,* Wheeler (1936) suggested that this species leads a nomadic life, moving its nest from one worked-out termite district to another.

Mounting evidence shows that group raiding occurs in at least some of the members of each of several ponerine or ponerine-related genera beside the three mentioned above, including *Simopelta* (Borgmeier, 1950; Gotwald and Brown, 1966), *Onychomyrmex* (R. W. Taylor, pers. comm.), and *Termitopone* and *Ophthalmopone* (Wheeler, 1936). Wilson's field results (1958a) demonstrated that colonies of *Leptogenys purpurea* emigrate from one temporary nest to another. These movements evidently had some connection with raiding as the ants carried booty as well as brood in their columns.

Systematically studying the behavior of these ants in genera belonging to what William L. Brown, Jr. (1954), myrmecologist at Cornell University, calls the "poneroid complex" should bring important results. As Wilson (1958a) suggests from evidence on ponerine legionnaires, their predatory-nomadic patterns, although rudimentary, may cast light on how the complex group behavior characteristic of dorylines arose. He lists four hypothetical evolutionary stages: (1) group predatism, perhaps with extracolony feeding of brood at booty sources, as in *Phyracaces* (Wilson, 1958b); (2) more specialized group predatism by nomadic and somewhat larger colonies, on the level presumed for most existing legionary ponerines; (3) more efficient group predatism by larger colonies against more varied prey (no examples known); and (4) large-scale group predatism by great nomadic colonies as in existing dorylines.

Characteristics of raiding in Wilson's stage (2) are exemplified by a raid of *Termitopone laevigata* that I observed in Panama.

> The shiny black ants moved along at a slow, meandering gait, all in one direction and strictly in single file, one closely after another. The column was headed by one or two ants that moved steadily forward, indicating that the trail may have been made before. I counted exactly 576 of the nearly monomorphic workers in close file; then the procession ended abruptly. Arriving at a stump, the ants bunched up on one side, then spread slowly around the perimeter in both directions, at the same time entering stump and ground by degrees with evident excitement. Soon raiders began to emerge carrying termites and termite brood, which they piled in caches under leaves in three or four locations around the stump. After nearly two hours, the attack seemed to have ended; then the ants filed back over the same trail, at the same pace although nearly all were now laden. The return column, unbroken from beginning to end, contained only 543 ants.

These legionaires—in their stereotyped modes of exodus and of attack after contact with the nest has been broken, in their way of concentrating forces and piling the booty in caches until the assault is ended, then returning 20 to 50 m to their base in a body—illustrate archaic features of raiding that may have typified pre-doryline ancestors.

By contrast, doryline raids are much more complex and highly developed affairs,[1] which usually advance by stages. Contact with the base nearly always continues throughout the raid. These ants commonly store part of their booty in caches and carry it to the rear by degrees, then end the operation by gradually returning to the bivouac or by emigrating to a new site.

---

[1] Types of group predatory behavior observed in legionary and army ants should be distinguished from forms of collective raiding carried out by ants of other subfamilies, e.g., the myrmicine *Harpagoxenus americanus* (Creighton, 1927) and the formicines *Polyergus lucidus* (Forel, 1899) and *Formica sanguinea* (Dobrzanski, 1961). These are convergent to the patterns of ponerines and dorylines (i.e., independently evolved) and rather different, commonly involving "slave-making" or a social parasitism with adults raised from unconsumed brood (Wheeler, 1910, 1928).

Borgmeier (1955) and Rettenmeyer (1963) have summarized the literature on raiding by New World dorylines, Cohic (1948) and Raignier and Van Boven (1955) that on Old World dorylines.

Army ants in raiding columns are typified by their rapid gait and meandering course with antennae arched forward and downward with tips barely skimming the ground, a veritable "scanning" process essential to following the trail, discovering prey, and keeping in touch with other workers. All dorylines possess certain resources essential to raiding which include an acute contact-odor sense used in dextrous trail following and trail communication: chemical products used in laying a continuous trail; strong mandibles and (in many) potent stings used in quick, vigorous attacks; stout (and, in many, long) and strong legs used in rapid trail running; and a strong body frame used in the body bracing and tenacious pulling and holding actions common in attacks on prey.

All dorylines raid on trails laid by the ants themselves, probably akin to those of hindgut products indicated for *Eciton* (Blum and Portocarrero, 1964) and *Neivamyrmex* (Watkins, 1964). In running these trails, the ants display a remarkable group cohesion. Their quick responses to one another are aided by an acute sensitivity to contacts and odors. As raiders advance on a trail, each of them in effect keeps approaching an attractive chemical that lies just ahead. As a test, the removal of a leaf over which the column runs breaks this trail; then workers gather on both sides of the gap or backtrack (Fig. 4.1). One can see that the ants increase in tension

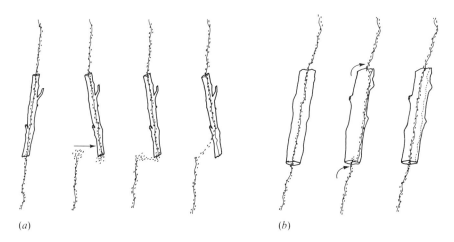

(a)    (b)

FIGURE 4.1.
Trail following by workers of *Eciton hamatum*. (a) In a field test, turning a stick disrupts the column of army ants but only at the displaced end. The column is going again within a few minutes and, as indicated, even shortens the route. Such tests show that the ants follow a continuous chemical trail made by them. (b) Rolling a log sideward a short distance disrupts a column of army ants only briefly; they soon resume their progress over the same trail although now at an incline. They travel best on a level surface, however, and soon have formed a new section of trail, shifting the old one (*dotted line*) gradually.

and agitation as evidenced by their quick jerky motions and by rapid oscillations of their antennae as they retrace the trail again and again. But before long some of them probe their way forward short distances from the break, others then go farther, and soon the column flows once more (Schneirla et al., 1966). If, as a control, the leaf is simply lifted and replaced at once, after a brief disturbance the column proceeds as before. When wind, rain, or passing animals displace ant viaducts over vines or leaves, the track is restored with surprising speed. Trail-interruption tests block progress similarly wherever they are tried along the route, showing clearly that the army ants make and follow a continuous chemical track in their raiding.

In first laying a trail, army ants behave much as they do in repairing a trail break. In the forward zone of raiding, a relay trail-laying process goes on that is essential for the assault, mopping up, transport of booty, and other actions of raiding that follow. Ants in the advance are not scouts in the human sense but temporary pioneers since trail pushing is done by any and all raiders that enter new ground (Schneirla, 1933). Each pioneer reacts at the trail end as though she had received a little shock there, then crawls forward with her body close to the ground in a tense, agitated way, antennae oscillating and tapping the substratum. These ants in the forefront usually advance only a few centimeters. Each of them rubs her abdomen against the ground, releasing thereby a substance that extends the trail; then she retreats. Slight as this trace must be, ants coming up behind her follow and extend it a little before they in turn are forced to reverse. In this way the raid advances, each group of ants taking its turn at trail laying as it reaches the front. All army ants in their forays, whether advancing on the surface or along underground channels, move in masses, columns, or small groups that leave scent traces, marking routes for use by others coming up in the rear. But, as circular column tests show, army ant trail running is not simply a matter of slavishly following an odor track.[2]

---

[2] Running in a column with nestmates is not simply a matter of following a chemical trail although descriptions (Wheeler, 1900; Gallardo, 1915) might have suggested this. Analysis (Schneirla, 1944c) shows that the ants change the scent trail according to numerous conditions that affect them as they run. The column may begin by following around the edge of a square object in the nest (Fig. 12.4), then become circular as the ants speed up. This suggests the effect of momentum on the course of the ants. Increasing the temperature speeds up the rate of rotation and also increases the diameter of the circular column. Indenting the column on one side with a card causes the circle to bulge out on the opposite side; dampening the ground or applying heat on one side has a comparable result. Each ant both follows a chemical and moves with reference to tactual and other stimuli from the movement of others (Fig. 2.4).

The success of army ants in getting their prey under a variety of conditions suggests humanlike abilities to many observers. Robert C. Wroughton (1892), for example, a forestry official watching the raids of *Aenictus* in India, was prompted to write:

> I have seen a strong column, marching on a white ant heap, detach. . . columns right and left, and the several detached columns enter the heap from different points of the compass. The notion irresistibly forced on anyone, watching these manoeuvres, is that they are either the result of preconcerted arrangement, or are carried out by word of command. (P. 178.)

In their raiding, army ants of diverse species often converge forces upon booty places, but this is done, as study shows, by dorylinelike methods Wroughton did not consider.

Operations in the raids, diverse and efficient as they usually are, call upon a fine repertoire of sensitive and reactive assets not related to "tactics" in the human sense of "planned actions." When the ants are active in the forest in groups, these resources are used to the fullest. The simplest type of "flanking" procedure usually arises when a small end group of a column-raiding species comes upon the nest of a booty insect. The ants spread to both sides around the border of the ring of home-colony odor that encircles the nest entrance and thus form two prongs moving to right and to left. Then as new forces press up in column from the rear, the ants pour into the victim's quarters from all sides, without, however, any such tactical plan or signals in the human sense as may have been implied by Wroughton's description. In comparable ways, a wasp's nest is soon covered and torn open (Fig. 4.2), and trail divisions are formed in column raids.

The masses of swarm raiders force their way in much higher numbers and more complex organization (Chapter 12). In the forays of *Eciton burchelli*, for example, great crowds of raiders sweep along the ground while auxiliary columns assault ground nests or at times mount high into the trees where they attack insects living in plant masses aloft.

The initial advance in all army ant raids is slow or rapid, steady or erratic, according to the numbers of participants, the degree of their excitement, and the situation. Because the mobile and highly sensitive antennae of the foragers play constantly over everything just ahead, raiding advances are influenced by the nature of the terrain and especially the location of booty odors. The advance is fast—often 20 m an hour in end groups of *Eciton hamatum* streaming along the tops of log or other canalized routes. The ants move forward slowly through

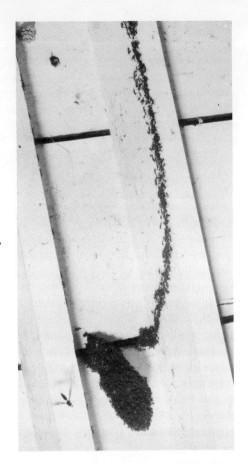

FIGURE 4.2.
Raiders from a nomadic colony of *Eciton hamatum* have thickly covered a small bulb nest of wasps and, after having expelled the adults, are ransacking the brood cells, leaving in column with quantities of pupae.

brush or booty-rich areas but push on rapidly over smooth ground. Topography influences their progress. Thus column raiders follow the contours of hills or the grooves of peccary trails; swarm raiders often follow the courses of ravines. Correspondences between the slope of the ground and the course of the raiders—e.g., nearly straight upward in mounting a vertical bank or with a deviation according to the angle of rise in passing up slopes (Crozier and Stier, 1928, 1929)—indicate the role of movement sense (proprioception) in the advance. Once a trail is well formed, column movement on it is rapid and steady.[3]

Distribution of booty notably affects where the raiders turn. They

---

[3] On moderately light, humid mornings in the forest, we timed workers in their rates of running in columns over level ground. Intermediate workers of *Eciton hamatum* outbound and unladen ran at rates between 19 and 33 sec/m; those inbound and laden ran at rates between 19 and 25 sec/m. Minor and major workers clocked under the same conditions were slower. Running in column is in general more rapid and variable in the raids than in the emigrations.

usually continue their forays into booty-rich localities but soon leave bare spaces or raided-out areas. Thus, comparably, nomadic colonies starting a new raid generally develop new routes, entering only briefly those used in the previous raid. Statary colonies, similarly, take different directions on successive days so that a map of their forays through the phase resembles a cartwheel figure (Fig. 4.3). This and comparable changes in the successive raids of driver ant colonies in their long nest stays (Cohic, 1948) bear an important relation to changing colony condition (Chapter 7).

Army ants contrast strongly with many other ants in their odor-track mode of reaching food sources from the nest. Foragers of *Formica* and other highly visual ants master long individual routes between nest and food by using light direction and other cues plastically (Brun, 1914; Schneirla, 1929; Jander, 1957). By contrast, foraging army ants utilize light only in minor ways as by withdrawing in groups from bright light in a fallen tree area or at the forest edge. Light, although forcing changes in their local movements, affects the main direction of the raid very little. Light, wind, rain, and ground moisture or dryness comparably influence the frequency of surface raiding in different species as well as the behavior of local groups.

Differences are notable in the raiding zones of army ants, even among species in the same genus. *Eciton burchelli* and *E. hamatum* are able to carry out colony raids by day but generally keep within the forest. In the tropical dry season, however, when surface conditions are often too harsh for them, these ants operate under surface cover, along ravines, or in shady humid areas. *E. mexicanum* and *E. dulcius,* which often use surface cover and underground channels even in the rainy season, expose their columns on the surface only in full shade. In dry weather they forage under even deeper cover or at night (Bruch, 1934; Schneirla, 1947; Borgmeier, 1955; Rettenmeyer, 1963).

*Neivamyrmex* is more restricted in surface raiding than *Eciton* (Borgmeier, 1955). *N. nigrescens,* for example, raids aboveground only at night—except on overcast, humid days—and operates under surface cover when the air and ground are especially dry. The sympatric species *N. opacithorax* is still less active aboveground, foraging well under cover or underground except in humid weather or when the ground is very wet. Species ecological differences (Chapter 2) clearly affect typical differences in the occurrence of surface raiding.

Nomadic colonies of surface-raiding *Eciton* species successfully carry out their daily forays on or near the surface despite natural interferences. In rainy weather, these ants manage to collect their daily hauls of booty despite even long downpours. Torrential rains bring surface action to a

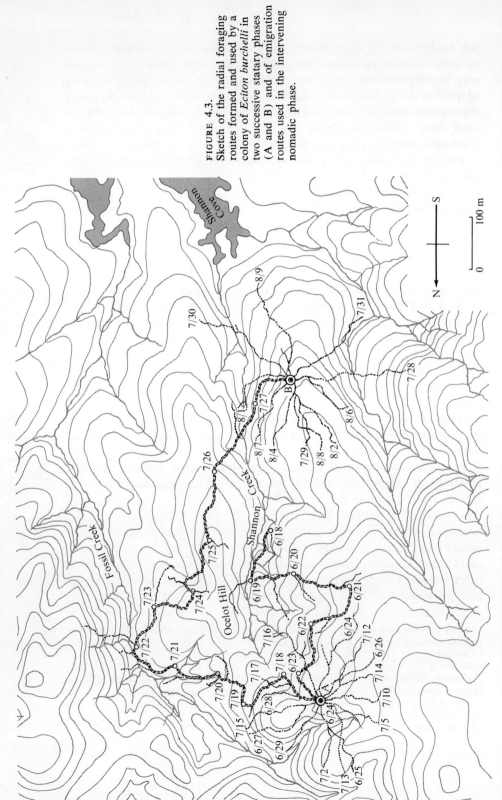

FIGURE 4.3.
Sketch of the radial foraging routes formed and used by a colony of *Eciton burchelli* in two successive statary phases (A and B) and of emigration routes used in the intervening nomadic phase.

halt as the ants disappear under leaves and other cover, but with a letup the ants are soon on their trails and in action. Light rains stop the raid only briefly; the ants first stop under surface cover but soon become adapted to the pelting and proceed to raid even in the open. Comparable disturbances from wind are met equally well. Doryline odor trails do not wash away readily (Schneirla and Brown, 1950), and where puddles cover them detour columns are formed.

Two types of army ant raiding systems are recognizable: column raids and swarm raids (Fig. 4.4). In column raids, one or more treelike systems of branching trails arise through the repeated division of small terminal foraging groups; in swarm raids, the advance groups generally become much larger. The many end groups in a column raid work over narrow strips of terrain in which the ants ferret out their booty; the main body in a swarm raid, in contrast, acts as a dragnet, flushing out a far greater variety of prey along a much broader front. Before discussing these types of raids, let us consider some general aspects of predatory behavior.

As a raid proceeds, stimulative effects from potential booty strongly influence the ants in their route and behavior.[4] The daily bag of prey remains high, even in dry weather, when booty is relatively inaccessible on the surface and must be captured in remote recesses (Schneirla, 1949; Schneirla and Brown, 1950). Versatile reactions of the raiders to odors and to slight movements enable them to reach those parts of brush masses, thickets, and leaf-strewn ground where booty is concentrated. Stirrings of wood roaches or centipedes attract the swarmers; they rush in; then the struggles of prey and odors from both attackers and victims attract still other raiders. Column raiders, not as responsive to movements of prey, dart to odors, then motion as potential victims are neared. To a bursting forth of prey or to stiff resistance at a discovered insect nest, the worker rises, antennae waving and mandibles opening and closing, then darts forward or retreats according to conditions. Army ants may even become so conditioned to familiar odors as to dash ahead or jerk back before the object appears and to catch the ground traces of their victims' nests from a distance. I have seen workers of *Aenictus laeviceps* and other column raiders turn sharply and dive quickly toward a small nest hole when still 2 or 3 cm away.[5] Conversely, raiders often withdraw from

---

[4] The same colony of *Eciton* can follow its own odor trails readily when they are reencountered after many weeks (Schneirla and Brown, 1950), and these trails seem to be chemically very stable (Blum and Portocarrero, 1964). From other evidence, colonies of the same species and even of different species use one another's trails (Schneirla and Brown, 1950; Watkins, 1964), and it is likely that any tropical forest bears a network of trail remnants guiding army ants to rich booty sources.

[5] Workers of *Camponotus* and other ants that are common victims of *Neivamyrmex nigrescens* give startle reactions to the surface trails of the army ants at distances of 1 to 2 cm.

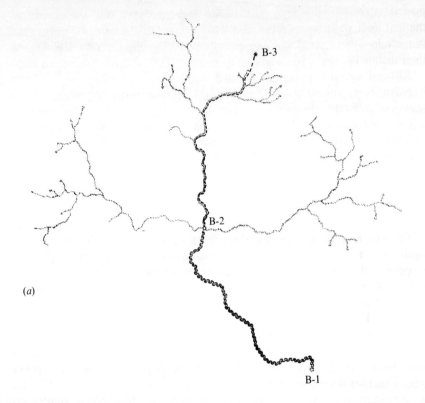

FIGURE 4.4.
Comparison between patterns of column raiding and swarm raiding. (*a*) A typical three-system column raiding pattern of *Eciton hamatum*, as sketched late in the afternoon. B-2: current bivouac, reached over an emigration route from B-1 on the previous evening; B-3: bivouac site reached in a nocturnal emigration over a former raiding route from B-2. Distance B-2 to B-3 approximately 250 m. (*b*) A typical swarm raiding system of *Eciton burchelli*, as sketched late in the morning. Biv: current bivouac; P: principal trail; F: fan; S: a network of columns behind the swarm; $S_1$: a new swarm formed by division of the main one about 11:00 A.M. Distance Biv to S about 95 m.

beetles or bugs at a distance in response to the odors of their noxious secretions. A familiar sight when army ants probe thickets is a large tick posted at the tip of a leaf, with its rear end pointed toward the base of the leaf and, perhaps a half-centimeter back, a row of raiders posted in a semicircle—their antennae stretched forward and vibrating—keeping their distance.

All surface-raiding dorylines, and, doubtless, subsurface raiders also, in advancing, surmount or bypass a variety of obstacles. Thus, when forays of different colonies happen to cross paths, there is little fighting as a rule, and no booty taking at each other's expense (Schneirla, 1938). Instead, workers of the two camps stand facing each other tensely with quivering antennae along a border of opposed picket lines behind which the two raids continue on opposite sides. At times, one raid ricochets from the other as the weaker colony or species gives ground. The swarm raiders generally stand fast, raiding as before, while the column raiders detour from them over vine bridges or under cover.

Blocked raiders often simply backtrack to branch trails on which they again advance until stopped once more, repeating the process as often as stoppages may require. In this way raiders of normally subsurface-active species (e.g., *Neivamyrmex opacithorax*), blocked below by seepage in rainy weather, often turn to the surface in their raids, even if not also attracted there by unusually humid air.

Communicative resources—actions, related secretions, and their by-products that can influence the behavior of colony mates—are vital to doryline raids. Contact usually combines with odor so intimately in group reactions that the passage of excitement from ant to ant along a column or across a swarm often brings results with lightning speed. Workers are attracted forward when they make frequent antennal contacts with returning ants that are doubtless bearing traces of booty odor combined with excitement odors.[6] This happens when advance foragers have found a new source of booty and return excitedly along a trail. But when booty is sparse, contact-odor stimuli are weak along the column, and newcomers drift away. A doryline raiding group thus advances steadily so long as it taps areas rich in booty—then its liaison runners are numerous and excited—but lags when little booty is found. Or a new line of advance may be abandoned as raiders retreat from strong resistance, reversing others as they meet them along the route.

As another resource of group predation, workers on nomadic raids, ex-

---

[6] Brown (1960) found that workers of *Eciton hamatum*, *Nomamyrmex esenbecki*, and *Labidus praedator* gave similar "alarm" reactions of disturbance and attack when presented in tests with the crushed heads of workers of their species (probably exuding a mandibular-gland pheromone). Such substances, in low concentrations, may excite and attract.

citable and strong in numbers, often smooth out their routes—and so improve the flow of traffic—with their own bodies. This occurs when the front of a swarm arrives broadside at the top of an overhanging bank, its hosts pressing ahead so strongly that strands and ropes of clustered bodies soon hang over the edge (Fig. 3.7). Workers crawl down over their fellows and hang, becoming immobilized in the lengthening cluster as others in turn hook downward from them. Finally the structure reaches the ground and serves as a ladder for traffic as the swarm presses on below the bank. Where traffic continues, such a ladder may serve for hours as a link in the trunk route behind the swarm. Let traffic on this structure slacken, however, and ant units in it soon stir, free themselves, and run off. With flanges widening paths along narrow vines, roadways smoothing rough sections of trail over leaves, and viaducts leveling the way over rocks in a stream bed (Fig. 12.1), these ants readily transform traffic obstructions into traffic-speeding devices (Chapter 12).[7] Similar formations arise in highly excited colonies of *Eciton hamatum* and other column-raiding species but less frequently and in less developed forms than in the swarm raiders.

Raiders of *Labidus, Dorylus,* and other subsurface-living army ants form walls of clustered workers on sections of their trunk routes aboveground, frequently with the addition of ragged walls or incomplete arcades of earthen pellets (Cohic, 1948). Such behavior, for reasons discussed in Chapter 5, arises more consistently in emigrations than in raids.

The column-raiding pattern is found in several doryline genera, both in surface-adapted species (e.g., *Eciton hamatum* and *Aenictus laeviceps*) and in hypogaeic species capable of surface activity (e.g., *Neivamyrmex pilosus* and *Nomamyrmex esenbecki*). Column-raiding species typically capture soft-bodied prey; they also attack less forcefully, with less excitement, and with less severe bites and stings than do swarm-raiding species, which take a much wider range of booty.

The simplest doryline foraging system yet studied is found in the column raiding of *Aenictus* (Schneirla and Reyes, 1966). In a developed foray of the surface-raiding *Ae. laeviceps* (Fig. 4.5*b*), a column from one to five ants wide meanders from the bivouac, ending in branches often more than 20 m out. Each branch column ends in a raiding group from a few centimeters to a few meters wide, depending on colony condition, stage of the raid, and the ground. These branches, as in other column

---

[7] Schips (1920) discusses a report of *Eciton burchelli* crossing a 12 m wide river by clustering into a thick, stiff rope that reached from one bank to the other and was used as a bridge. Savage (1847) mentioned occurrences of this kind in *Anomma arcens*. This might happen when a highly excited colony meets a water barrier that is not too wide and fast-flowing. I doubt that it occurs often as I have never seen army ants cross a running stream more than 2 m wide except by using logs, rocks, and the like.

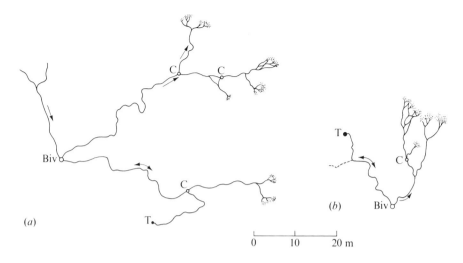

FIGURE 4.5.
Diagrams of raiding systems of *Neivamyrmex nigrescens* (*a*) and of *Aenictus laeviceps* (*b*), typical of the late nomadic phase. C: cache of booty; T: tree mounted in persistent raiding.

raiders, drop away as the raid moves outward, except for those invading rich sources of booty. Although *Aenictus* usually has only one such trail system, highly excited colonies may have two or more. The terminal raiding groups of nomadic colonies may be 5 to 6 m wide at the peak of the phase, with column networks that are swarmlike in appearance and behavior. These large raiding groups can spread, divide, and redivide rapidly. In the few statary raids of *Aenictus,* terminal raiding groups are much smaller.

The most complex and specialized type of column raiding is found in *Eciton hamatum.* A nomadic raid in this species usually develops three systems of trails, each with its base column from the bivouac, with successive branch trails, and with terminal raiding groups (Schneirla, 1933, 1938). One of these trail systems, when well developed, resembles a diagram of a river system (Fig. 4.4*a*). In pressing forward, the advance groups of a column raid probe the surface and low vegetation, prying beneath surface cover and into insect burrows and other niches. By contrast, colonies of this species in the statary phase usually build up only one relatively slender trail system (Fig. 4.6*b*). *E. mexicanum, E. vagans,* and *E. dulcius* also have similar trail systems, which, however, usually run at least in part under surface cover or through underground channels.

Among species-raiding patterns of the branching-column type, those of *Eciton* seem to be the most complex and those of *Aenictus* the simplest,

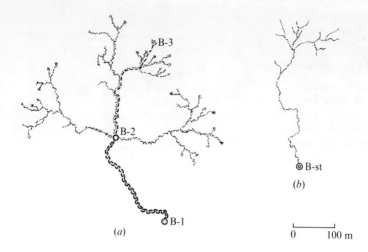

FIGURE 4.6.
Diagrams of raiding systems typical of *Eciton hamatum* (*a*) in the nomadic phase and (*b*) in the statary phase of a functional cycle. B-2: bivouac of one day in the nomadic phase from which a three-system raid developed; B-st: statary bivouac site.

with *Neivamyrmex* intermediate (Fig. 4.5). *N. nigrescens* uses surface cover much less often than *N. opacithorax*, also a nocturnal forager. Among species that raid both by day and by night in tropical forests, transitional forms are represented by *N. pilosus* and *N. gibbatus*, the former more like *N. nigrescens* in its patterns, the latter more like *N. opacithorax*. When colonies of both *Aenictus* and *Neivamyrmex* become highly excited, their end groups often break into small-scale swarm patterns, the extent and frequency of which differ with the species.

Colonies of *Eciton mexicanum, E. dulcius, E. vagans,* and others raid in branching-column systems similar to those of *E. hamatum* (Schneirla, 1947), as do *Nomamyrmex* spp. The difference is that in *Nomamyrmex,* a subsurface-nesting species, the basal columns shift underground as the raid advances. As a result, local forays of this species (probably from the same colony) are often seen in the same area at successive times on the same day but in well-separated places.

The swarm forays of a few species in the genera *Eciton* and *Labidus* and of most *Dorylus* species are the most spectacular raids of all. These ants are highly excitable, muster large numbers quickly, and are very energetic on the attack. They all have strong bites. The New World swarmers also have potent stings, but the driver ants have relatively weak stings. All of the swarmers take a much wider range of prey, including many hard-bodied arthropods, than do the column raiders.

The best-known swarm raider of tropical America, *Labidus praedator,* though its swarms are not the largest, is notorious for its stinging, biting masses that often invade human dwellings at night. Of this army ant, Belt (1874) said:

> A dense body of the ants, three or four yards wide, and so numerous as to blacken the ground, would be seen moving rapidly in one direction, examining every cranny, and underneath every fallen leaf. On the flanks, and in advance of the main body, smaller columns would be pushed out. These smaller columns would generally first flush the cockroaches, grasshoppers, and spiders. . . . The greatest catch of the ants was . . . when they got amongst some fallen brushwood. The cockroaches, spiders, and other insects, instead of running right away, would ascend the fallen branches and remain there, whilst the host of ants were occupying all the ground below. By-and-by up would come some of the ants, following every branch, and driving before them their prey to the ends of the small twigs, and nothing remained for them but to leap, and they would alight in the very midst of their foes, with the result of being certainly caught and pulled to pieces. (Pp. 18–19.)

The advance bodies of *Labidus praedator,* much smaller and shorter-lived than those of *Eciton burchelli* in the same forests, usually vary more in direction and group behavior than the latter. They are only small models of the great safaris of the African driver ants, whose colonies mass far greater forces over far greater distances. Although the swarms of *L. praedator* often cover the ground, it is generally a one-ant-thick carpet in contrast to the masses of ants pouring in layer upon layer that are mustered in driver ant raids (Vosseler, 1905; Cohic, 1948; Raignier and Van Boven, 1955). The drivers and *L. praedator* begin their raids by day or night when heavy columns pour out of cracks in the ground, from insect burrows, or from other avenues enlarged by digging. Behind the main body in a looser column network than in *E. burchelli,* local groups capture and tear up the prey. The subsurface-nesting swarmers, however, soon carry their booty underground along routes to which their heavy trunk columns shift as the main body advances. There are ample signs that these ants also do much raiding below the surface.

Of all swarm raiders, *Eciton burchelli* is most regular in its internal organization and the timing of its raids (Schneirla, 1940, 1944b). Usually, a large swarm moves ahead, with a fan-shaped network of columns trailing behind and in the rear a base column connecting with the bivouac. As such a raid near its height, after midmorning (Fig. 4.7), we see a great rectangular body of a few hundred thousand excited reddish-black workers of all sizes moving along and columns and masses eddying about. These ants often swarm in a single phalanx more than 15 m wide and

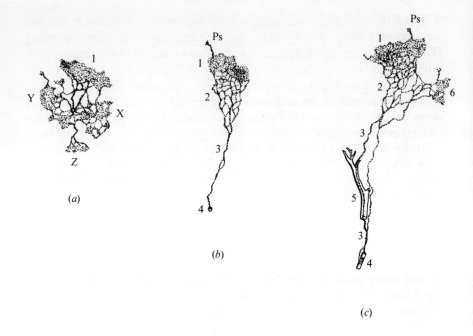

FIGURE 4.7.
Stages in the development of a swarm raid of *Eciton burchelli*. (*a*) About 90 minutes after dawn, ants spreading from the bivouac are now forming a swarm (1) advancing 8 m out and abandoning other sections (X, Y, and Z) of the expansion. (*b*) Sketch of the same raid at about 9:00 A.M. with the swarm now about 30 m from the bivouac. Ps: pseudopodic column; 2: fan of columns; 3: basal trail connecting with the bivouac, 4. (*c*) Raiding system of *Eciton burchelli,* sketched at 11:00 A.M. 5: alternate basal route over large log previously raided; 6: new swarm just formed by splitting from the large one.

one to two meters deep—and I have seen them as wide as 25 m. This complex of scurrying ants moves along broadside, covering the forest floor and acting as a great net from which columns probe here and there into brush, ground vegetation, and even high into the trees—with surprising steadiness and good direction despite all the complications.

Swarms of *Eciton burchelli* are better organized and hold one direction of advance much better than those of *Labidus praedator* or *Dorylus*.[8] This we see in the regular flanking movements of the body, alternately to right and to left, carried out with one wing serving as pivot while the other wheels broadly (Fig. 12.2*b*). These occur at intervals of ten to twenty-five minutes, depending upon the condition of the colony and

---

[8] Comparisons of swarm behavior in *Eciton burchelli* and *Labidus praedator* lead me to attribute the greater variability in organization and direction of progress in the latter to the less specific direction of progress of recruits arriving from underground (Chapter 12).

stage of the raid and upon such occasional interruptions as a massive invasion of a fallen tree thicket or a great tree. Such flanking operations (Chapter 12) reveal a superior organization to that in other swarm raiders, in which they are much less distinct. But what the swarms of driver ants may lack in mass organization, they make up for in their ability to concentrate great numbers quickly and attack large prey with more power than can *E. burchelli*—which, to be sure, is formidable enough in its own right.

To most of the harried small life of the forest, which responds by running, hopping, climbing, flying, or standing—each in its own manner—as the raiders pass, the broad front of the swarmers present a greater hazard than the slender tentacles of the column raiders. The swarmers owe their effectiveness to their wide sweeps and greater numbers, also to their workers' vigorous and persistent attacks, their more potent bites and, in *Eciton burchelli* and *Labidus praedator,* stings. Scorpions and other large victims given to strong resistance are first pinned down by raiders anchored firmly by their tarsal hooks, then spread-eagled by oppositely pulling groups, torn apart, and finally carried off to the rear in pieces.[9] The swarmers, pouring into cracks and crannies, catch their victims both by overrunning them and by darting at sources of motion and odor. Few insects escape them consistently, not even those nesting in plant growths high in the great trees.

Doryline swarm raids, although impressive as predatory operations, do not "consume everything in their path," as the fable writers have it. They do not harm the vegetation and, except in their most excitable forays, may miss up to half of the potential victims in their path (Schneirla, 1949; Rettenmeyer, 1963). Not many vertebrates seem to be killed by the New World swarmers although I have seen them kill snakes and lizards that they have trapped between the buttressed roots of trees and bite and sting to death nestling birds intact that were left behind. Swarmers of *Eciton burchelli* may subdue smaller vertebrates that they overrun or trap in sleep, killing the victim by mass attacks in response to its struggles, death coming either through shock or asphyxiation or both. For others like the giant toad of Central America *Bufo marinus* that always seems to escape, being routed from daytime retreats is still hardly a minor affair as it involves being stung and bitten in the bargain. It is the

---

[9] For a laboratory test of gang pulling, I put several hundred workers of *Eciton burchelli* into a shallow observation nest which had strips of doubled cloth loosely cemented to the tops of the four walls under the glass; then I began to pull gently outward on one of the strips, back and forth, with tweezers. At once the motion attracted some of the ants to grip the cloth and pull inward. This soon brought many others, and in a short time a row of them was pulling at the cloth all along the wall. Before long the cloth was ripped loose by the strong work of many chains of ants, formed by jaws gripping legs in series, all well anchored to the floor and pulling steadily.

driver ants of the Old World, however, with their cutting, shearing mandibles and great numbers that have earned the reputation of killing and tearing to bits large and small forest animals and domestic stock unable to escape them (Savage, 1847; Loveridge, 1922).

Swarm raids can be located through the calls of antbirds attending them; impressive solo effects stand out against the audible background rustle made by the ants as they run over leaves and vegetation and capture prey and by potential victims struggling or escaping. Among the familiar sounds of tropical American forests often heard across ravines and through tangles are: the excited whistling of the spotted antbird; the sharp accelerating cheer "wheeta, wheeta" of the bicolored antbird; the soft but penetrating whistled crescendo series of the ocellated antthrush; then, at closer range, the distinctive low "churr-r-ing" and higher "chirr-r-ing" of some of these same birds. The chorus at its height involves a great variety of piping, twittering, whistling, and churring delivered in notes or in short phrases by the avian swarm attendants.

Spotted and bicolored antbirds are common followers of ant swarms, flitting about in the low vegetation close to the eddying masses. At intervals one or another of them flashes down, snatches up some scurrying insect flushed out by the ants, then perches again. Its booty swallowed, the bird utters an excited call, preens itself a little, and is ready for another foray. These birds generally feed on arthropods run out by the ants, rarely taking the raiders themselves except by accident or when booty is scarce.

Swarm raids of nomadic colonies of *Eciton burchelli* are so readily seen and even heard and drive out so many insects that they attract a whole host of birds.[10] Stages in the study of the antbirds are represented

---

[10] Johnson (1954) reported that such birds as the spotted and bicolored antbirds are given to following ant swarms that pass through their areas whereas others including the ant tanager, woodcreepers, and cuckoos attend the swarm usually for short periods only and may be attracted socially to the flock. He observed one large swarm raid which at its height was attended by twenty-one birds of thirteen different species. Willis (1960, 1967) has carried out long-term studies on Barro Colorado Island, banding several individuals in every study for records from year to year. He describes the following birds as "professional followers" because of their constancy in attending swarms:

| | |
|---|---|
| Spotted antbird | *Hylohylas naevioides naevioides* |
| Bicolored antbird | *Gymnopithys leucaspic bicolor* |
| Ocellated antbird | *Phaenostictus moleannani moleannani* |

Each pair, he finds, has a central area (usually around the nest) in which it "dominates" other birds attending raids, generally taking the topmost position in perches and flights over the swarm, chasing others "with a puff and a snap." Their frequent calls probably play a part in these behavioral relationships.

Many birds in the New World family Formicariidae are common among those that follow or visit swarm raids. Certain African birds are comparably associated with driver ant swarms (Bequaert, 1922).

by the studies of the American ornithologists Frank Chapman (1929), Robert Johnson (1954), and Edwin Willis (1960, 1967), carried out to a large extent on Barro Colorado Island. Some of these birds must have evolved in close relation to the swarm raiders, so intimately are their lives bound up with them. The spotted and bicolored antbirds, for example, appear so regularly at scenes of swarm action and around bivouacs, where their soft calls are often heard soon after dawn, as to suggest their having learned a variety of cues from the ants. They are most in evidence when the swarm is highly active and the booty heavy, thus with nomadic raids, but much less with statary raids and more often with the raids of *E. burchelli* than with those of *Labidus praedator*. Rarely does one appear, even briefly, at a column raid. They respond to the masses of the ants themselves, to signs of swarm activity, and to cues of booty with an alacrity surely based on experience with food in the presence of these stimuli. Their music and lively actions are an unforgettable part of this forest phenomenon.

Another striking feature of the swarm raids is a buzzing cloud of flies; some types hover, spiral, or dart over the main swarm or its fringe, others over the network of columns behind the swarm (Bequaert, 1922; Curran, 1934). Often variable notes of higher pitch are heard as flies swoop down here and there upon insects that burst suddenly into view. They respond to visual effects from the movements of harried prey as in tests they dart upon small pieces of light sponge jerked about quickly among the ants. After a swift descent, each fly rises rapidly, having barely touched the apparent target. But in this instant it can release an egg, which may be carried into the bivouac on booty with a chance of developing among the doryline brood. The life cycles of these flies and of other insects found with colonies may be regularly intermeshed with those of the ants (Akre and Rettenmeyer, 1966). Certain other flies at times also capture booty by swooping down on the columns, as we have observed in raids of *Aenictus laeviceps*.

Doryline raids in different species differ greatly in the ground they cover. In nomadic raids, the usual distance reached from the bivouac is 15 to 35 m in *Aenictus laeviceps,* 25 to 55 m in *Neivamyrmex nigrescens,* 20 to 50 m in *Labidus praedator,* 70 to 140 m in *Eciton burchelli,* and 140 to 350 m in *E. hamatum*. But distance is not very reliable as an indicator of raiding strength. I have observed, for example, strong raids of *E. hamatum* that moved out less than 40 m in fallen tree thickets rich in booty, and weak statary raids of this species that advanced more than 350 m from the bivouac when the same trail was used on more than one day (Schneirla, 1949). The distance of a raid depends especially upon the terrain, the distribution of booty, and the condition of the colony.

The raids of group A dorylines differ greatly in the nomadic and the statary phases of the colony cycle (Chapter 7). Colonies of *Eciton, Neivamyrmex,* and *Aenictus* are in their most excitable condition in the nomadic phase when they carry out their heaviest raids. Moreover, as this phase goes on, the raids increase in vigor, the workers attacking more and more persistently and in larger numbers and retreating less often from resistance. Results summarized in Table 4.1 for three colonies of *E. hamatum* indicate heavier traffic in the nomadic phase than in the statary phase.[11]

All of these dorylines carry out successive statary raids in different directions from a fixed site (Fig. 4.3). Raids by statary colonies of *Eciton hamatum* usually have just one slender column with shorter branches (Fig. 4.6b) in contrast to the multiple trail systems of many branches and strong columns typical of the nomadic phase. Nomadic colonies nearly always begin their raids at dawn; statary colonies begin more slowly or do not raid at all. Colonies of *E. burchelli,* when statary, usually develop small, relatively unexcited raids that begin slowly or at times soon retire into the bivouac, attract few or no birds, and gather much less booty than when nomadic. The greatest difference is seen in colonies of *Aenictus,* which from a highly aroused condition when nomadic lapse into a statary condition usually with no surface raids at all except on the first few and the last few days of the phase (Schneirla and Reyes, 1966). The basis of these differences is discussed in Chapters 6 and 7.

A doryline raid is an exodus in which booty is captured; it is followed

TABLE 4.1.
*Raiding traffic (ants/min) between 10:30 A.M. and noon in three colonies of* Eciton hamatum, *through intervals of 6 days or more in each activity phase*

| | | | Outgoing | | Incoming | | | |
| | | | | | Unladen | | Laden | |
| Colony | Phase | No. of days | Range | Avg. | Range | Avg. | Range | Avg. |
| --- | --- | --- | --- | --- | --- | --- | --- | --- |
| '52 H-C[a] | Nomadic | 12 | 65–242 | 130 | 27–171 | 106 | 27–166 | 80 |
| | Statary | 11 | 5–143[c] | 58 | 2–147[c] | 43 | 2–86[c] | 27 |
| '52 H-O[b] | Nomadic | 11 | 26–122 | 60 | 14–80 | 50 | 2–60 | 27 |
| | Statary | 6 | 3–58[c] | 15 | 6–28[c] | 12 | 3–26[c] | 8 |
| '49 H-35[b] | Nomadic | 8 | 24–108 | 61 | 19–146 | 67 | 15–126 | 55 |
| | Statary | 17 | 18–92[c] | 28 | 24–74 | 26 | 3–54[c] | 20 |

[a] An especially large colony for the species.
[b] Near normal population size for the species.
[c] Days without raids or not visited.

---

[11] In this field study, carried out in 1952 at Barro Colorado Island, I was assisted by Carl W. Rettenmeyer, now professor of biology at Kansas State University.

by a return to the nest or an emigration, depending upon colony condition. There are important generic differences in the timing of raids (Fig. 14.2). Raiding has its most distinct diurnal routine in surface-adapted *Eciton* species whose colonies start raiding at dawn and end at dusk (Schneirla, 1938); raiding has its most distinct nocturnal routine in *Neivamyrmex nigrescens* whose colonies start surface raids at dusk and end them before dawn (Schneirla, 1958). Surface-adapted *Aenictus* species, although at times influenced by dawn and dusk, may, when nomadic, begin raids either by day or by night (Schneirla and Reyes, 1966). Hypogaeic dorylines that are surface-active are also variable in their times of raiding. For example, *Dorylus (Anomma) wilverthi* in the Congo begins raiding most frequently at dusk, less often at dawn, and never in midday (Raignier and Van Boven, 1955).

The hypogaeic dorylines *Labidus praedator* and *Nomamyrmex esenbecki* of tropical America are known to raid both day and night. Less well known is *L. coecus,* of which Bates (1863) wrote:

> The armies move . . . as far as I could learn, wholly under covered roads, the ants constructing them gradually but rapidly as they advance. The column of foragers pushes forward step by step, under the protection of these covered passages, through the thickets, and on reaching a rotting log, or other promising hunting ground, pour into the crevices in search of booty. (Pp. 364–365.)

This species seems to raid around the clock and to be active mainly underground except on dark, humid days or when termite nests or other sources of booty attract its columns upward (Borgmeier, 1955). Its activity schedule should be interesting to study in comparison with that of *Labidus praedator.* (Generic differences in the timing of raids are considered in Chapters 5 and 7 in relation to schedules of emigration.)

Surface-adapted *Eciton* species, whose raids are closely related to the daily march of environmental events, offer the best opportunity to study the development of mass forays. I shall now summarize a study conducted on Barro Colorado Island of a representative nomadic raid carried out by a colony of *E. burchelli* (Schneirla, 1940). Measures in the growth of this raid are plotted in Figure 4.8 (main stages are sketched in Fig. 4.7).

Quiescent in the cool hours before dawn, ants on the bivouac surface began to stir with the first faint light. By sunrise, at 6:00 A.M., they began a radial expansion on the ground, increasing with rising light intensity and temperature.[12] The excitement and spreading grew until at

---

[12] This was a cooperative study in which Dr. Orlando Park, zoologist of Northwestern University, and his associates took environmental records and notes on

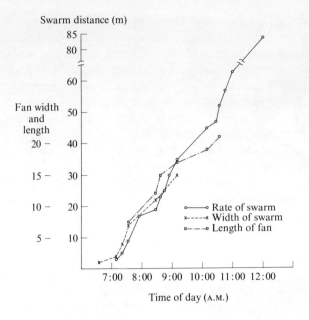

FIGURE 4.8.
Rate of development of a typical swarm raid of *Eciton burchelli* from dawn to noon (measurements by pacing). Measurements of swarm width and fan length stopped after complications (i.e., events leading up to swarm division) made them unreliable.

6:33 A.M. a small mass had pushed out 3 m on one side of the bivouac. The expansion mounted with excited returns of ants to the bivouac. By 7:00 A.M., the southwest part of the expansion had gathered forces both from the bivouac and from areas under evacuation. At 7:15 A.M. the foray had become a distinct swarm, advancing broadside outward, and pushed by masses surging up from the rear. Shortly after 7:30 A.M., as recruits flooded into the swarm, the ants began to attack prey and the first antbird was heard churring nearby.

vertebrate activities as I worked with the ants. Light readings at dawn were as follows:

| Time (A.M.) | Light intensity (foot candles) | Temperature (°F) |
| --- | --- | --- |
| 5:50 | 0.1 | 75 |
| 6:05 | 0.5 | 75 |
| 6:08 | 1.0 | 76 |
| 6:21 | 2.0 | 76 |
| 6:33 | 3.8 | 77 |

FIGURE 4.9.
Basal column of a raid of *Eciton burchelli* issuing from the bivouac (under root mass) around midmorning.

At 7:30 A.M. the growing swarm was connected with the bivouac by a network of columns. An hour later, when the swarm had widened to about 10 m and was out 20 m from the bivouac, a fan-shaped network of columns formed in its rear now connected with the bivouac by a wide trunk column on which the ants mainly rushed outward into the raid. At 9:00 A.M. the swarm was 15 m wide and still growing—as were the fan and base column—while the exodus continued from the bivouac (Fig. 4.9).

As the sections of this raid grew in step with one another, the scope and vigor of raiding increased steadily until at 11:00 A.M. the swarm front was more than 60 m from the bivouac. The pattern of the foray at this time, typical of nomadic *Eciton burchelli,* is shown in Figure 4.7.

The giant army in such a raid, as it pushed outward, cuts a swath of increasing width and moves at a fairly steady rate of 10 to 12 m per hour while keeping to a main direction of progress. In the case shown in Fig.

4.7, the swarm made two wide swings within thirty-five minutes, the first slowly to the left through a booty-rich area, the second rapidly to the right across open ground.

Nomadic swarms of *Eciton burchelli* commonly advance steadily, at least until midmorning (Fig. 4.8) when signs of disorganization appear. This change (Chapter 12) is marked by variations in rate of advance and in internal swarm behavior attributable chiefly to excessive numbers in the swarm and to extreme fluctuations in mass pressure from the rear. Loss of swarm organization is marked by a growing tendency of the flank sections to diverge. Finally, they split away from each other and continue raiding independently except for columns joining them in the rear.

As the forefront of the raid advances, the subswarms divide again until by midafternoon the raid is led by numerous bodies of foragers penetrating well-separated areas. Because these various advance groups are connected only by columns in the rear, nomadic raids of *Eciton burchelli* in their late stages resemble the branching-column raids of *E. hamatum* except that their end groups are much larger.

The basal columns are communication lines between the raiding front and the bivouac; hence they are good posts for reviewing changes in the main raid.[13] The truth of Bates' statement (1863), that in this column one sees two trains of ants running oppositely, depends upon the time. In a nomadic raid of *Eciton burchelli,* after the early exodus the principal trail usually is dominated by outgoing traffic until about 9:30 A.M. (Fig. 4.9) Returning ants have difficulty bucking this stream and are squeezed to the sides. The majors, particularly, after repeated collisions, are shunted to the column borders where standing or moving they serve (incidentally) a protective role. The majors, sometimes spoken of as "officers" (Bates, 1863; Belt, 1874), do control traffic in a sense though only incidentally as by simply getting in the way at junctions, thereby contributing to important changes in the raid (Chapter 5).

A period of mixed traffic then begins in which more and more returning ants, laden and unladen, get through to the bivouac. The homeward-bound stream now takes over the route, first behind the fan, then closer and closer to the bivouac. One now sees long trains of foragers with individual loads, and larger ants running in tandem, straddling

---

[13] A chance observation shows how effectively the basal column of a nomadic raid of *Eciton burchelli* holds its own against disturbances. Early one afternoon in a Panama forest I came across a large boa constrictor, coiled and asleep, with its nearest part only about 30 cm from a wide trunk column of *E. burchelli*. From the circumstances, this snake, having engulfed an agouti or other small mammal, had been moving downhill toward a ravine but had settled in place where the procession of ants stopped it. Thirty minutes later it was still asleep in the same place as the column proceeded.

FIGURE 4.10.
Submajor workers of *Eciton burchelli* carry in tandem a sizable piece of booty, the tail of a scorpion. (Photo courtesy of C. W. Rettenmeyer.)

grasshopper legs or carrying other sizable burdens (Fig. 4.10). They come from the front of raiding, from mopping-up zones, or from caches in the area of the fan and from brush masses or trees in the rear where raiding continues. Nearly always the reversal of traffic is still in force at 11:00 A.M.

These changes become clear through studies of subgroup relations within the raiding system. The initial exodus in nomadic raids of this species at its height is a 2 to 3 cm wide ribbon of hurrying ants, which exerts a steady, broad pressure on the base of the swarm into which it flows. The main factors leading to the reversal of this exodus after midmorning are mounting booty-laden returns, also the breakdown of swarm unity as the advance body grows very large. Thus, in nomadic raids of *Eciton burchelli,* both the morning exodus and its reversal are long and complex but highly predictable occurrences (Schneirla, 1944b).

Nomadic *Eciton hamatum,* as diversified in behavior as *E. burchelli,* also carries out a lengthy morning exodus and a complex reversal stage in its expansive column raids (Schneirla, 1933, 1938). In its case, usually, there is first a radial outpouring through which a column network spreads from the bivouac and gradually gives rise to three trunk columns—less often two[14]—each with many terminal branches and raid-

---

[14] This is a predictable matter, depending especially on the numbers and excitement of the early expansion and the presence of trails of the previous day's raid leading from the bivouac. Near the end of the nomadic phase, when colony excitation is high or more generally in large colonies, the expansion mounts rapidly and is likely to result in three trunk lines from the bivouac. Between the second and fifth nomadic days (N-2 to N-5) or in smaller colonies, only two trunk lines may arise.

ing groups (Fig. 4.6a). The exodus, as in *E. burchelli,* is kept going by a mounting excitement within the bivouac. This entire movement is usually strongest on the trunk line of greatest booty return.

In these great column raids, many conditions determine which branch trails drop away and which persist, e.g., booty haul, momentum, and ease of advance. Trails to trees and to brush heaps, rich booty sources as a rule, are likely to remain longest in use and attract recruits best both through excitatory stimuli from returning raiders and through recruits drawing others in after them.

In a nomadic raid of *Eciton hamatum* around midmorning, widely separated raiding groups head far-flung columns at distances of 150 m or more from the bivouac. On one raiding system, inbound laden traffic becomes dominant; on the others, traffic may head first one way then the other or may be mixed. The traffic seems chaotic with much crowding and bumping, but usually the strongest forces hold the center of the odor path. The smaller travelers are often overrun; the majors, through frequent buffeting by faster runners as in *E. burchelli,* are often forced aside.

In these massive raids of *Eciton,* traffic on a basal route does not always indicate what transpires in the raiding. Light traffic or currents of unladen ants on a base trail may or may not be a sign of a light capture of booty in that area of raiding. As an example, heavy booty returns in one part of an area could lead to a redistribution of forces that blocked bivouac-bound traffic at many junctions, in which case the booty would go mainly into caches; then, for some time, only a light, sparsely laden procession would be reaching the bivouac.

What makes a forager turn back toward the bivouac when she picks up booty? The process may be complex as laden ants often run about for a time before they become clearly directed toward the bivouac. Several types of cue may influence the behavior of ants that are laden. On picking up booty, the load, through contact, odor, and taste, may introduce a set toward running *with* other laden ants and *against* unladen ants. At junctions, one cue useful to these ants may be spots of *Eciton* chemical left at the bivouac-ward turn by the first excited laden ants to pass; for example, most booty carriers first to arrive at trail-division points in the morning take the bivouac-ward turn, perhaps turning to ant chemicals alone. A second cue may be the scenting of this turn with booty odor; for example, when pieces of booty are rubbed lightly along the false turn at a junction, returning ants hesitate, and many of them take this turn. Both the results of trail-switching tests and the behavior of laden ants after "wrong" turns indicate that odor cues exist. Several factors seem to favor homeward progress by laden ants.[15]

---

[15] Results of my trail-orientation tests with workers of *Eciton* both laden and unladen may be summed up as follows. Workers displaced to one side from the

Throughout the morning build-up of raiding and even into early afternoon, laden ants of *Eciton hamatum* generally turn correctly at junctions. Then wrong turns begin to increase. The reason may be that the afternoon exodus so disrupts incoming traffic that laden ants arriving at junctions often turn into a branch leading *away* from the bivouac. Odor differences may thereby become obscured at junctions where many traffic reversals have occurred. Thus, after midafternoon anyone following the turns of booty carriers may be led as often away from the bivouac as toward it.

Raiding in the known surface-adapted species of *Eciton* is diurnal and ends at dusk. Prior events in raiding have a typical schedule which is clearly related to the sequence of environmental events. After the morning exodus, raiding slackens through midday in what may be called the "siesta effect" (Schneirla, 1949).[16] With activities then reduced through the lethargizing effects of strong light, heat, and often drought, most ants get under cover, and traffic on base trails falls off greatly. With the prey also inactive and secluded, predatory action is then much reduced.

A recovery from this effect early in the afternoon brings a strong exodus from the bivouac and a resurgence of raiding. In *Eciton burchelli* there is then a notable increase in the size and rate of advance of swarms operating far apart at trail termini. In all surface-active *Eciton*, the crescendo in afternoon raiding—with sharp increases in cache storage of booty and in traffic complications—is basic to the transition from raiding operations to emigration. With falling light intensity in the forest, the ants attack prey less often and less vigorously until at dusk the general assault has stopped. In *Eciton*, these events (Chapter 5) are prerequisites to emigration.

---

trail return often enough by circling within 10 to 15 cm (30 to 40 cm in *E. burchelli*) to indicate a lateral chemical gradient from the trail outward. Tests involving end-to-end turning of sections of trail up to 1 m long are negative (provided section ends are smoothly joined), indicating no longitudinal polarization of the trail. Other tests, though, in which widely separated sections of trail are exchanged, suggest chemical differences between sections of trunk trails near the bivouac and far out. For example, queen odor and booty odor may be heavier at the bivouac end.

Cues differentiating turns at trail junctions include the angles of approach, most of which are Y's facing outward, thereby helping returning raiders into the bivouac-ward turn by momentum. One indicated chemical cue is greater concentration of attractive odor (e.g., booty) on the bivouac-ward turn. That a distinctive odor attracts returning laden ants to take this turn is indicated by their quick reversals after going just 1 to 2 cm into the "wrong" branch and by results from trail-exchanging tests in which laden workers usually take the former bivouac-ward turn on whichever side it is placed (as a cut-out section of leaf).

[16] This stage of reduced action is particularly marked in the dry season when traffic on the base trails of statary raids of *Eciton* at times stops for long distances from the bivouac (Schneirla, 1949; Schneirla and Brown, 1950).

As mentioned above, raiding is greatly reduced in the statary phase; in *Eciton* it is low even in the first few and the last few days of this phase when the afternoon exodus is weak and a general return to the bivouac begins around midafternoon. This reduction is marked in *E. hamatum* by just one trail system in contrast to the multiple systems of the nomadic phase (Fig. 4.6b). In the early part of the statary phase, in both *E. hamatum* and *E. burchelli,* this system spreads from the bivouac outward; later the ants use old trails and do not expand much as a rule until they reach relatively untapped territory farther out. A comparable condition arises in the driver ants (Cohic, 1948).

Although, in the early statary days of *Eciton,* the return of raiders extends well into the night, by midphase the raids end much earlier, at times even before dusk. This change is also indicated by the beginning of raids more slowly and later in the morning, sometimes not until early afternoon, and sometimes not at all (Schneirla, 1938, 1949). These signs have their parallels in other dorylines. The dropping out of raids is more frequent in *E. burchelli* and *Neivamyrmex nigrescens* (Schneirla,

TABLE 4.2.
*Statary phase raids in colonies of three genera* [a]

| Species Colony | Day in phase | | | | | | |
|---|---|---|---|---|---|---|---|
| | 1 | 5 | 10 | 15 | 20 | 25 | 30 |
| *Eciton hamatum* | | | | | | | |
| '36 H-A | r r r r r | r r r r r | r r r r r | r r r | | | |
| '46 H-B | r r . r . | . . . – r | . – . . . | r r r . r | r | | |
| *Eciton burchelli* | | | | | | | |
| '46 B-I | r r r r . | r r . . r | . . . r r | r . r r r | r | | |
| *Neivamyrmex nigrescens* | | | | | | | |
| '56 N-Br.[b] | r r r r r | r . r r r | r r r r . | r r | | | |
| '56 N-Cr.[c] | . . r r r | r r r . r | r r – r – | r r – r r | | | |
| *Aenictus laeviceps* | | | | | | | |
| '61 Ae. 1.-XXVIII | . r r . . | . . . . . | . . . . . | . . . . . | . . . r r | . | |
| '62 Ae. 1.-IV | r . . . . | . . . . . | . . . . . | . . . . . | . . . r r | r r r | |
| '62 Ae. 1.-VIII | r r . . . | . . . . . | . . . . . | . . . . . | . . . r r r | r r | |
| '63 Ae. 1.-I | . r r r r | . . . . . | . . . . . | . . . . . | . . . r r r | r r | |
| *Aenictus gracilis* | | | | | | | |
| '61 Ae. g.-16 | r r r r . | . . . . . | . . . . . | . . . . . | . . . . . | . . . | |
| '62 Ae. g.-5 | r r r r . | . . . . . | . . . . . | . . . . . | . . . . . | . | |
| '62 Ae. g.-10 | r r . . . | . . . . . | . . . . . | . . . . . | . . . . . | . . r | . . r |

[a] Each r represents a day with raids; each dot, a day without raids; each hyphen, a day not visited.
[b] Bridge colony.
[c] Creek colony.

1958) than in *E. hamatum* and in *Aenictus* is so extreme that colonies seldom raid at all except in the first few and last few days of the phase (Schneirla and Reyes, 1968). Results on generic differences in the frequency of statary raids are summarized in Table 4.2.

The reduction of statary-phase raiding is marked in *Eciton burchelli* by smaller, slower moving swarms limited in their booty to grasshoppers, ant brood, and other easily taken booty. Individual attacks on prey are less lively and energetic, and group communicative behavior is then much more sluggish than in the nomadic phase. These differences, significant for the basis of army ant behavior, are discussed in Chapters 6 and 7.

A traditional explanation offered for army ant nomadism has been booty scarcity. In a somewhat different approach, Wilson (1958a) has suggested that the dorylines may have evolved through primitive stages in which group predation became specialized in the type and size of booty taken. Before discussing these matters, we will consider the subject of booty in existing dorylines.

All army ants are carnivorous in that their main or sole food consists of proteins and other substances obtained from the tissues of captured animals (Brues, 1930). There are notable species differences in dietary range and types of food taken, influenced chiefly by the following factors:

1. Species habitat and vertical levels of nesting and of action.
2. Worker structures, e.g., size and strength, types of mandibles and mouth parts, length and stoutness of legs, potency of sting, hardness of exoskeletal armor.
3. Workers' chemotactic thresholds, speed of action, neural coordination.
4. Species' raiding pattern and level of excitability, readiness to attack, rate and numbers of recruitment.
5. Colony population, ranges and types of colony conditions, degree of worker polymorphism.

The first point has much to do with typical booty but not because dorylines are prone to take prey close to their nests.[17] Rather, nesting behavior is related intimately to group predatism and nomadism in many existing army ants. Hypogaeic dorylines, for example, have so evolved as to subsist to a great extent on termites, also on leaf-cutter ants and their brood, and through these affinities often nest in the preempted quarters of their victims (Wheeler, 1936; Borgmeier, 1955). The evo-

---

[17] Most dorylines at the beginning of a foray from their bivouac are much less excited (and less responsive to booty stimuli) than later when the raid has advanced farther out.

lutionary relationship of points 1 and 2 is suggested by the slender bodies, weaker stings, shorter legs, and thinner exoskeleton more common in hypogaeic than in epigaeic species.

Bearing on point 2, swarm raiders of the Old World have cutting, shearing mandibles and thereby take vertebrates as prey whereas *Eciton burchelli* and other swarm raiders of the New World are largely restricted to an arthropod diet. Differences in factors listed under points 3 and 5, and especially in the ability to muster great numbers rapidly, also aid the driver ants to overcome large prey.

*Dorylus, D. (Anomma)*, and *Labidus praedator*, by virtue of their patterns of mass attack, capture a wide range of booty that Brues characterized as a "veritable arthropod chop suey" (1930). Factors basic to this pattern include a high level of excitability and resources for speedy recruitment. Colonies of *Eciton burchelli* form their largest swarms and capture their greatest range and amount of prey at the peak of their excitement in the nomadic phase. At a comparable time, *E. hamatum* captures a much larger proportion of adults than in the statary phase, then overcoming large colonies of *Atta* able to fend them off at other times. Colonies of *Neivamyrmex nigrescens* at their nomadic peak attack colonies of the harvester ant, *Pogonomyrmex* spp., ordinarily unassailable by them. Corresponding differences, dependent on colony condition, arise in all army ants.

Column-raiding dorylines, which are generally slower in combat than the swarmers and have smaller end groups and slower recruitment, capture a much narrower range of prey than the latter—mainly the soft-bodied brood of ants, wasps, and other social insects. Colonies of the hypogaeic column raiders *Eciton vagans* (Mann, 1916) and *Nomamyrmex* spp., weaker on the attack than *E. hamatum*, commonly take such easier ant prey as *Camponotus*.

Certain column-raiding species in the hypogaeic *Neivamyrmex* (e.g., *N. pilosus* and *N. gibbatus*) resemble *Eciton hamatum* in their types of booty, which, however, are generally smaller in proportion to the lower size range of their workers. There are exceptions, however, as several smaller dorylines (e.g., *N. nigrescens* and *Aenictus* spp.) whose raids approximate the column type (Schneirla, 1958; Schneirla and Reyes, 1966) take a wide range of booty, including hard-bodied arthropods and termites. Despite similarities in raiding pattern to *E. hamatum*, these dorylines are quicker and have sharper stings, and in their range of prey and often also in their raiding groups they seem to be transitional between the column- and the swarm-raiding patterns.

In the dorylines no clear correspondences appear between taxonomic status, types of booty taken, and typical raiding patterns, hence Brues' generalization (1930) that dietary specializations "have determined to a great extent the evolution and differentiation of insects" needs validation in their case. *Eciton burchelli,* for example, resembles *Labidus praedator* both in its swarm pattern and its wide range of booty but differs greatly in both respects from its close relative *E. hamatum. Eciton,* with wide species differences in behavior and booty range, may be more specialized than either *Dorylus* or *Aenictus* with their greater species similarities.

Food may have played very diverse roles in evolution. This may be true of termitophagy, i.e., preying and feeding on termites. Attacks on termites seem related to the degree of an army ant's subterranean adaptation as the dominantly hypogaeic genera *Neivamyrmex* (Borgmeier, 1955), *Nomamyrmex, Labidus* (Luederwaldt, 1926), and *Dorylus* (Cohic, 1948) are generally termitophagous in contrast to such strongly epigaeic types as *Eciton hamatum* and *E. burchelli* which take no termites. We have studied two species of *Aenictus,* however, which are both distinctly surface-adapted, yet take termites freely. The role of booty specializations in doryline evolution remains to be clarified. We will consider the general role of colony food supply in nomadic behavior in later chapters.

Although most dorylines are almost entirely carnivorous, many of them are known to feed in part on nonanimal substances. As examples, *Dorylus (Alaopone) orientalis* feeds frequently on potatoes and the tubers of plants (Green, 1903); other driver ants feed on carrion and fruit oils (Savage, 1847), also manioc and the debris of fruits and palms (Cohic, 1948); *Labidus coecus* feeds on carrion, nuts, and seeds (Wheeler, 1910; Borgmeier, 1955). Both *L. praedator* (Borgmeier, 1955) and *Aenictus gracilis* (Schneirla and Reyes, 1966) carry away cooked rice and similar substances in large quantities. For the most part, dominantly hypogaeic species seem to be the ones attracted to many protein, starchy, and fatty substances of nonanimal origin used as foods.

Although at times attacks by swarm raiders on grasshoppers and other solitary victims and on ants, bees, and wasps in their nests seem very one-sided, the conflict involved in many cases is hardly without loss to the raiders.[18] To be sure, many of the victims' defensive measures do not injure the army ants but only reduce their booty. Grasshoppers, roaches, and others fly, hop, or run to safety from the swarmers; merely standing

---

[18] The brood of many wasps is a common prey of army ants, especially of *Aenictus* and *Eciton.* As attackers of *E. hamatum* pour over a wasp nest, the wasps fly excitedly around the entrance, even crushing or beheading some of the army ants. But once their nest is lost (Fig. 4.2), they form an open cluster on a nearby bough or palm leaf where they spend the night, then disappear the next morning.

Few ants are immune from doryline attacks. Even *Nomamyrmex* is formidable in numbers. Borgmeier (1955) reports a running battle between *N. esenbecki* and *Atta laevigata,* in which an area of 8 sq m was covered with the large and small workers of both; meanwhile numbers of the army ants carried off the leaf-cutter's brood. Raiders of *Nomamyrmex* often do considerable damage in attacks on beehives (Luederwaldt, 1920).

still or moving slowly about saves stick insects, phalangids, and others. The heavily armored workers of *Cryptocerus* species walk about near the bivouacs or across the columns, seemingly immune from attack. Workers of *Pheidole* and many other ground-nesting ants burst forth carrying their brood and scatter to refuges under leaves and other cover even before the invaders have entered their nests, thereby reducing the shock of combat and saving much of their brood and adults. No army ant raid, even by the swarmers, ever makes a clean sweep of potential booty.

Because many of the common victims resist very strongly, it is likely that army ants often suffer as heavy losses in their raids as do their prey. Swarm raiders must lost large numbers to the repellent (and noxious) secretions of the many beetles and bugs they flush out (Roth and Eisner, 1962); subterranean raiders must suffer heavily from the secretions of nasute termites whose nests they invade (Allee et al., 1949). Many attackers must be crushed in the mandibles of ants capable of strong resistance, as in assaults on strong colonies of *Atta* or from the bites, stings, and chemical repellents of myriads of small workers in the colonies of *Pseudomyrmex, Crematogaster,* and other tree-nesting ants. To doryline invasions, *Camponotus* and other ants in sizable colonies often are able to stand their ground at the nest entrance, catching many of the raiders in their jaws and crushing them as others slip past into the brood chambers. Workers of the ponerine *Odontomachus* kill many attackers by severe blows from snapping mandibles; workers of formicine species kill many with jaws and acid spray. Heavy combat often arises between *Aenictus* and the tree ant *Polyrachis,* whose leaf nests they invade (Fig. 13.3). The dorylines fight a constant battle with the myrmicine *Pheidole* all over the world, driving its colonies from their pillaged nests which are taken over as bivouacs. But often, as in conflicts of *Pheidole* with raiders of *Aenictus* or *Neivamyrmex,* the dispossessed victims remain in marginal galleries from which they plague the intruders in return sallies during their long, vulnerable statary phase.

The army ants have many enemies: the countless species of ants that do them battle, the termites many of them commonly raid, even jumping spiders that sit beside their columns and snatch out small workers. Their predatory way of life clearly holds much risk, for each doryline species has its chief foes as well as its chief victims. But the army ants, in their turn, take a heavy toll from the arthropod and other small life of the forest; also, as a rule they recoup their own losses well.

To all dorylines, therefore, large-scale raiding is a way of life. Foraging operations bear significant relations to bivouac formation, population, diet, and daily life; to emigration (Chapter 5); to raising broods (Chapter 6); and to cyclic colony function (Chapter 7). Studying the relevance of food to the adaptations of army ants requires an investigation of all their life activities.

# 5

# The Emigrations

Nomads are beings that have no permanent home but move from place to place. The traditional reputation of the army ants as nomads was verified when Bates (1863) first distinguished nest-changing columns from raiding columns. He saw what may well have been an emigration of *Eciton:* a "dense column [in which] all were moving in one and the same direction [and] all the small-headed workers carried in their jaws a little cluster of white maggots which . . . might be young larvae of their own colony. . . ." (Pp. 359–360.)[1]

A conventional explanation for the emigrations is typified by Savage's statement (1847) that the African driver ants move from place to place "as the wants of the community may require" and Belt's (1874) that colonies of *Eciton* he observed in Nicaragua "move on from one place to another, as they exhaust the hunting grounds around them."

Although this hypothesis of food shortage may suggest an ancestral basis for nomadism in army ants, there is no evidence that a paucity of booty around the nest is a regular cause of emigration in existing army ants.[2]

Even so, in starting my work on *Eciton* in 1932 I found this concept dominant, especially on the strength of Vosseler's studies of African driver ants (1905). One colony, he said, moved away after a "sojourn of more than 11 days in which it had raided over an area larger than 10,000 sq m. . . . When the booty is no longer sufficient for the feeding of the

---

[1] Sumichrast (1868), from his observations on *Eciton drepanophorum* in Mexico, made the same distinction, noting that the columns were "sometimes expeditions of pillage, sometimes changes of domicile." Recent reviews of emigration in New World (Borgmeier, 1955; Schneirla, 1957a; Rettenmeyer, 1963) and Old World dorylines (Raignier and Van Boven, 1955) bear out this distinction for all of them.

[2] At least a few members of other ant subfamilies, however, move their nests at times in response to local nonoptimal conditions, as appropriate tests show. For example, colonies of both the dolichoderine *Iridomyrmex humilis* and the myrmicine *Monomorium pharaonis,* nesting in building walls in wintertime, move—as I found —with brood and queens into humid, food-baited traps placed a few meters from their current locations. These ants, through such responses to food and ecological conditions, aided incidentally by man, have spread as introduced species around the world.

colony, the colony must again settle upon an emigration." No alternatives to this assumption were considered.

My research on Barro Colorado Island made me doubt this idea. On the first day I found and studied a colony of *Eciton hamatum* in a large raid and traced its columns to the bivouac. In the late afternoon the ants seemed to be returning mainly to the center, so I left for supper. But when I returned, the colony was gone. Even with a good headlight, no traces of it could be found.

From other evidence that colonies moved off nocturnally when (as usual) they left much potential prey behind, a different idea arose, namely, that of a nomadic condition or a statary condition, depending on the stage of the brood. I tested this idea, directly at odds with the concept of food scarcity, in 1933 and 1936 (Schneirla, 1938).

Actually, the classical literature offered more than one idea to explain nomadism. In working up my 1932 results, I found that some of my own cases—colonies of *Eciton* that changed from the statary to the nomadic condition when pupal broods emerged—were like one reported by Müller in 1886. He watched a colony of *E. burchelli* that for twelve days occupied the same hollow tree. His attempts to expel the ants with smoke brought only short moves to similar sites nearby. To explain this behavior, Müller suggested that "at the time, a queen was among them . . . with great gaster, heavy and difficult to transport." Brauns (1901) offered a similar hypothesis for the behavior of colonies of *Anomma*, which, he thought, must remain in fixed nests when they were raising queens and males—large developing forms that must be difficult to transport. Wheeler (1900) suggested, comparably, that colonies of *Neivamyrmex* species he observed in Texas did not move until they had completed the development of "their annual broods." All three writers seemed to think that an "encumbrance," i.e., something in the colony that was difficult to move, prevented emigration.

Another hypothesis concerned the weather. Sumichrast (1868), from observations of army ants in Mexico, pointed to the "inroads or migrations which they undertake at undetermined epochs, but in relation, it appears to me, with the atmospheric changes." Von Ihering (1894) pointed out that certain dorylines appeared on the surface so often just before rainy weather that a common one in Brazil, *Labidus praedator*, was known to residents as the "rain ant." This idea is relevant to the conditions under which given species seem especially active but does not specify a cause of emigrations.[3]

---

[3] Raiding sorties of *Labidus praedator* appear more often on the surface on overcast than on clear days (von Ihering, 1894), perhaps mainly in response to increased humidity and to dim light aboveground. This condition may apply similarly to emigration columns.

The notion of food scarcity in the colony area was generally the most favored, however. Müller (1886) used it to account for the eventual departure of his colony with a newly emerged brood of workers which, he thought, "must have had a strong need for food." This idea appealed to theoretical writers on animal migration, some of whom (Heape, 1931; Fraenkel, 1932), specifically on Vosseler's authority, classed the dorylines as nomads but not as migrants. Later in this chapter we will consider the validity of this distinction.

The more I studied emigration in *Eciton,* however, the more evidence mounted against the hypothesis of food scarcity and other hypotheses on army ant emigrations. Rather, colonies of *E. hamatum* (Schneirla, 1938) and *E. burchelli* (Schneirla, 1940) emigrate regularly (i.e., are nomadic) or do not emigrate for days (i.e., are statary) according to the condition of their brood (Schneirla, 1949; Schneirla and Brown, 1950).

Significantly, in both of these *Eciton* species and in others as well (Schneirla, 1947), the colonies always carry out day-long raids before they emigrate in the night. These great raids, I thought, had to be analyzed to understand the causes of emigrations. As it proved, raids and emigrations, both instances of colony exodus, differ greatly in their patterns of behavior but are significantly interrelated.

The daytime exodus of *Eciton* is absorbed in raiding. *E. burchelli,* for example, starting at dawn, creates a great swarm system with a heavily traveled liaison trail between advancing masses and bivouac (Schneirla, 1940). Booty-laden workers returning over this route arouse successive waves of recruits, which expand the vast raiding front by stages. Even the midday siesta letdown, although retarding the flow of events for a time, opens the way for a new stage. Environmental changes in early afternoon (Chapter 4) revive both the exodus and the raiding operations that maintain and increase it in feedback fashion.

This exodus is not an emigration, but in nomadic colonies of *Eciton,* through chains of behavioral events, it usually leads to an emigration. Three conditions make this result almost inevitable: (1) Excitement continues high within the bivouac and keeps up the exodus strongly (Chapter 6); (2) increasing traffic complications prevent a general return of raiders to the bivouac; and (3) toward dusk environmental changes reduce all ants operating outside the bivouac to actions other than raiding.

While daylight lasts, the exodus usually is absorbed in expansions of raiding. Meanwhile, traffic becomes complicated, both as a result of collisions between the exodus and incoming caravans of workers at trail junctions and of interference between these forces with trains of ants shifting from one raiding zone to another over connecting trails. Usually, after midafternoon, the route of exodus becomes one way because the outgoing procession forcibly reverses traffic at junctions successively

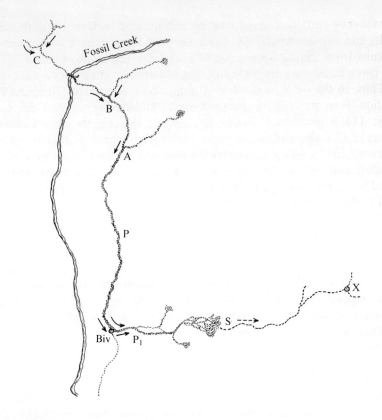

FIGURE 5.1.
A sketch to show the case of a colony of *Eciton burchelli* in late afternoon, with the afternoon exodus (*lower right*) forcibly diverted from the main raiding route of the day (P) by a heavy return of booty-laden traffic from across Fossil Creek. A, B, and C: major trail junctions. On the principal trail ($P_1$) a swarm (S) now advances 30 m from the bivouac. Dotted line shows route of eventual nocturnal emigration; X: site of new bivouac.

farther from the bivouac. At length, a stream of traffic pours from the bivouac that first includes only booty carriers but later also carriers of brood. This exodus, if it prevails, is the emigration.

How *Eciton hamatum* reaches a comparable result, despite its several trails from the bivouac, is suggested by an event that often occurs in maximal nomadic raids of *E. burchelli*. In such raids, heavy returning traffic often breaks up the exodus. When this happens, the outgoing movement—sustained by mounting excitement within the bivouac—shifts into another route (Fig. 5.1). The new foray, pushed on strongly by recruits from the bivouac and by forces from the main raid that bypass the bivouac, advances quickly and later becomes an emigration.

Thus, in the daily operations of nomadic *Eciton,* raiding leads to emigration. Four main causes that enter after midafternoon insure the transition: (1) a plethora of booty that excites workers to carry it (somewhere); (2) the waning daylight that reduces excited raiders to trail runners; (3) a rising excitement within the bivouac that sustains the exodus; and (4) traffic congestions that favor travel radially and outward but oppose travel converging on the bivouac.

To illustrate the final transition in *Eciton,* a colony of *E. burchelli* (Schneirla, 1940) at 4:00 P.M. had a large raid of several subswarms and trail-end foraging groups still in action. But as light intensity fell, the raiding groups decreased in size and attacks on booty. At 5:30 P.M., with just 4 meter-candles of light, the largest subswarm was only about 2 m wide; at 6:30 P.M., with light down to 0.2 meter-candle nearly all raiding had stopped. Finally, at 6:44 P.M., with no light, the exodus pushed strongly outward from the bivouac for 60 m in a column 2 to 3 cm wide.

Dusk simplifies the problem of emigration vastly for surface-adapted *Eciton.* As light and temperature fall, workers lose their daytime reactivity to the movements and odors of booty and begin to run the trails in lock-step fashion, much as in their circular columns under monotonous stimulative conditions in the laboratory (Schneirla, 1944c). The ants now respond to a world narrowed down to tactile and odor stimuli from nestmates moving close by from the trail and from booty or brood (Figs. 5.2 and 5.3) held firmly as they run along. All types of workers run more regularly in the emigration than in raiding columns, and when laden they run even a little faster.[4]

Thus, at the day's end, *Eciton* changes from raiding and from handling and feeding brood in the bivouac to routine trail running with brood. Observations indicate that at dusk with environmental influences greatly reduced in variety and potency, three types of stimuli assume control. These are brood or booty to carry, a chemical trail to follow, and nestmates as sources of sensory influences on running (Fig. 5.4). We may postulate this effect: The great brood, responding to day's end changes in its own condition and in worker behavior, undergoes an increased excita-

---

[4] Groups of twenty workers of *Eciton hamatum* recorded in an emigration ran at rates (seconds to go one meter) ranging mainly between twenty-two and thirty seconds and were less variable when laden than when unladen. Progress is generally a little slower and steadier in columns of emigration than in those of raiding.

FIGURE 5.2.
Emigration columns of *Eciton burchelli* and *Aenictus laeviceps* carrying nearly mature larvae. (*a*) *Eciton burchelli*. Note the flange of workers bordering the column on its lower side. (Barro Colorado Island, Panama Canal Zone.) (*b*) *Aenictus laeviceps*. Note that the burdens swing outward a little as the carriers round the turn. (Negros Island, Philippines.) See Plate III, following page 138.

FIGURE 5.3.
Larva-carrying column in a nocturnal emigration of *Eciton burchelli*.

tion that causes workers to pick up the larvae and carry them forth.

Thus, through prerequisite steps in raiding, emigrations develop in *Eciton* (Schneirla, 1933, 1938). By definition, the exodus becomes an emigration when, in response to intercolony excitation, workers in column begin to carry brood from the bivouac (Figs. 5.3 and 5.5). In *E. hamatum* this change often starts in late afternoon and is well under way at dusk, with the last ants leaving the old bivouac site and building the new bivouac about one hour earlier than in *E. burchelli*.[5] In both species, these events begin earlier in the first days of the nomadic phase than they do after midphase; then they are complex and variable and especially in *E. burchelli* with its large colonies, may go on past midnight (Schneirla, 1949; Schneirla and Brown, 1950; Rettenmeyer, 1963).

---

[5] In two continuously studied colonies of *Eciton* (Schneirla, 1949), 18 of 21 bivouacs were started by *E. hamatum* between 6:00 and 8:00 P.M., and 20 of 24 bivouacs were started by *E. burchelli* between 7:30 and 9:30 P.M.

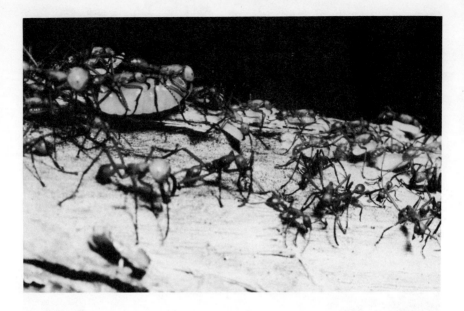

FIGURE 5.4.
Section of column in a nocturnal emigration of *Eciton hamatum* passing along a vine. Most of the ants carry their larvae; a few carry booty. (Photo courtesy of C. W. Rettenmeyer.)

Once brood carrying is well advanced, the dorylophiles begin to appear, the larger ones running with the ants in column,[6] the smaller ones skipping along in the column or even riding on the ants themselves. Their appearance in numbers often indicates the approach of the queen.

The passage of the queen with her entourage, an impressive event first described for *Eciton lucanoides* (Reichensperger, 1934),[7] usually occurs when transport of the brood is more than half completed. Early in the nomadic phase this is most often between 7:00 and 8:00 P.M. in *E. hamatum* and between 8:00 and 10:00 P.M. in *E. burchelli;* later in

---

[6] Among representatives of one genus (*Ecitomorpha*) of staphylinid beetles observed (Akre and Rettenmeyer, 1966) in an emigration of *Eciton burchelli,* 26 were running in the centers of the column, 12 were riding on booty, and 3 were riding on brood. During the nocturnal emigrations of *Neivamyrmex nigrescens* in southeastern Arizona, Topoff (1969) found as many as 15 carabid beetles of the species *Helluomorphoides latitarsis* running either within or alongside of the emigration columns. The beetles fed primarily on the army ants' larvae.

[7] The Costa Rican collaborator who reported this observation to Reichensperger (1934) stated that he had also seen a nocturnal emigration of *Neivamyrmex pilosus.* and he believed that the emigrations of these ants occur almost exclusively at night. R. Z. Brown (pers. comm.), who observed an emigration of the hypogaeic *N. gibbatus* at night, noted that the movement of the queen involved a long entourage with excited workers crowding around her, much as in *Eciton hamatum.*

FIGURE 5.5.
Bivouac of a colony of *Eciton hamatum* during a nocturnal emigration just as the queen and her worker entourage are leaving the site. The queen is now hidden under a mass of workers on the vine at right of bivouac. (Photo courtesy of R. E. Logan.)

the phase the event is more variable in both species and is often long delayed.

As the queen approaches, the column first thins to a file and then is broken by workers that rush back and forth on the trail. Then, watching back over the route, one sees a solid phalanx approaching, a long, broad mass of turbulent workers, under the peak of which now and then one glimpses the queen (Fig. 5.6). The entourage, which in *Eciton* is usually several centimeters wide just back of the queen and converges to column width a meter or two behind, prominently includes numbers of major workers and also groups of workers that are darker than the others and unladen (members of the queen's guard). At times of high colony arousal, the entourage may progress literally in layers studded thickly by the white, bobbing heads of the majors.

The queen runs along steadily, despite the workers rushing around and over her and even riding upon her. When she hesitates, as in mount-

(a)

(b)

FIGURE 5.6.
Two views of nocturnal emigrations of *Eciton hamatum* at the time the queen passes with her entourage. (*a*) In this lateral view the queen, at the head of her wedge-shaped retinue, cannot be seen as she is covered by workers. (*b*) In this view from above, only the head of the queen (*arrow*) can be seen as workers move over her.

FIGURE 5.7.
Section of a nocturnal emigration column of *Eciton burchelli,* widened at the passing of the queen. But the queen, suddenly disturbed by a photo-flood light as she passed over the stick and leaf at the right, now is held within a bolus of workers below this leaf (*arrow*).

ing a vine incline, they at once cluster over her. After slight interruptions she moves on; but in the longer delays introduced by rain or gusts of wind, the workers form a thick bolus over her, which may remain for minutes with the continuing procession eddying around it. Then finally the queen burrows her way from beneath this mass to rejoin the column, the worker cluster marking the spot until it has melted away. Often, at the height of the nomadic phase, and particularly in *Eciton burchelli* (Fig. 5.7), the queen progresses by tunneling her way through clusters that form at many places on the trail where she passes. Almost certainly the queen's passage, similar in *Neivamyrmex* (Schneirla, 1958) and *Aenictus* (Schneirla and Reyes, 1969) although on a smaller scale than in *Eciton,* is marked by excitatory odors from secretions of both queen and workers, attracting the workers to her in large numbers and thereby guaranteeing her a safe passage.

In the wake of the queen and her entourage, the regular procession of the main emigration soon resumes (Fig. 5.3). The appearance of this column indicates a uniform stimulative guidance throughout as any slowing of progress, e.g., over a rough log or around an obstacle, is forecast by a slackening of the column just ahead; and any speeding of traffic, e.g., along a smooth vine, is marked by an acceleration ahead. The lock-step running of workers in column involves responses to stimulus cues from the trail and from nestmates as circular column studies show (Schneirla, 1944c). The regularity of this type of behavior is basic to doryline life.

Factors determining the route of an emigration by *Eciton,* never simple, seem to complicate as the nomadic phase runs on. Then raids become larger, and through the day traffic problems mount at central trail divisions and successive outlying forks. Watching the outward-rushing exodus as it reaches a trail division, one often thinks that it might take either route. This happens at junctions where traffic snarls continue as they do in the first emigrations of the nomadic phase, which are often delayed by callows that reverse direction en bloc at each local disturbance. Their actions increase difficulties, which, in the event of rain or high wind, may greatly prolong the emigration. Later in the phase, worker behavior related to the advanced larval brood complicates the movement as new behavioral relationships arise between brood and workers.

One such relationship involves a mass feeding of larvae during the emigration itself. As the larvae grow, this operation enlarges through more vigorous colony raids and the larger and more numerous booty caches that result. The logistics of transporting the brood then increase greatly. Laden workers in the procession, interrupted at each further trail junction, are attracted to the cache nearby and soon cover the booty with a layer of feeding larvae. One large colony of *Eciton hamatum,* during a late nomadic emigration, at one time had seven larva-feeding booty caches in the route of its exodus, located at points between 60 and 220 m from the bivouac. At this stage in the cycle, every cache of any size reached by the exodus offers another opportunity for the feeding, resting, and exercise of larvae. The major workers, incidental blockers of traffic at junctions, are contributors to these useful results.

The time required for emigrations in *Eciton,* although variable, generally increases as the nomadic phase wears on. Within the first few nights the callows fall into line better, and with the larvae still small and largely transported in packets, the eventual bivouac may begin soon after dusk in *E. hamatum* and before 9:00 P.M. in *E. burchelli.* Then as more and more of the larvae are carried individually by workers, the brood-carrying procession lengthens and must trace a longer and more complex route. As a result, after midphase the emigrations of *E. hamatum* usually last two or three hours longer than before and those of *E. burchelli* need even more time and often continue until after midnight. Durations of these movements also vary according to colony size, interruptions by rain or wind, roughness or regularity of the terrain, and amount of booty carried.

Doryline emigrations often differ greatly in the speed of travel and the width and evenness of columns characterizing the species (Borgmeier,

1955; Rettenmeyer, 1963). The colonies of *Eciton burchelli*, generally much larger than those of *E. hamatum*, need more time to emigrate. Nonetheless, by forming wider, more densely crowded columns, they carry out their emigrations in proportionally less time. *Neivamyrmex nigrescens*, with smaller ants in smaller colonies, generally needs not much more than half the time required by *E. hamatum* (Schneirla, 1958). The still smaller colonies of the diminutive *Aenictus*, early in the nomadic phase, take little more than two hours (later only four or five hours as a rule) for the complete operation (Schneirla and Reyes, 1969).

For colonies of *Eciton*, despite possible delays, one night is generally ample time to change bivouacs. In the latter part of the nomadic phase the emigrations, now usually longer than before, may at times be greatly protracted. But notwithstanding intracolony complications and the weather, colony action continues somehow. I have often watched for the reappearance of a column of *Eciton* after the ants had been driven under cover by rain, only to find, as the downpour lessened, the heavy column already under way on an alternative route under surface cover or on elevated paths along limbs and vines.

Doryline resources for emigrating under adverse conditions are extensive. If worse comes to worst, as when a long, heavy rain strikes in late nomadic days, emigrations may continue in the hours after dawn. The workers, then not as light-shy as they are earlier in the phase, seem to adapt to the morning light as it increases. Sometimes, a colony of *Eciton* that has had its emigration interrupted for some time, e.g., by heavy rain or by crossing paths with another doryline, may break off the movement at dawn, though still divided. In this case, the second section, as a rule, combines forces with the first in the day's raid, then moves out promptly in the early evening as the column of a single emigration.

These "consolidation emigrations," as we call them, seem much more common in certain *Neivamyrmex* species than in *Eciton*. *N. nigrescens*, which begins nearly all its nomadic raids at dusk, usually starts emigrating within two or three hours after the raid begins and completes the movement on the same night. This species, with smaller colonies, smaller ants, and raids of smaller scope than any *Eciton*, operates a correspondingly more contracted schedule. But as the brood grows, protracted emigrations occur when sections of the colony cluster with larvae at caches formed (e.g., under rocks) in series along outlying raiding trails. At such times, the movement either continues underground after dawn or is completed on the next evening. This "leap-frog method of emigrating is common in *N. nigrescens* (and in *Aenictus* also) late in the nomadic phase.

In most dorylines, as in *Eciton*, routes of emigration generally follow raiding trails and thus are influenced particularly by the distribution of booty. The point may be impressed by a simple rule: To follow emigrations of these ants without losing the colony, one should trace out the column step by step, not taking short cuts even to avoid fallen tree thickets and other obstacles. Booty is likely to be heaviest in such places; hence the raid and finally the emigration commonly pass through them. The same rule applies to anticipating the final stopping place; for the ants, in dorylinelike ways, often pass sites that an observer considers very good but end in places he would not have chosen.

An emigration of *Eciton* that operates regularly and rapidly usually opens the way for stable clustering in early evening at some advanced trail junction. As mentioned, the ants commonly begin clusters in more than one place before one of these becomes the bivouac. The chances that such a place will be the new bivouac increase greatly when most of the brood has been carried in and the queen has entered. Late in the nomadic phase, however, the exodus may go on further even after most of the colony has settled. This is one aspect of the process of bivouac "selection" (Chapter 3), by which the colony increases its chances of securing a site of spatial and ecological properties appropriate to its current condition. A similar gain may result from the series of clustering stops often made on the same night at corresponding times by colonies of *Neivamyrmex nigrescens*.

Doryline emigrations differ greatly in scope according to species and conditions. Colonies of *Eciton hamatum* emigrate over distances ranging between 100 and 350 m, averaging about 100 m. I have followed emigrations of *E. hamatum,* especially with large, highly active colonies, over paced-out courses as long as 450 m; but at other times, I have traced columns winding among areas of concentrated booty in the same fallen tree thicket that ended only a few meters in a direct line from the start (Schneirla, 1949). According to how the scope of raids varies with the terrain, the distribution of booty, colony size, and colony condition in the functional cycle, emigrations over very different distances may occur in series (e.g., Fig. 5.8).

Distances of emigration differ broadly with the size of the ants and their colonies. Thus *Aenictus laeviceps* ranges from about 10 to 80 m and averages about 20 m (Schneirla and Reyes, 1969) whereas *Neivamyrmex nigrescens* ranges from 15 m and averages about 40 m (Schneirla,

FIGURE 5.8.
Eastern part of Barro Colorado Island, Panama Canal Zone, mapped to show the itinerary of colony '46 H-B (*Eciton hamatum*) through a period of 112 days, February 12 to June 6, 1946. Small circles indicate nomadic sites; double circles are statary sites; double broken lines are successive routes of emigration; dotted lines are the principal routes of some of the daily raiding systems; contour interval 6.1 m; scale: length of N-S arrow at lower right 300 m.

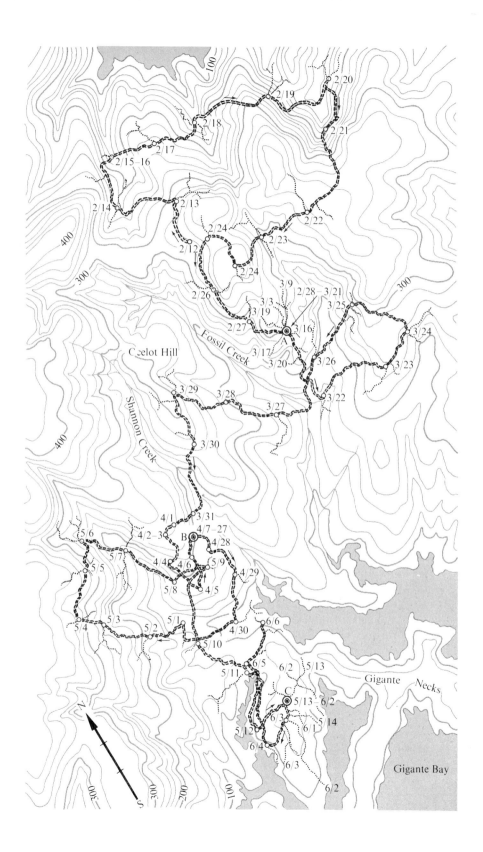

1958). These results reflect smaller zones of raiding than in *Eciton* although *Aenictus* differs from the others in emigrating at times well beyond the limits of that zone (Chapter 13). The longest emigrations thus far reported for dorylines are those for *Dorylus* (*Anomma*) *wilverthi,* in which twenty-six emigrations recorded in one study averaged 226 m (Raignier and Van Boven, 1955). Further consideration shows, however, that these results must be evaluated in view of generic differences in the relationship of raiding to emigration (Chapter 7).

The dismantling of surface bivouacs in *Eciton* differs with time in the nomadic phase. Early in the phase, the outer bivouac wall usually retains its form until near the end. Then after the queen has left and as workers carry out the last of the brood, the outer wall melts away from the bottom upward. At this stage, the entire bivouac may be gone and a new one constructed far away by 7:00 P.M. in *E. hamatum* and by 8:00 P.M. in *E. burchelli.* But later in the phase bivouacs tend to be dismantled more variably, the walls often collapsing early as workers carry brood into wide caches at the base of the exodus trail. For example, in *E. hamatum* at times the old bivouac may dissipate before 9:00 P.M. with the entire colony on the move at 10:00 P.M., spread out in column or in caches with brood and booty but with the new bivouac not yet begun. It is instructive to compare this pattern with the much simpler one of *Aenictus* (Chapter 13).

*Eciton* offers a good illustration of how emigrations end. Early in the nomadic phase the queen may run near the tail of the procession; usually, however, she comes earlier, after the emigration is about two-thirds completed and when brood carrying is nearly finished (Schneirla, 1938, 1949). Further on in the phase she may appear early or late, depending notably on the type of bivouac and her station in it. As the last ant strands of the old bivouac are unraveling, the end of the procession as a rule is marked by three features. One of these is a number of dorylophiles running in the ragged and thinning column of ants; another is the backtracking of workers at the borders of the procession; and finally, after the main column has ended, a few scores of workers bring up the rear, making short runs back and forth on the trail.[8]

Normally, when these actions have stopped, no ants return. Seldom are any members of the colony seen later at the old bivouac site, except for a few crippled workers that may have straggled back to it late from the raid. Usually no part of the colony returns except under conditions to be discussed in Chapter 8.

---

[8] For reasons given in Chapter 9, emigrations terminate very differently when a sexual brood is present in the colony.

Emigrations of colonies in surface-adapted *Eciton* species generally end before midnight, and thereafter the entire colony stays within its new bivouac until raiding begins at dawn. Although this bivouac seems quiet to outside view, inside it is very active, except for workers forming the bivouac framework. Most of the others are busy feeding and tending larvae and queen, shifting larvae about, and at intervals feeding by themselves. Comparable brood feeding and internal reorganization seems to occur in colonies of other species after their emigrations. Postemigration behavior may be most variable in *Aenictus,* in which emigrations themselves are most variable (Chapter 13).

Genera in our group A differ greatly in the scheduling of their emigrations. In surface-adapted *Eciton* there is just one emigration a day in the nomadic phase; this is a continuous operation that follows the raid directly; hence it is mainly nocturnal. In *Neivamyrmex nigrescens* also there is as a rule just one emigration in each twenty-four-hour interval, beginning sooner after the start of raiding than in *Eciton* (Schneirla, 1958). *Aenictus,* in contrast, through the intermediate and largest part of each nomadic phase, emigrates both by day and by night, usually more than once in each twenty-four-hour interval.

The described contrasts in daily schedule seem referable to generic differences in the relationships of raiding to emigration. In most surface-active species these relationships are linked to the twenty-four-hour march of events in the environment. Of particular importance is the pattern of worker reactions to the brood. Thus, the emigrations of *Eciton* arise much more gradually through more complex and specialized processes of raiding than those of the relatively simple *Aenictus* (Chapters 7 and 13).

Emigration schedules in hypogaeic group A species may be similar to those of surface-living species but seem more variable. Emigrations starting at dusk after raids have been observed in *Eciton mexicanum* and *E. vagans* (Schneirla, 1947) or later in the night as in *E. dulcius* (Rettenmeyer, 1963). Nocturnal emigrations with behavioral conditions resembling those of *Neivamyrmex nigrescens* have been observed in *N. pilosus* (Borgmeier, 1955) and in *N. gibbatus* (Schneirla and Brown, 1950) in Central America and also in *N. opacithorax* in Arizona (Schneirla, 1958). Borgmeier's review (1955) of this genus indicates that great numbers of its species are virtually eyeless and may carry out all or most of their emigrations underground. At present, their emigration schedules are unknown.

Army ant colonies also carry out exceptional movements called "shifts," which as a rule are clearly distinguishable from emigrations. Shifts are changes of bivouac site that occur not at expected times of emigration[9] but evidently in response to the effects of rain, sunlight, or other environmental disturbances. Such movements, usually very short and seldom more than a few meters long, occur now and then in *Eciton* (Schneirla and Brown, 1950) but more often in *Neivamyrmex* (Schneirla, 1958) and *Aenictus* (Schneirla and Reyes, 1969). Müller (1886), using smoke, forced a short movement in his statary colony of *E. burchelli*. I have obtained similar results by slicing open the bivouac with a machete or by reflecting sunlight on the ants with a mirror at intervals (Schneirla, 1949). From the results, shifts may be aroused much more readily in nomadic than in statary colonies and more readily in the first and the last part of the statary phase than in the intermediate days. Other common causes of shifts are encroachments by another ant. Colonies of *Pheidole*, ousted from their nests by *Neivamyrmex* and *Aenictus*, often remain at the margin, attacking repeatedly and forcing the army ants to shift. At times, doryline dispossesses doryline in this way. As an example, I saw a statary colony of *E. hamatum*, clustered near one end of a hollow log, forced to shift its position a few meters after repeated conflict with a colony of *Labidus praedator* bivouacked close by in what had been the base of the tree (Schneirla, 1949).

Driver ant schedules of emigration differ greatly from those of *Eciton* and other genera of group A. Colonies of group A dorylines emigrate frequently when nomadic—*Eciton* nightly, *Aenictus* often several times in twenty-four hours. By contrast, colonies of *Dorylus* move at irregular intervals from a few to many days apart. Also, their emigrations are lengthy operations and continuous except for breaks during rain or when the sun is high at midday. Twenty-eight movements observed in colonies of *D. (Anomma) wilverthi* varied roughly between 34 and 38 hours and averaged close to 57 hours in duration (Raignier and Van Boven, 1955). Causes of this remarkable difference between dorylines of groups A and B are discussed in Chapter 7.

At the end, the millions of workers of a driver ant colony, with their brood and queen (Raignier and Van Boven, 1955), are all transferred to a new underground site usually more than 200 m distance from the previous one. They run in a broad procession often several ant-layers thick in a trench (Fig. 5.9) that is steadily deepened both by the heavy traffic and by the energetic digging of the smallest workers. Here and there the trench has walls or even a partial roof of clustered workers. Generally, however, only limited sections of the emigration take to the surface, and most of the route follows underground channels.

---

[9] Contrary to the "encumbrance" hypothesis discussed earlier in this chapter, shifts of bivouac occur at times in statary colonies of *Neivamyrmex* and *Aenictus*, even with physogastric queens (Schneirla and Reyes, 1969). Movements of this type in *Aenictus* involve exceptional clustering behavior of workers (Chapter 13).

FIGURE 5.9.
Sections of emigration routes in the driver ant *Dorylus* (*Anomma*) *wilverthi* on the surface become trenches through which the procession runs. The trenches are walled by earthen pellets removed from beneath the hurrying column. (Photo courtesy of A. Raignier and J. Van Boven.)

There are notable similarities between these driver ant emigrations and those of *Labidus praedator* of the New World, whose colony movements, seldom observed, may be infrequent and mainly subterranean. The only four emigrations of this species I have observed (Schneirla, 1957a) all involved a broad column four to ten ants wide, in which thousands of callow workers ran in a nearly continuous procession and great numbers of worker cocoons were carried. The column is usually flanked on both sides by walls of earthen pellets,[10] roofed over here and there into arcades, and pieced out at intervals with rows of clustered or standing workers. The stretches aboveground, generally short, seem to be portages across streams and other places at which underground progress is blocked. One colony, in emigrating, emerged from a hole in one bank

---

[10] In *Labidus,* the walls and arcades, as well as formations in which workers cluster or stand (usually facing outward) at both sides of the column, resemble those described for *Dorylus* (Cohic, 1948; Raignier and Van Boven, 1955). The smaller workers, trodden upon and rolled underfoot in the column, do most of the building. They pick up tiny pellets of earth from beneath the procession and with these scramble out to one side. Working from inside as a rule each carrier mounts the wall and pushes her pellet into place, often with little shakes of her head and forebody. This behavior, with moisture on the particles, gives even the arcades an appreciable resiliency. *Anomma* makes arcades when the earth is damp but only ragged walls when it is dry (Cohic, 1948).

of a ravine, crossed the stream over rocks—with clustered workers filling the gaps—and, after a course of 15 m on the surface, disappeared under a log on the opposite bank. In two days on this route, traffic stopped or changed direction only during midday hours and at times of heavy rain. Carl Rettenmeyer (1963), entomologist at Kansas State University, reported in detail a colony emigration of this species observed in its course aboveground for a stretch of 4 m. In this movement, which continued for more than a day, he estimated traffic at between 300 and 1000 ants per minute.

During several investigations in tropical areas where colonies of *Labidus praedator* are common, I have seen many of their raids, both by day and by night, but few of their emigrations. This experience, a typical one (Borgmeier, 1955), may mean that emigrations are infrequent in this species unless the colonies often confine their movements altogether to underground passages. Various authors (Sumichrast, 1868; von Ihering, 1894) have reported long nesting stays in colonies they observed but without any mention of emigrations. Colonies of *Labidus,* which seem to be very large, may carry out single lengthy emigrations at well-separated intervals, as do colonies of driver ants.

*Labidus coecus,* although more hypogaeic than *L. praedator,* is often seen on the forest floor in trenchlike passageways with walls and occasional short arcades like those of *L. praedator.* One rarely sees a few meters of undirectional column in which callow workers run with booty-laden adults—probably a section of emigration. Bates (1863) reported following tunneled trails of this ant for more than 100 m on occasion in Amazonian forests. The emigrations of this ant are still unsolved.

*Nomamyrmex,* also common in tropical forests and often seen in surface raids, is rarely observed in emigration. I have seen hundreds of surface raids by these ants but only three of their emigrations. In these movements, which continued day and night for thirty hours or longer (Schneirla, 1957a), two of the colonies crossed open spaces of a few meters between hills, and one emerged from a hill, crossed a stream on a log, and entered the opposite bank. Outstanding in all three cases were the great numbers of callow workers running and carried in the procession. Rettenmeyer (1963) reported a notable case in which a colony of this species, first observed in a surface raid, changed after dusk into an emigration which was still going on when observation stopped thirty hours later. Shortly after 10:00 P.M. on the second night, excitement increased greatly, and rows of "guard workers" along the column borders grew denser and more agitated, standing with front legs raised and antennae vibrating rapidly. Then, after twenty minutes, as these changes intensified and passed along the column in waves, the queen (contracted) was spotted moving along more slowly than the workers. "Except for a slight increase in speed and an increase in excitement shown by the workers near the queen, no retinue could be distinguished." To date, this is the only report of the queen of this species observed in an emigration.

To take stock, significant differences exist among the doryline genera in the timing, regularity, and other aspects of their emigrations. The species of *Eciton, Neivamyrmex,* and *Aenictus* that we have studied, included in our group A, are all distinguished by regular emigrations carried out in series through long nomadic phases. In sharp contrast to these, species of *Dorylus, Labidus,* and others in our group B all seem to carry out movements of long duration irregularly and at well-separated intervals rather than in series. These generic differences in nomadism seem related to other differences in adaptive patterns (Chapters 12 and 14).

Army ant genera also differ in the daily timing of their emigrations and in the relationship of these movements to raiding (Fig. 14.2). In species of *Eciton* and *Neivamyrmex* studied, the colony movements always follow complex raids and have species-typical relationships to day and night; but in the *Aenictus,* as well as in the driver ants and others studied, emigrations seem to vary in their relationships to raiding and also in their timing with respect to night and day.

From the evidence cited, the status of army ant emigration in at least group A genera seems clear. The classification of these colony movements as mere "emigrations" forced by local food scarcity, offered by the English biological theorist Walter Heape (1931) and others, cannot stand against evidence from many well-studied colonies (Schneirla, 1949, 1958; Schneirla and Brown, 1950; Schneirla and Reyes, 1969) favoring a theory of intracolony causation.

Actually, army ant nomadism, evaluated on this evidence, conforms to Heape's own definition of animal migration. Important criteria are: The colony movements (1) depend upon cyclic reproductive processes, (2) occur in close relationship to corresponding reversible biological conditions, (3) enable the colonies to nest in situations of contrasting ecological properties, and, therefore, (4) have the aspect of directionality. These points, all fundamental, are discussed in Chapter 6 and others that follow.

I contend that *each movement* of an army ant colony from one bivouac site to another is an emigration because in it the colony actively changes its abode. Furthermore, evidence presented in Chapters 6 and 7 indicates that reproductive processes play the necessary casual role in these operations. These results, with other evidence (Chapter 3), support my interpretation of the series of emigrations in a nomadic phase of group A colonies and the separate emigrations of group B colonies as each being one phase of a *migration* in which the colony reverses habitats of contrasting ecological properties. These alternating conditions of army ant migrations are based on phase-specific physiological states that arise reciprocally and repeatedly in the same individuals of each colony (Chapter 7). The patterns of migration well known for desert locusts (Kennedy, 1951), many butterflies (Williams, 1958), and other insects (Schneider, 1962), although broadly equivalent to these in their cyclic physiological basis, usually involve different individuals in their successive phases.[11] They have arisen independently (i.e., are convergent) and so differ in their contributions to species survival (Chapter 14).

Although the hypothesis "food scarcity in the environs" is not supported as a necessary cause of emigration in *existing* army ants, conditions of the type it suggests are discussed in Chapter 14 in view of their possible importance for the evolution of nomadism in these ants. The role that alimentary conditions *in the colony* play in setting off emigrations can be studied to advantage in a comparison of *Aenictus* and *Eciton* (Chapter 14).

Army ant colonies do not reverse their directions of movement as do migratory birds; yet their nomadic movements have two aspects of directional change. The first of these depends upon the behavior described in this chapter for group A colonies, which in emigrating do not backtrack but move in some other direction. That successive emigrations thereby generally move forward (Fig. 5.8) is indicated by evidence from abnormal cases in which backtracking of colonies does occur (Chapter 8).

As another aspect of directionality, army ant colonies, by emigrating, reverse their microclimate from that of the preceding phase (Chapter 7). Colonies of surface-adapted species, in the nomadic phase, emigrate into similarly exposed sites successively but at the end of this phase move into a sheltered site (Chapter 3). Colonies of hypogaeic species, correspondingly, move downward to deeper and more stable sites when statary and toward the surface and more variable sites when nomadic.

In these ways, the mechanism of emigration permits army ant colonies to maintain a food supply commensurate with their condition and to reverse their responses to the general environment as their reproductive condition changes. The nature of these reversible processes can be examined in studies of brood effects on the colony (Chapter 6) and their bearing upon cyclic colony behavior (Chapter 7).

---

[11] Desert locusts, after being crowded for some time, raise a physiologically active generation (*Shistocerca gregaria*) capable of successions of mass flights into verdant, rainy areas. When sparsely distributed, however, these locusts raise individuals of an opposite phase (*Shistocerca solitaria*), phyiologically depressed, which reside locally as separate, nonmigratory individuals (Kennedy, 1951).

# 6
# The Broods

The army ant brood—the mass of developing young present in the colony—is vital to colony function. All normally operative colonies of doryline ants contain developing broods. These broods are *holometabolous,* i.e., they pass through the stages of egg, embryo, larva and pupa,[1] emerging on maturity as callow adults. These consecutive stages have held in the social insects from the time of ancestral solitary wasps. Army ants diverge in their development from other insects not so much in their individual embryology as in their collective aspects, i.e., in the general make-up of the brood and in the changing relationships of the young to colony behavior.

The importance of the brood in the colony life of social insects had been overlooked even into the early twentieth century when the larvae were still being considered little more than passive food receptacles. But W. M. Wheeler (1928), extending arguments first advanced by the French biologist, Emile Roubaud (1910), from studies of social wasps, emphasized in his concept of trophallaxis the general importance for social insects of interrelationships between brood and adults. This concept featured the attractions of *both* the brood for adults and of adults for brood, effected especially through exchanges of nutrients and a variety of exudates and secretions.

Wheeler assigned to mutualistic relationships between brood and adults a basic role in the evolution of social life in insects, and in his classic

---

[1] George C. Wheeler (1943), biologist at the University of North Dakota, has described the developmental stages of army ants. From the minute egg (about 0.3 mm) an embryo grows, from which, when the chorion or egg membrane breaks, the microlarva (first instar larva) appears. The doryline larva is elongate and cylindrical with a head and thirteen segments, of which three belong to the thorax and develop one pair of leg discs each on the ventral side, and the other ten belong to the adult petiole and gaster. The mature worker larva of *Eciton* and of most *Labidus* spins a cocoon (that of *Neivamyrmex, Aenictus,* and *Dorylus* does not), then straightens and transforms into the inert prepupa, which, after constricting behind the epinotal segment, develops its legs, antennae, and other parts beneath the cuticle. As the pupa nears maturity, pigmentation appears gradually, first anteriorly and then backward over the body. The mature pupa ecloses as a callow (i.e., young adult).

treatise on the subject (1928) he vigorously advanced the evidence and theory centering on the concept of trophallaxis. This was a good step toward understanding group behavior and colony organization in social insects, but it was only a step, for this concept meant to Wheeler a "pleasure-principle" basis for colony unity and not an active role of the brood in colony function.

Presumably, because of this distinctly Freudian and subjective view of the role of trophallaxis, i.e., to keep the individuals of the colony together as a social group, Wheeler overlooked the causal relationships between the developmental condition of the brood and changes in colony behavior. Thus, when Wheeler (1921) encountered an emigrating colony of *Cheliomyrmex* during his studies of army ant taxonomy, he described it merely as "a great army moving with its brood." On another occasion, when Wheeler was concerned with the first description of the queen of *Eciton hamatum,* he described the highly active colony with its larval brood in an open bivouac as "great numbers of nearly full grown larvae" within the cluster of ants hanging in the open (1925). In neither case did Wheeler remark about the relevance of the developmental condition of the brood to the activities of the colonies he was observing. Müller (1886), despite confused interpretations, at least gave brood condition some relevance to changes in colony behavior.

Army ant broods were certain to receive notice because of the following reasons: (1) They are large (as Müller, 1886, and Wheeler, 1925, noted for *Eciton*), much larger than the broods of most other groups of ants. (2) The entire brood is at nearly the same stage of development (Eidmann, 1936); in most other ants, if you look into the nest, you will see all stages of development, including eggs, larvae, pupae, and young—or callow—adults. And (3), there is a consistent relationship between the stage of development of the brood and the behavioral condition of the colony (i.e., nomadic colonies have larval broods while statary colonies have pupal broods).

Results of my own first study led me to describe (Schneirla, 1933) two colony conditions that were correlated with *different* brood stages: with larval brood, a nomadic condition of high colony activity, and, with pupal brood, a statary condition of low activity. Working out the relationship between the great raids and the emigrations typifying nomadism supported the thesis that active broods promote a highly active condition in the colony and, conversely, that inert pupae somehow are related to the reduced action of colonies in the statary condition (Schneirla, 1938).

Because army ant broods are large, discrete populations (or subpopulations) that arise frequently in the colony, they offer exceptional material for research in their bearing on colony function. The large all-worker broods distinctive of army ants are the main subject of this chapter. (The sexual broods of these ants, which contain potential queens and males only, are discussed in Chapter 9.)

All of the workers in colonies of army ants are sexually female. The mechanism for this sexual determination is suggested by the so-called "Dzierzon rule," first stated for honeybees (cf. Butler, 1962), by which females develop from fertilized eggs and males from unfertilized eggs. Although this is an acceptable generalization for most social insects, the more accurate concept is that of "haplodiploidy" (Rothenbuhler, 1967).

A striking feature of many species of army ants is that although they are all sexually female, they often differ markedly in size and morphology. This differentiation of individual types within a sex is termed "polymorphism." Thus, although both workers and queens of ants arise from the same types of fertilized eggs,[2] they result from different conditions of development (Chapter 2). It is probable that in army ants, just as in honeybees, queens are produced from larvae that receive much food, and workers from larvae that receive very little food. Similarly, differences in morphology among the workers may also result from their having been fed different amounts of food early in larval life.[3]

One of the marked characteristics of all dorylines is the periodic scheduling of their broods. This is clearest in group A genera (Schneirla, 1965), in which all-worker broods normally appear regularly in a sequence with rarely a break by a sexual brood. Possible reasons for this state of affairs are discussed in Chapter 9, when we take up the different effects these sexual broods have on the colony. Chapter 7 considers similarities in the relation of sexual broods to colony function.

All-worker broods are immense in all dorylines and are among the largest broods of all ants. The mass of these broods, seen in a torn-open bivouac (Fig. 6.1), makes very difficult any estimate of their numbers.[4] Current approximations, given in Table 6.1, range from one or two million for certain driver ants to about 30,000 for *Aenictus*. The largest army ant brood populations may be roughly fifty times greater than the smallest ones.

General estimates are undependable, and accurate census taking for

---

[2] Evidence for ants supports the view that in these insects sexual dimorphism is determined by the genetic mechanisms of haplodiploidy (i.e., dependent on fertilization) but that polymorphic difference among workers are largely determined trophically (Ezikov, 1922; Wesson, 1940; Gregg, 1942; Brian, 1957a).

[3] The English myrmecologist Michael V. Brian (1957b) found with controlled larval feeding tests in *Myrmica ruginodis*: "that a strong basis in favor of large larvae operates in the distribution of food (and perhaps other services)." (P. 333.)

[4] Although my own early estimates (Schneirla, 1934) of brood magnitudes in these ants were much too low, almost certainly others (e.g., Eidmann, 1936) were far too high. To test the subjectivity of such judgments, I presented a jar containing a known total of 24,005 pupae of *Eciton burchelli* to several zoologists. Their estimates ranged from 8500 to 110,000!

FIGURE 6.1.
The bivouac of a colony of *Eciton hamatum* late in the nomadic phase with its wall opened by a stroke of the tweezers. The nest contains a nearly mature larval brood massed near the outer wall.

TABLE 6.1.
*Approximated "average" populations of workers and of brood in normal colonies of doryline ants*

| Species | Workers | All-worker brood | Reference |
|---|---|---|---|
| *Dorylus (Anomma) wilverthi* | 15,000,000 | 1,500,000 | Raignier and Van Boven, 1955 |
| *Labidus praedator* | 4,000,000 | 1,000,000 | Schneirla, 1957b |
| *Eciton burchelli* | 900,000 | 300,000 | Schneirla, 1957b; Rettenmeyer, 1963 |
| *Eciton hamatum* | 250,000 | 80,000 | Schneirla, 1957b; Rettenmeyer, 1963 |
| *Neivamyrmex nigrescens* | 120,000 | 50,000 | Schneirla, 1958 |
| *Aenictus laeviceps* | 100,000 | 30,000 | Schneirla and Reyes, 1966 |

these broods is difficult. One procedure is to collect an entire brood and then count a measured fraction in order to calculate the whole. A simpler procedure, necessary for large colonies, is to trap several measured sections of emigration columns carrying numbers of brood judged to be typical, then to count the samples and multiply for the whole in view of the duration of brood transport in the movement. In Table 6.1 the first procedure was used to give the approximations for broods of *Eciton burchelli* and *E. hamatum* and for *Aenictus laeviceps;* the second procedure gave the approximations for *Dorylus (Anomma) wilverthi, Labidus praedator,* and *Neivamyrmex nigrescens.* All of these approximations involve sources of error for correction in systematic census studies (Rettenmeyer, 1963).

Within group B, whose brood populations may be the largest in the dorylines, census counts indicate that the broods of *Dorylus (Anomma) nigricans* are larger than even those of *D. (A.) wilverthi* (Raignier and Van Boven, 1955). Colonies of *Labidus praedator* and *Nomamyrmex esenbecki,* from studies of their emigrations, also have immense broods. Brood numbers in group A species seem to be distinctly smaller on the whole (Schneirla, 1958). The largest of those assayed, that of *Eciton burchelli,* is calculated to be roughly three times that of its close relative *E. hamatum* and ten times larger than that of its distant relative *Aenictus.* But the magnitude of an army ant brood, although important (Chapter 13), is only one of its characteristics that influences colony function (Schneirla, 1938, 1945).

The all-worker broods of *Eciton* and *Neivamyrmex,* beside being *regular* in their timing (Fig. 7.1), are *polymorphic,* each consisting of a smoothly graduated series of individuals differing in size and in structure (Figs. 2.1 and 2.2); *synchronized,* each of them with all its members developing roughly in step with one another (Fig. 6.2); and *coordinated,* for successive broods in the series, i.e., they overlap in time and are interrelated (Fig. 7.1).

The all-worker broods of *Aenictus* are like those of the other two genera of group A in their regularity, synchronization, and coordination. They differ from the others, however, in being virtually *monomorphic,* with all individuals of each brood closely similar in size and structure. Representatives of a monomorphic brood of *Aenictus* and a polymorphic brood of *Eciton* are contrasted in Figure 6.3. Implications of this difference are discussed in Chapter 13.

In addition to brood synchronization, the coordination of broods[5] is

---

[5] The phenomenon of overlapping, interrelated broods, first reported for *Eciton* (Schneirla, 1949), was named "brood coordination" by Brian (1957c), who investigated it experimentally in myrmicine ants. In *Myrmica ruginodis* he found two overlapping brood cycles in one season. These were interrelated through the effects of such conditions as handling and feeding by workers that influenced brood development.

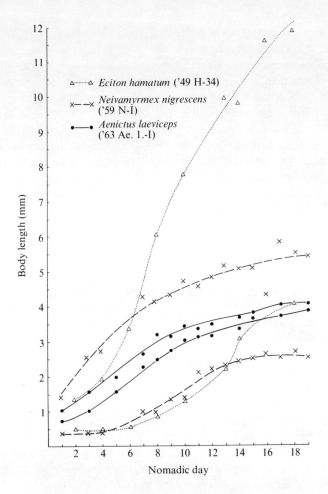

FIGURE 6.2.
Series of brood samples from colonies of three doryline (group A) genera, representing the course of growth in major and minor larvae through the nomadic phase.

significant to colony function in all army ants. To understand why this is so, it is necessary to study features of development within the broods as well as among the broods of different dorylines.

In a broad sense, biological factors controlling development are alike in all social insects (Wigglesworth, 1965). There are differences however in the rates of development. One type is represented by examples of three subfamilies of ants, including two dorylines, all surface-active, running insects; another is represented by honeybees. Honeybees evidently

(a)

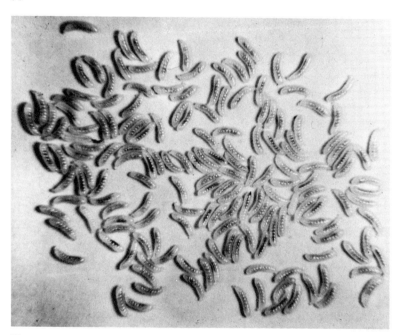

(b)

FIGURE 6.3.
Representatives of polymorphic and monomorphic army ant broods.
(a) A graduated polymorphic series of a nearly mature larval brood of
*Eciton burchelli*. The largest larvae are potential major workers, the smallest
ones potential minor workers. (b) An essentially monomorphic brood of a
colony of *Aenictus laeviceps* nearing larval maturity. Dark streaks are
partially digested food.

develop at a distinctly faster tempo than the ants, perhaps because they are flying insects which live a faster-paced life centered on foraging by flight.

It is interesting to note, however, that the estimated developmental times for all-worker broods of *Dorylus* (*Anomma*) *wilverthi* (Raignier and Van Boven, 1955) are considerably shorter than those for the other dorylines and for other subfamilies of ants. We will discuss later (Chapter 7) this difference among the dorylines, which may be ecological and related to colony size (Chapter 2); developmental times for these broods, however, may be considerably longer in the more hypogaeic *D.* (*A.*) *nigricans* (Raignier and Van Boven, 1955). We will just note here that members of any one army ant brood differ in their developmental rates in each of the stages. The ranges of these individual differences, significant for collective function, are greater in *Eciton burchelli* than in *E. hamatum* and much greater in *Dorylus* than in any group A doryline we studied (Fig. 6.4).

In group A army ants, each brood enters the colony as a great series of eggs laid by the queen midway in a statary phase (Fig. 7.1). Hence, the members of this brood are synchronized: They develop roughly in step with one another; yet they differ individually in a smooth series from minor to major workers. The successive broods of each colony take approximately the same time to complete the corresponding stages of their growth. These features suggest that broadly equivalent developmental conditions prevail in colonies of the three genera (Chapters 3 and 4).

Figure 6.2 shows an important difference among the broods of the three genera during their larval stage: the widening ranges between smallest and largest individuals at any one time in *Eciton* and *Neivamyrmex* as compared with the limited range throughout in *Aenictus*. Broods of the first two genera are polymorphic, i.e., they present graduated series of individuals differing in size and in structural details from the smallest to the largest individuals. These individual differences seem to be correlated with differences in the order of egg laying. The brood series of *Aenictus,* in contrast, is quasimonomorphic, i.e., its members differ only a little in size and in structure. Most dorylines have polymorphic broods and polymorphic adult worker populations (Chapter 2). Why *Aenictus* alone produces monomorphic broods, though with a schedule of egg laying much like that of other group A genera, is as yet unanswered (Chapter 13).

We find a remarkable similarity between successive broods in these dorylines, illustrated in Figure 2.3, which compares proportions of individual size types in brood and adult samples from the same colonies of two *Eciton* species. In both populations the smaller intermediates are most frequent, the minors are next, and the majors are least frequent.

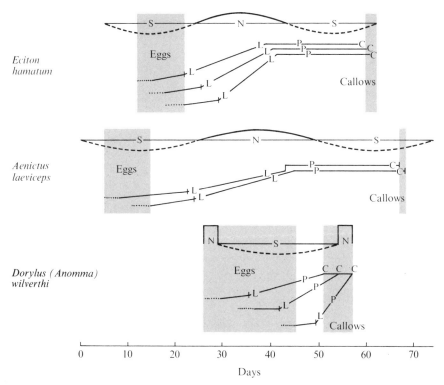

FIGURE 6.4.
Schema of brood developmental stages in *Eciton hamatum, Aenictus laeviceps,* and *Dorylus (Anomma) wilverthi,* representing the principle of developmental allometric convergence—differential growth in the brood decreasing the time range at maturity over that at egg laying. N: nomadic; S: statary; L: larvae; P: pupae; C: callow.

Since the adult population is made up of survivors of two or more earlier broods, the close resemblance of the current developing brood to its pattern indicates that conditions of development must be very stable and similar from brood to brood. What are these conditions?

Results of studies on development in various ants have generally been interpreted to mean that all workers develop from genetically equivalent fertilized eggs. Several investigators (Ezikov, 1922; Wesson, 1940; Gregg, 1942; Brian, 1956) working with numerous species have found that varied nutritive and other conditions affect larval growth with the result that polymorphic patterns are changed correspondingly. Worker polymorphism in ants, therefore, seems to have a trophogenic basis and to depend mainly on differential feeding and handling of the larvae by workers (Chapter 2). This statement may be accepted as a reasonable hypothesis for the dorylines, notwithstanding the known complexity of conditions affecting polymorphism in social insects (Weaver, 1966; Topoff, in press).

With doryline polymorphism considered nongenetic, the chief types of factors that may influence individual development are: (1) changes in the queen's physiological condition as she matures and lays her long series of eggs; (2) stimulogenic effects, i.e., tactile, mechanical, and chemotactic effects introduced by workers; (3) trophogenic effects, i.e., food and other nutritive and secretory effects; and (4) incubative conditions. Effects of queen and brood mates on development, though incompletely known, doubtless are complex. As our results (e.g., Fig. 2.3) indicate, developmental conditions are species-typical and so are relatively stable for each brood. Even so, from colony to colony and within the same colony from time to time, these conditions vary somewhat according to the existing situation. This impression is enforced by evidence concerning the brood and cycle (Chapter 7) and the colony queen (Chapter 8).

The graduated polymorphic brood series of most dorylines may be attributed theoretically to fine differences in the effects of these four main factors on individual larvae during growth. The first individuals in the egg series may be retarded the least—as oöcytes—through condition (1) above (see Chapter 8); they are also the first to begin feeding as microlarvae (i.e., first instar larvae) and are the largest and most attractive to workers, hence the ones with the highest stimulative and nutritive advantages. With a head start on all others, these individuals maintain their advantages, finally becoming major workers (Fig. 6.5). At the other extreme, the last individuals in the series presumably lose the most through condition (1) and begin their larval development latest and at the lowest ebb of all. With the lowest quota of stimulative and nutritive advantages, they attain the reduced size and form of minor workers. Presumably, the differences that prevail for individuals intermediate in the series correspond to the order of their initiation as oöcytes in the queen. Incubative differences also, as suggested in Chapter 3, are roughly correlated with sequence in the brood developmental series.

The all-worker broods of army ants are immense populations, in which food and related developmental advantages are severely restricted for all members. As species-typical brood ranges suggest (Fig. 6.2), standard and highly rigid limits must prevail for the total of these advantages

FIGURE 6.5.
Polymorphic range of *Eciton hamatum*. (*a*) Workers and mature larvae in a laboratory nest. (*b*) The same group of workers and brood a few days later after enclosure of the larvae in cocoons. Note mite on gaster of minor worker near center of photo. (See Plate IV, following page 138.)

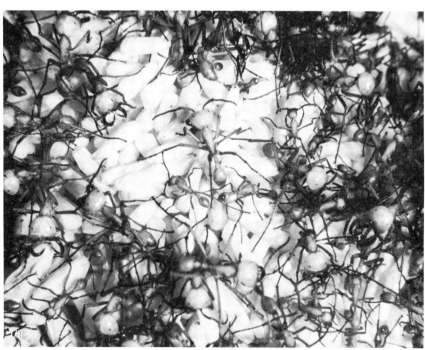

available to any one brood. Because of the competition that prevails among great numbers for any given condition (e.g., food and space) available only in limited amounts, these broods are notable examples of insect "starvation series" (Emery, 1895c; Wheeler, 1928) in which all members are underprivileged at all stages. As an outcome, mature all-worker larvae of *Eciton hamatum*, for example, range only from about 3.0 to 11.5 mm in body length whereas the mature larvae in sexual broods of this species, from an equivalent start, have far greater advantages in all developmental conditions and so attain a much greater bulk and a very different form (Fig. 9.3). In the all-worker series of *Neivamyrmex* species, comparable individual differences prevail in factors aiding development, though not as marked as in *Eciton* (Fig. 6.2). In contrast with certain myrmicine and other ants, in which differentiations in feeding and related conditions may enter later in the larval stage (Weaver, 1966), in the polymorphic army ants there is every reason to believe that these differences are involved from the start and continue throughout growth. Thus, in brood samples of *Eciton* taken at any stage, we find species-typical proportions of individual size types, with the largest individuals fewest and smaller individuals more numerous. This condition is illustrated in Figure 6.6 for a brood nearing pupal maturity. These differences, critical for colony function, have been studied to advantage in allometric growth research.

In Chapter 2 we discussed the Huxley allometric formula as first used (Huxley, 1927, 1932) in comparing whole/part dimensions through the range of individuals in adult populations of crabs and driver ants, and as further applied (Wilson, 1953; Hollingsworth, 1960) to the study of adult series of various doryline ants.[6] In our program we have found this method a very useful means of studying doryline broods in the larval stage, which is probably the interval of their most plastic development.

---

[6] Allometry is of course much more than the size-frequency distribution of individuals in a population as it expresses relative differences among these individuals, e.g., in the whole-to-part or the part-to-part structures of members of the series. In these first studies (Tafuri, 1955; Lappano, 1958) the Huxley formula was modified to emphasize the exponential function (appropriate to a developing series) rather than the power function (appropriate to an adult series). These alternatives were expressed in a later study (Schneirla et al., 1968) as $y = ax^b$ (power function) and $y = ab^x$ (exponential function), in which $x$ stands for larval body length, $y$ for leg-disc area, and $a$ and $b$ are constants. The logarithmic form of the latter formula is:

$$\log y = (\log b) x + \log a$$

in which $\log b$ represents the slope of the growth curve when viewed in a semilog coordinate system (e.g., Fig. 6.7). (My collaborators in this study were Rosamond Gianutsos, candidate for the Ph.D. in psychology at New York University, and Professor Bernard Pasternack, Department of Environmental Medicine, New York University School of Medicine. The research was aided by grant GB-5105 from the National Research Foundation.)

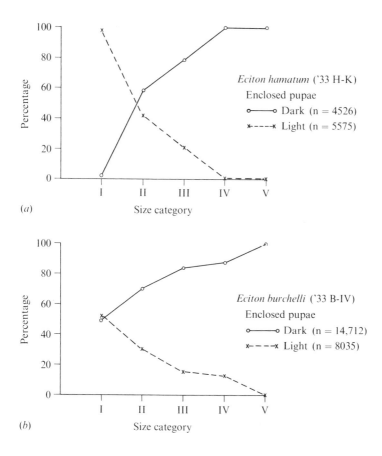

FIGURE 6.6.
Developmental condition of different size groups of army ant pupae. (a) Conditions in a large sample of an all-worker brood of *Eciton hamatum*, preserved at a stage near pupal maturity. Specimens nearly mature (dark cocoons) were then separated from those less mature (light cocoons); then each series was sorted into size groups ranging from smallest (minor workers) to largest (major workers). (b) Conditions in a large sample of an all-worker brood of *Eciton burchelli*, preserved at a stage near maturity and treated identically with the sample of *Eciton hamatum*.

Our purpose has been to investigate individual differences in the (synchronized) brood of the same doryline colony at successive times in its larval stage. Thus, John Tafuri (1955), working with *Eciton hamatum* in his doctoral research, used a formula for expressing ratios between the body length of larvae and the dimensions of one "walking set" of imaginal leg discs to differentiate these values through the main interval of larval growth falling in the colony's current nomadic phase. Compared in terms of these ratios, the smallest larvae in the series of samples had the steepest curves, the largest larvae the flattest curves, and the intermediate larvae fell in between. These results indicate that the smallest individuals in the broods of these species grow at the fastest rate, the largest individuals at the slowest rate, and intermediate individuals at corresponding transitional rates. In her doctoral study, Eleanor Lappano (1958) found equivalent differences in the larval series of *E. burchelli*.

In a further investigation of the same type (Schneirla et al., 1968), we compared larval series of *Eciton hamatum, Neivamyrmex nigrescens,* and *Aenictus laeviceps*. The differences are represented in Figure 6.7, in which a brood of *E. hamatum* and one of *Ae. laeviceps* are compared with respect to the slope of regression lines expressing the growth rates of their largest and smallest larvae at successive stages. The lines for major and minor workers of *E. hamatum* (and of *N. nigrescens*) are reliably different in slope, but those for the corresponding size types of *Ae. laeviceps* are not (although a small difference is apparent by inspection of the figure). This research, therefore, offers a means for comparing rates of development between genera as well as species in the dorylines.

Allometric growth studies also bring out the adaptive advantages for the colony of differential growth rates within a brood during the functional cycle. When the brood is delivered in the statary phase, as eggs, a difference of about one week separates the first- and the last-laid eggs in *Eciton hamatum* and of about ten days in *E. burchelli*. In both species, on the first nomadic day (N-1) the most advanced individuals are microlarvae and are feeding[7] although the least advanced are still embryos. As Figure 6.4 shows, the smallest individuals in both *Eciton* and *Neivamyrmex* begin their larval development at a substantially later time than the largest. Comparisons based on our allometric growth results (Fig. 6.7) show, however, that larvae then advance relatively more rapidly the smaller their size. They seem to "catch up" so well that at pupal maturity the major workers are separated from the minor ones by an interval of only two or three days in *E. hamatum* and of four or five days

---

[7] Feeding can be observed under a lens in the larvae of *Eciton hamatum* and other group A species studied on the night of the first emigration. This behavior may begin within the first day they hatch late in the statary phase.

FIGURE 6.7.
Series of brood samples from colonies representing the army ant group A genera *Eciton* (*a*), *Neivamyrmex* (*b*), and *Aenictus* (*c*), indicating the general course of growth in larvae of major and minor size groups through the main part of the larval stage (i.e., the nomadic phase of the colony).

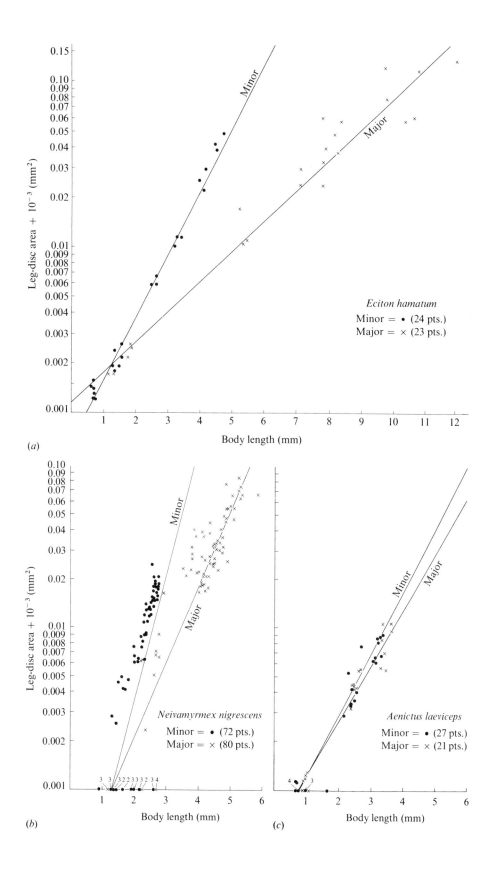

in *E. burchelli*. This condition, which seems to hold for all polymorphic army ants, I call "developmental convergence." The concept is significant for brood-stimulative theory of doryline cyclic function.

By virtue of developmental convergence (Chapter 7), the excitatory effect of an emerging callow brood of *Eciton* or of *Neivamyrmex*, delivered by many more individuals within a shorter time, builds up into a much stronger mass impact on the colony than if the time range at egg laying were to persist. But, as Figure 6.4 illustrates, *Aenictus* achieves a maximally summated brood-stimulative effect by virtue of the practically monomorphic, highly synchronized character of its brood throughout development. There may be an even larger differential between the egg-laying and the callow-arousal intervals in the great broods of the driver ants, whose overall developmental time has been estimated (Raignier and Van Boven, 1955) at 20 to 25 days as compared with a median of 46 days for *E. hamatum* (Schneirla, 1957a). As Figure 6.4 suggests, the brood range of *Dorylus* in early stages is much wider, in proportion to that of later stages, than in group A species studied.[8] The degree of developmental convergence in the driver ants may be greater, therefore, than in group A. Scattered reports for *Labidus coecus* (Weber, 1941) and *L. praedator* (Luederwaldt, 1918) indicate the occurrence of coordinated broods in that genus, the young brood with a much wider range (e.g., eggs to midlarvae) than the advanced one (e.g., advanced pupae). The significance of this condition is discussed in Chapter 7.

Among army ants in general the broods of each species normally are found to have predictable distributions of individuals at each cross-section from early to late stages of growth. To account for these circumstances, the hypothesis is reasonable that major workers develop from first-laid eggs, minor workers from last-laid eggs, and intermediate individuals from eggs laid at times in the series corresponding to their respective sizes. Consistent with this assumption, broods representing the polymorphic group A genera have wide ranges of size types at each point in larval growth, and related differences in growth rate are indicated in Figure 6.2.

To study population conditions within a brood presumably related to growth rate, the broods of colonies of *Eciton hamatum* and of *E. burchelli* were sampled when they were close to pupal maturity (Schneirla, 1934). Each large sample of cocoons was first divided into "light" (unpigmented, less advanced) and "dark" (pigmented, more advanced) specimens; next, each of these two groups was sorted into five size classes, from smallest

---

[8] Comparable differences may exist among the species of a genus corresponding to their degrees of surface adjustment. For example, we have found that in *Neivamyrmex* brood ranges are appreciably smaller in the surface-active *N. nigrescens* than in the more hypogaeic *N. carolinensis*.

# Color Plates

PLATE I.
Curtain- (or half-cylinder-) type bivouac formed by a nomadic colony of *Eciton hamatum* between the buttressed roots of a tree.

PLATE II.
Test of species interactions based on odor. *Left*, workers of *Eciton burchelli* hanging next to the front wall of a tall, narrow cell, separated by cheesecloth from a similar cell of workers of *Eciton hamatum*. For several days, these groups are exposed to each other's odor. *Right*, workers of *Eciton hamatum* were released into an arena. Within an hour the workers had formed this circular column. *Below*, workers of *Eciton burchelli*, then admitted to this arena, join the circular mill, running smoothly in it for about 20 minutes. *Below right*, as workers of the two species become disturbed (presumably by the other's odor), fighting begins and the mill breaks up. Preexposed workers of the two species are longer together, without fighting, than are unexposed control groups.

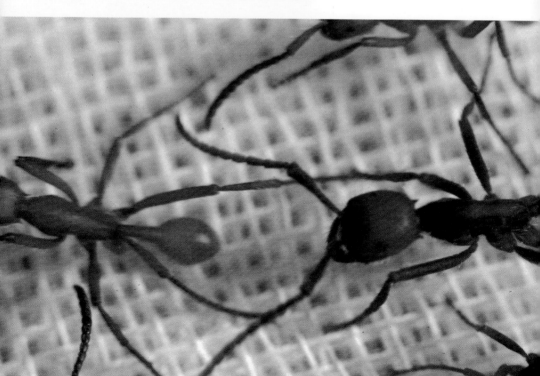

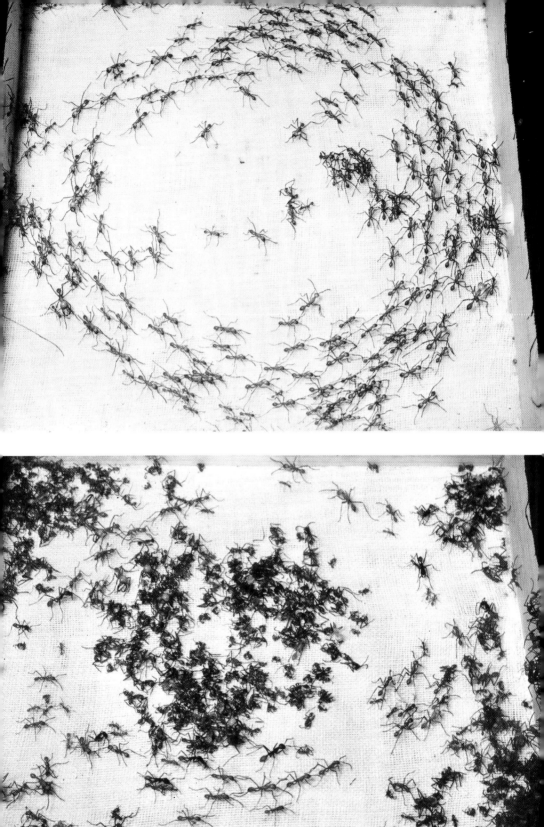

PLATE III.
Emigration columns carrying nearly mature larvae.
*Above, Eciton burchelli* (Barro Colorado Island, Panama Canal Zone).
Note the flange of workers bordering the column on its lower side.
*Below, Aenictus laeviceps* (Negros Island, Philippines). Note that
the burdens swing outward a little as the carriers round the turn.

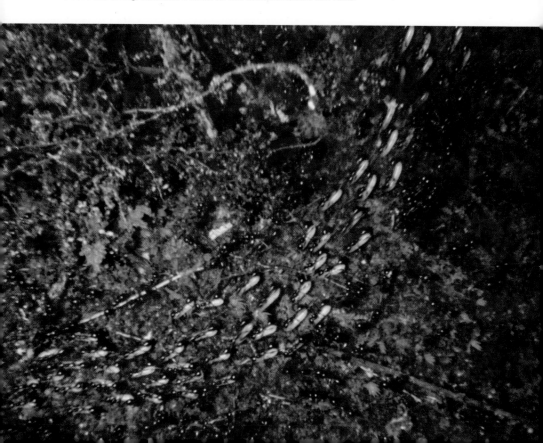

PLATE IV.
Polymorphic range of *Eciton hamatum*.
*Above*, workers and mature larvae in a laboratory nest.
*Below*, the same group of workers and brood a few days later
after enclosure of the larvae in cocoons. Note mite on
gaster of minim worker near center of photo.

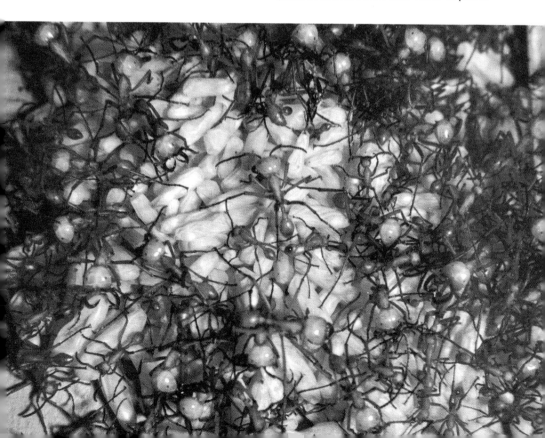

PLATE V.
Alate males of *Eciton burchelli* (and other species), started at the base of a narrow incline in laboratory tests, readily mount on the path. Note that the hairy tip of the gaster is in contact with the substratum, and that the wings are functional.

PLATE VI.
Callow queen and alate (pre-flight) male of *Eciton burchelli* in mating test. The male has the odor of the queen's colony; hence she responds to him; he has not had his flight, so does not respond to her.

PLATE VII.
Monomorphic brood of a colony of *Aenictus laeviceps*, as the colony is about to enter the statary phase. This brood, just entering the prepupal stage, has just beeen removed from the colony's new bivouac in a hollow log.

PLATE VIII.
A platter-type bivouac of the surface-adapted *Aenictus laeviceps*, formed under dry leaves and a rock, with exposed clusters on its downhill side. The ants are beginning a raid by spreading from the bivouac.

to largest. The results were clear-cut for all four broods. As Figure 6.6a shows for a sample of *E. hamatum,* all minors were light, and proportions of light pupae decreased regularly toward the large extreme; and all majors were dark, and proportions of dark individuals decreased regularly toward the small extreme. Figure 6.6b gives comparable results for a brood of *E. burchelli.* These results point to species-typical developmental conditions (e.g., bivouacs) that affect the broods of all functioning colonies similarly.

As they develop, army ant broods pass through series of changing relationships with the workers, important for colony behavior. In *Eciton,* for example, in the midstatary interval the still unpigmented pupal brood in cocoons exerts no evident excitatory effect and may even have a quieting effect on workers that stand or hang in the bivouac holding cocoons in their jaws. In the colony, a highly reduced schedule of raiding then prevails (Chapter 4).

But toward the end of the statary phase, this relationship changes greatly. The more the pupae darken and begin to stir within their cocoons, the more their tactual and odorous effects excite the workers. In study nests, we see the workers pick up the dark cocoons and shift them about more frequently than the light ones. When the pupae are virtually mature, the excitatory effects are obviously great, as indicated by what happens when a dark cocoon is put down near a few workers. The workers are plainly attracted to the cocoon, and especially to its anterior end, where they soon catch hold. In their ensuing counterpulling and twisting there, the case is soon torn open. Mature pupae nearly always emerge from the slit-open anterior ends of those cases, clearly an outcome of the greater attraction of workers to the head end of the cocoon than elsewhere. Two kinds of effect may be inferred: mechanical stimuli from the frequent twitching of antennae and legs by handled mature pupae, and greater odor from them anteriorly than elsewhere.

The callows, as soon as they emerge, become the objects of much licking and carrying about by the workers. Almost without exception (Fig. 6.6) the majors are the first to be removed from cocoons and to be so treated, the intermediates next, and the smallest members of the pupal brood last. This sequence clearly parallels the order in which individuals of different sizes become pigmented and begin to stir within their cocoons.

In *Eciton* and in *Neivamyrmex,* when the new processes of reciprocal stimulation between workers and mature pupae reach a high level, the colony begins a nomadic phase. Although the pupae of *Neivamyrmex* are naked, in this genus much as in *Eciton* the workers pull the remnants of pupal tissues from callows, lick their bodies energetically, and stroke them with antennae and palps. Workers rush about, excitedly touching antennae to brood or other workers, and many leave the bivouac and join the raid. This great wave of increased colony excitation and activity, evidently initiated by the brood (Schneirla, 1938), clearly is sustained by an almost incessant mechanical and chemotactic arousal of workers by brood and of brood by workers—and probably also by pheromones. Although there are indications of a comparable arousal effect based on pupal maturation in *Aenictus,* from the behavioral circumstances it is weaker than in *Eciton* and *Neivamyrmex* (Chapter 13).

Within the first few nomadic days in *Eciton,* the callows seem to excite adult workers less and less the more they resemble them in behavior (Chapter 7). Meanwhile, a new type of brood-worker relationship has emerged, involving the oncoming young brood. Bivouac and laboratory studies of this relationship are best started in midstatary phase, when the eggs are delivered (Fig. 7.1). In the bivouacs, the eggs are usually massed in a single bolus held in strands of workers hanging below the queen's position. As a rule, this mass is penetrated only by the smallest workers, who tunnel their way through it. As eggs transform into embryos, these form a thickening "shell" in the outer layers of the mass, where both minor and smaller intermediate workers are now busy (Schneirla, 1949). When the embryos hatch as microlarvae, these are fed by the small workers, probably with eggs and other tiny bits of food they can ingest. Army ant larvae may even consume eggs of their own brood, as investigators (Le Masne, 1953; Freeland, 1958; Weir, 1959) have observed in other ants.

As the nomadic phase begins, worker- and brood-stimulative relationships proceed both outside and inside the bivouac. In the early emigrations, the young brood is transported in many tiny packets. The workers, in forming a new bivouac, rearrange this brood in a central position which it occupies next day. But as the brood grows and its smallest members also become larvae, its distribution in the bivouac changes (Fig. 1.1). Usually, in *Eciton,* the smaller larvae are kept in the center of the bivouac, reached mainly by the smaller workers, and the larger ones are held in strands farther out, where they can be reached by larger workers. In relation to these differences in brood placement, the scope of brood feeding and of stimulative relationships between larvae and workers expands throughout the phase.

In the nomadic phase then the workers are increasingly occupied in grooming and feeding the brood. Army ant workers deliver food to their larvae much as do ponerines (Wheeler and Bailey, 1925). Morsels of booty, macerated as they are rolled about and squeezed between the jaws, tongue, and palps of the adult workers, become soft pellets which are laid upon larvae or larvae are laid upon them. The process of food preparation often begins with workers pulling against each other in pairs with a piece of booty held between them in their jaws, evidently extracting juices with each new grip and pull. In keeping with the workers' degenerate crops (Eisner and Brown, 1956) and small gasters, their own food may be the fluids more often than the tissues. The larvae, in contrast, feed voraciously and are often seen with their mouthparts applied to morsels or—increasingly as they grow—to whole pieces of booty.

By and large then the larvae of polymorphic army ants are treated differently according to their size and developmental stage; in handling and feeding the brood, the smallest workers can deal most readily with small larvae, the larger workers with larger larvae. The worker's handling of larvae depends not only on such properties as her mandibular and antennal stretch but also on her physical size. A large worker, likely to be carrying a larger piece of booty, as a rule cannot pass as far toward the center of the bivouac mass as a smaller worker. The latter, in contrast, threads her way farther inward where channels are mostly narrower. As a result of such conditions, differential feeding prevails among polymorphic army ants.

As the nomadic phase advances in *Eciton,* larvae are heaped increasingly in small pockets or held individually by workers in the bivouac strands. Their positions may often be changed according to time of day, stage of the raid, and amounts of food available. The more the larvae grow, the more the workers seem to be occupied in feeding and handling them. At length, even the major workers are often seen near brood-feeding groups, opening their great mandibles widely at intervals, and applying their mouthparts to larvae or to booty much as in drinking or in touching the queen. Most of the time in brood feeding, larvae are given morsels roughly proportional to their size. But there are many exceptions late in the nomadic phase.

Our observations indicate that the larvae, through both their movements and odors, attract the workers who, in their turn, both feed the larvae and stimulate them (Schneirla, 1938, 1949, 1965). It is a reasonable premise that army ant colony functions center on a variety of stimulative, trophic, and nutritive processes kept in force by brood growth and brood coordination. The mutualistic concept of trophallaxis hardly suggests the summative and progressive aspects of reciprocal arousal and of the feedback effects through which doryline broods can arouse and maintain colony action. Further research may be expected to support the concept of the brood as a collective initiator of complex, expanding stimulative relationships *necessary* to the cyclical colony functions of these ants (Chapter 7).

Army ant workers are involved increasingly with their larvae as the brood grows. For the workers, larvae and booty belong together, food being dropped more and more often on larvae in the bivouac, larvae more and more often on food in cache feeding. Although the larvae are not motile, they play active roles in colony life. When touched by workers, they bend their anterior ends ventrally, as do many other ants (Le Masne, 1953); at times, according to the stimulus pattern, they also turn their foresegments to one side or even extend them. Gently breathing over larvae excites a mass wriggling as a rule like that seen when workers move over unfed larvae. The prompt, vigorous reactions of unfed larvae to stimulation contrast with their relative sluggishness when satiated. Accordingly, we may say that thresholds of sensitivity and reactivity vary according to larval alimentary condition, dropping low with lack of food but rising with satiation. To larval stimuli, workers respond in a variety of ways from feeding or grooming the larvae to joining a raid.

In studying larvae of *Eciton* in laboratory nests with workers (Fig. 6.5), I have observed that from time to time they exhibit stirring movements in which the twisting and ventral bending of anterior segments is prominent (Le Masne, 1953). These movements, which seem attractive to workers, and which may involve both mechanical and chemical stimuli, often lead to feedings (Brian, 1965). From series of such observations with the broods of numerous colonies, I conclude that on the whole the larger larvae of any sample attract the workers most and are likely to receive more food in any one feeding than the smaller larvae. On the other hand, larger larvae never seem to feed as often as the smaller larvae. This result suggests that the smaller larvae receive more food in relation to their size than the larger larvae (see footnote 3).

From tests in which workers of *Eciton* cluster readily on paper discs where larvae of their colony have rested for a time (but not on clean discs) and likewise turn in far greater numbers into the arm of a Y tube through which larvae have been drawn rather than into the clean one, the strong attraction of larval odor for workers is evident. The results indicate, moreover, that the attraction of larvae for workers increases through the nomadic phase (Schneirla, 1945). Although in *Eciton* this effect falls off sharply after the larvae spin their cocoons, in *Neivamyrmex* (Schneirla, 1958) and *Aenictus* (Schneirla and Reyes, 1969) it increases for a few days beyond larval maturity into the prepupal stage (Chapter 7). As the larvae are then inert, the basis must be chemical.

Functional studies are needed on the possible sources of larval stimulation. As an example, Lappano's histological results (1958) suggest that in *Eciton burchelli* the labial glands are ready to function in the largest larvae at the start of the nomadic phase, and the intermediate and small larvae at about the third and sixth nomadic days, respectively. These and other secretory tissues, especially those having to do with ingestion and feeding, may well contribute to raising the level of colony excitement as the nomadic phase advances.

A different type of change arises in the broods of *Eciton* as larval maturity approaches, namely, a transformation of labial gland tissue into the precursor of the spinning material. These changes too appear first in the largest larvae (a few days before the nomadic phase ends), next in the intermediate, and last in the smallest larvae (Lappano, 1958). Comparable changes arise in maturing larvae of *E. hamatum*. Significantly, laboratory observations show that at corresponding times larvae of the respective size types begin to decrease their feeding and at length cease to feed, but without any signs that their excitatory effect on the workers has lessened.

Changes are apparent, however, in the behavioral relationships of workers and larvae. The workers still seem attracted to the maturing larvae but when near them behave differently than before. They pick up the larvae, now distinguished by frequent waving motions of their anterior ends, but instead of feeding or shifting them about within the cluster as they would have done a day or two earlier, the workers form broad columns and carry them from the bivouac to nearby places where there is wood dust or other detritus. Here the larvae are dropped and start to spin their cocoons, an event described vividly by the American naturalist, William Beebe (1919) for a colony of *Eciton burchelli* that had settled under the roof of a building near his laboratory in British Guiana.

These are busy scenes indeed. The ants form platter-size spinning aggregations of larvae in places near the bivouac, in areas between buttressed roots, within logs, or in similar places. Here the workers constantly shift larvae about, placing them on particles of detritus or dropping the particles on them. Watching a spinning larva under a lens, one sees that with each extension it touches its mouthparts to a nearby surface and there extrudes a fine thread that hardens at once into a fiber. Through alternate extensions and contractions in which it passes its mouthparts first along and then across its body, much as in other ants studied in this type of behavior (Wallis, 1960), the larva gradually spins the fabric of its cocoon. At intervals, partially enclosed larvae are snatched up by workers who turn them or shift them about. Now and then a worker, after mouthing the thin envelope of an enclosed larva and picking off loose particles, returns the individual to the bivouac where it completes spinning its cocoon (Chapter 7).

In *Eciton,* while the maturing larvae are spinning, they continue to excite and attract the workers but at close quarters now seem to disturb the workers. Theoretically, we may think of a worker's responses to spinning larvae as a modified form of the treatment she normally gives mildly repellent objects (e.g., spoiled food) that are carried from the bivouac and discarded in a refuse dump nearby. Instead the mature larvae seem to lose their repellent effect once they are within thin envelopes.

There are striking contrasts in the occurrence of cocoon spinning among the dorylines, both in all-worker and in sexual broods, as Table 6.2 shows. Mature worker larvae of *Eciton* always spin cocoons, but those of *Neivamyrmex* and *Aenictus* do not; mature larvae of *Labidus* spin cocoons, but those of *Dorylus* do not. The basis of these interesting differences remains to be investigated.

TABLE 6.2.
*Contrasts in larval maturation among the dorylines*

|  | All-worker broods | | Sexual broods | |
| --- | --- | --- | --- | --- |
| Genus | Cocoons spun | No cocoons | Cocoons spun | No cocoons |
| *Dorylus* |  | X |  | X |
| *Aenictus* |  | X |  | X |
| *Eciton* | X |  | X |  |
| *Neivamyrmex* |  | X | X |  |
| *Nomamyrmex*[a] |  |  |  |  |
| *Labidus* | X |  | X |  |
| *Cheliomyrmex*[a] |  |  |  |  |

[a] Unknown.

As this chapter has suggested, one important role of every young generation in doryline ants is to arouse and maintain colony nomadism. This is done in complex ways. Each brood, because it is synchronous and advances its members at much the same rate of development, delivers a virtually unitary stimulative and trophic effect on the colony at each stage from egg to callow. A coordination of successive broods can thereby arise, through which they not only overlap partially but are functionally interrelated: They undoubtedly influence one another physiologically. Advantages to the colony from this state of affairs may be glimpsed from preliminary results. A prominent illustration is the repressive effect evidently exerted by the new mass of callows on the early brood at the outset of a nomadic phase. The callow brood of *Eciton* and *Neivamyrmex*, through its strong attraction for workers, seems at first to reduce developmental advantages (e.g., food, worker handling) for the young brood. But in their few days of ascendance the callows raise the excitatory and trophic condition of the colony high and so in the end facilitate the development of the young brood (Schneirla, 1957a). Also, through their brief excitation of the colony, the callows may stimulate a reproductive change in the queen (Chapter 8). This is only one illustration of brood coordination, taken at one cross-section of the colony functional cycle.

In *Eciton* and *Neivamyrmex* of group A the described interrelationship between the callow and the young brood regularly follows the statary phase and insures the continuation of colony nomadic behavior once it begins. In *Dorylus,* however (perhaps also *Labidus*) and others of group B, the callows emerge and excite one long emigration, but only that for the time. Group A dorylines and *Dorylus* share the callow-stimulative effect that introduces nomadism though *Dorylus* lacks mechanisms of brood coordination that might continue nomadic behavior. These contrasting conditions in groups A and B are discussed in Chapter 7 in relation to cyclic behavior.

Other important functions of the brood include stabilizing the bivouac. In *Eciton,* for example, the colony centers its bivouac around the young brood and the queen, which thereby provide a basis for organization of activities within the nest. Later in the phase, although these relationships are much more complex and differently organized, a stabilizing effect of the brood is still recognizable (Chapter 3).

It is remarkable that although the army ant brood arouses and stimulates the colony, normally the workers lick and groom the brood but do not consume it to any great extent. We may postulate the basis of this immunity as residing in colony and queen odors which the brood bears, for tests show that larvae at the same stage but from other colonies of the same species are consumed with the booty. The queen odor is crucial for the normal differentiation of their own brood from foreign brood and booty since after a colony has lost its queen (Chapter 8), the workers begin to cannibalize their own brood.[9] They also consume their own brood when the colony supply of booty falls low (see footnote 6 in Chapter 1), suggesting that at times the brood may serve as a food reserve sustaining the colony through times of hardship.

Excitatory functions influencing colony behavior, clearly important in the army ants, may play a role in social insects generally. Their investigation should advance with further attention to seasonal and other changes in colony life as well as with increased emphasis upon problems of functional development. For example, in the army ants, characteristics

---

[9] A colony of *Eciton hamatum* deprived of its queen for two days or longer is likely to start feeding on its brood. After a queenless colony has merged with a colony of the species that has its queen, cannibalizing of brood occurs, but it is mainly or entirely the brood of the queenless colony that is consumed (Schneirla and Brown, 1950).

of function and behavior might seem to arise simply through maturation (i.e., the direct effects of tissue growth and differentiation to development). To be sure, in the social insects, properties of structure and physiology influence individual behavior patterns rather directly—much more so than in mammals; hence I have termed them "biosocial" (Schneirla, 1946). This consideration naturally does not force us to accept conventional "instinct" theory with its dogma of behavior patterns as "innate" (e.g., Lorenz, 1965; Schneirla, 1966). As our analyses of army ant behavior show, problems of species-typical behavior are not by any means that simple, even in insects.

It is a useful view, both for stimulating research and for species and generic comparisons, that a coalescence of processes of maturation and of experience (i.e., effects of stimuli and their combinations) forms the basis of behavioral development (Schneirla, 1956b, 1957c). From the evidence on army ants, these processes must arise and fuse in ways characteristic of insects but in ways very different from those effective in mammalian development (Schneirla and Rosenblatt, 1961). This fusion of maturation and experience may hold even until the new worker has ended its development as a callow. Such an approach is very helpful in the study of army ant behavior and function (Topoff et al., in prep.).

Some years ago Basil P. Uvarov (1932), then director of the Anti-Locust Research Centre in London, pointed to the importance of conditioned reflexes for insect behavior. There is evidence for conditioned learning in ants (Fielde, 1904; Heyde, 1924) as well as in other insects (Schneirla, 1953; Butler, 1962; Wigglesworth, 1965). Conditioning, although held to very simple forms in army ants, may be indispensable for the development of their species-typical patterns. To illustrate, when larvae feed in the bivouac, they are presented with the stimulus combination: (X) colony odor and (Y) booty odor. After many repetitions (i.e., experiences) of these stimuli together, the larva's developing nervous system may be so changed that it now responds more readily to booty with its own colony odor than to booty with a foreign colony odor. This process, closely meshed with the maturation of sensory and other mechanisms, may provide a basis for the individual's later approaches as a newly emerged callow adult to objects bearing its *own* colony odor. On a similar basis, perhaps begun in the larval stage, the new worker may be prepared to approach booty or brood when their odors are encountered.

There are indications that the development of army ant workers continues for at least a few days after they have emerged as callows. These new workers, distinguished in *Eciton* and *Neivamyrmex* by their light pigmentation, at first stay in the bivouac grouped usually around food. On the first nomadic day, only a few of the larger ones are seen on the raiding trails and not far from the bivouac. On the next day or two they appear on the trails in greater numbers and farther from the bivouac; but even now they take little part in the raiding. Although they do not at first seem to differentiate odors of mandibular gland secretions and other stimuli as do adults, they soon improve. More and more they join in carrying booty and in attacking prey. By the fifth day the callows, including the smallest individuals, are seen frequently on the trails, even at points far from the bivouac, and in raiding.

These behavioral changes are developmental. Observations suggest that early callows react first to food, then to the brood and nestmates, and at length to the trails.[10] In *Eciton burchelli* callows are seen carrying inert pupae even on the first emigration, their numbers increasing the following nights (Chapter 5). At first they are clumsy and easily reversed; after a few days they run along as the adults do. Observations indicate that in these first days these new workers accomplish: (1) a last phase of maturation (e.g., gains in sensitivity and in muscular strength); (2) a conditioning to colony odor, trail chemical, and odor differences at trail branches relative to trail running; and (3) experience with combinations of stimuli involving colony mates, booty, and brood—also with reactions of feeding brood, of attacking prey, and the like.

Development inside the bivouac thus provides a foundation for action outside. The callow worker when unfed first responds to nearby food or bustles about, next begins to feed larvae, and finally responds to brood stimuli by leaving on the raid. Thus, after their first few days, callows of both *Eciton* and *Neivamyrmex* are so much like the adults in appearance and behavior that it is difficult to tell them apart (Schneirla, 1938, 1958).

From these considerations, I propose that adultlike behavior in callow army ant workers arises out of the close combination of tissue maturation and "experience" (as defined) including early conditioning. Thus, through early callow self-feeding and feeding of brood, odors of colony, of brood, and of booty can become equivalent as stimuli to leave the bivouac in the raid. Trail following in the raid may be a specialized outcome of a developing behavioral process that begins with early adjustments to food and to brood. Because the doryline trail chemical comes from the hind gut (Blum and Portocarrero, 1964; Blum and Wilson, 1964; Watkins,

---

[10] A few hours after emerging, callow workers of formicine ants stand in place with excited movements of antennae and anterior body, attracting adults who feed them by regurgitating liquid food. Within a few days, however, the callows move about the nest, initiating feedings by approaching nestmates contacted with antennal movements alone. Meanwhile they have begun to feed the brood (Heyde, 1924). Their first foraging trips follow.

Honeybees, as has been known for some time (Rösch, 1925; Lindauer, 1961), normally pass through an early period of transitional activity after they have emerged from their cells as callow workers; they are first occupied in the hive with brood feeding and related activities, then begin an adult life of foraging with orientation flights and food-gathring trips of increasing length (von Frisch, 1950).

1964), the ability of the worker to follow it may arise through early feeding in the colony and may thus be considered a continuous approach to a food-related odor. The more the new workers follow this odor, the more readily they react to booty odor (i.e., food) with strong approach (i.e., attack). In these ways callow army ant workers finish their main development and become adult members of the colony.

Thus, the colony nurtures the brood, with its bivouac beginning as its incubator and feeding center and ending with an introduction to the trails. At all stages, there are buffering effects against extreme conditions; at first, when the larvae are small and vulnerable to cooling, they are kept in the center of the cluster; in the emigration they are carried in packets directly to the new site; later, when the larvae are larger and are carried individually, they are insulated by their own fatty tissues; they are also transported by feeding stages, which, as they increase in size, facilitate airing, warming or cooling, exercise or feeding, according to conditions. Throughout the nomadic phase, overexposure of the larvae is thus counteracted in ways appropriate to the stage.

The army ant mode of carrying larvae, slung back beneath the worker's body, ventrum close to ventrum the entire length (Wheeler, 1910), buffers the larva against exposure and against damage from bumping into objects. This way of carriage has several adaptive aspects. It insures an intimate contact between larva and worker, as carrier and burden are in virtually a mouth-to-mouth contact throughout. By grasping the larva in its most sensitive region, where the principal neural ganglion is located, the workers incidentally immobilize the young during the trip. This relationship, furthermore, permits brood and workers to experience a regular and lengthy reinforcement of each other's odors and oral stimuli. The process cannot but help increase prepotency of brood and nestmate stimuli in group situations. Multifaceted adaptive processes of the kinds described, common in the dorylines, must aid greatly in forming the social bonds that weld new individuals into their colony (Schneirla and Rosenblatt, 1961).

# 7

# Functional Cycles and Nomadism

Doryline predatory and nomadic operations are not the random events they were once presumed to be (Vosseler, 1905). The army ant cycle[1] might have been predicted from the observations of Müller (1886) on a colony of *Eciton burchelli* or even from the contrasting conditions of queens that were captured, but not from the food-scarcity hypothesis, which seemed reasonable (Heape, 1931; Fraenkel, 1932) only so long as evidence on the behavior schedules of dorylines was lacking.

Army ant behavior is always cyclic, i.e., it involves a periodicity in the occurrence of alternating sets of phasic processes in the colonies. Our investigations on surface-adapted species of *Eciton* brought evidence for: (1) two contrasting conditions in the colonies, the nomadic and the statary (Schneirla, 1933), which (2) recur in interrelated ways as the two phases of functional cycles (Schneirla, 1938, 1945) and which normally alternate through the year (Schneirla, 1949; Schneirla and Brown, 1950) on the basis of (3) the changing excitatory effects of successive broods on the colony (Schneirla, 1938, 1957a).

Stage (1) of the work on cycles, carried out on Barro Colorado Island, involved short-term studies with colonies of two species of *Eciton*. I found a high-level pattern of daily raiding that led regularly into nightly emigrations, characterizing the nomadic condition, and a contrasting low-level pattern that led regularly into a late-day return to the same bivouac (i.e., without emigration), characterizing the statary condition.

In stage (2) further research concentrated on the transition from one of these conditions to the other and the reverse (Schneirla, 1938) in the same two *Eciton* species; colony '36 H-A (*Eciton hamatum*), as an example, was found on August 5, 1936, bivouacked deep under a great log. After a large raid the next day, the colony emigrated in the late

---

[1] I use this word from the Greek meaning "circle," or "a return to the place of origin," to emphasize the periodic, repetitive alternation of phases in army ant nomadism.

afternoon and evening to an open site 160 m distant. Many thousands of light-colored callow workers crowded the column, cocoon opening continued, and a few unopened cocoons were carried along. On an empirical basis I judged this to be the first day of a new nomadic phase (N-1).[2] The next day from an exposed bivouac the colony developed a large raid on three extensive trail systems and then in the evening carried out an emigration of 150 m, which nearly finished at 11:00 P.M. This pattern of daily raids and nightly emigrations continued through a nomadic phase of eighteen days in which the colony moved through a winding course over an ant trail distance of nearly 3000 m. In the meantime, a large new all-worker brood passed through its larval stage.

Then, clustered far beneath a fallen tree mass, the colony passed through a low activity (statary) phase of nineteen days, during which its enclosed brood passed through the prepupal and pupal stages.[3] Within a few days at this site, the colony so decreased its level of activity that the forays usually had withdrawn into the bivouac by early evening. Daily raids, which early in the phase started slowly after dawn, began even more slowly later in the morning or not until early afternoon, and sometimes not at all. Raids often began on trails made earlier in the phase, with the ants frequently using the same trunk lines for distances up to 50 m before diverging into new paths. In this way the colony "boxed the compass" from its statary site, covering in its stay an area of roughly 300 m radius around the bivouac (cf. Fig. 4.3).

On the first day of the new nomadic phase, conditions of behavior and of the brood in this colony were like those that held on August 5 when the study began, i.e., a large-scale raid developed during the day, an emigration followed in the evening and early night, and a mature pupal worker brood had nearly completed its emergence from cocoons. On the following night (N-2) in the second emigration of this phase, a new brood of very young all-worker larvae was identified carried in packets by workers. The colony had come full circle in completing a functional cycle of thirty-six days.

Comparable results were obtained with a colony of *Eciton burchelli*, which finished a statary phase, then passed through a short nomadic phase, and had begun a new statary phase when the study ended (Schneirla, 1944b). These and other findings reveal a routine typical of this species in which colonies alternate regularly between the two phases, with concurrent changes in their behavior and in their broods.

---

[2] In most cases, colonies of *Eciton hamatum* enter a new nomadic phase with all or virtually all of the cocoons opened before the first emigration ends on N-1. In colony '52 H-C (Table 4.1), which was exceptionally large for the species, the last of the cocoons (the smallest ones) were opened on N-2.

[3] On a night early in the phase, rain forced a shift of less than one meter farther back under the brush ceiling of the bivouac (Chapter 5).

In stage (3) the linkage of the phases was clarified by means of a brood-stimulative theory of cyclic function in *Eciton* (Schneirla, 1938). Briefly, as each statary phase ends, the colony is aroused by high-level stimulation from an emerging callow brood and begins a nomadic phase. As the callow-arousal effect wanes, the nomadic phase is maintained by the stimulative effect of a developing larval brood. When these larvae become mature, the colony lapses into reduced function, and a statary phase then runs its course under low-level stimulation from this brood in its prepupal and pupal stages. The parallelism between colony behavior and the developmental condition of broods in the two phases is represented in Figure 7.1. Concurrent changes in the queen, also represented, are discussed in Chapter 8. Because the two phases involve several differences beyond the obvious ones in colony behavior, i.e., differences in brood condition (Chapter 6), in worker physiology, and in queen's condition (Chapter 8), this clearly is a functional cycle and not merely an activity cycle.

Evidence from cross-sectional colony studies of several *Eciton* species in southern Mexico (Schneirla, 1947) support this theory of cyclic function both for epigaeic and hypogaeic species of *Eciton*. Further research on Barro Colorado Island (Schneirla, 1949; Schneirla and Brown, 1950) involved studies of colonies of *E. hamatum* and *E. burchelli* over long periods under a variety of conditions. As an example, in a continuous survey of colony '46 H-B (*E. hamatum*) begun on February 12, 1946 (in the dry season), and ended on June 12 (after the first rains had come), successive phases occurred as listed in Table 7.1.

When the research ended, the colony was in the fourth day of a new nomadic phase. The behavior of colony '46 H-B followed the pattern of alternating phases given above, and the correlated changes of its brood are sketched in Figure 7.1. In this time it produced three all-worker broods, each emerging as a distinct population of callow workers at the end of a statary phase. In its fifty-two emigrations observed in three complete

TABLE 7.1.
*Cycles completed by colony '46 H-B* (Eciton hamatum)

| Cycles | Phase duration (in days) | | Days in the cycles |
|---|---|---|---|
| | Nomadic | Statary | |
| 1 | 16 | 21 [a] | 37 |
| 2 | 17 | 20 [a] | 36 |
| 3 | 16 | 20 [a] | 36 |

[a] All-worker brood produced as callows.

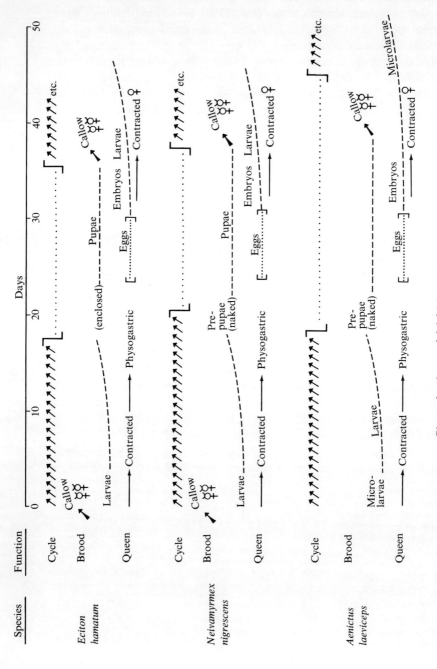

nomadic phases and in the start of a fourth, the colony covered a total paced-out distance of more than 8000 m, moving in a widely curved course that included two extensive loops (Fig. 5.8). The bivouacs were largely in open sites during the nomadic phase and always in secluded sites during the statary phase.

In this same project, beginning on February 7, a second continuously studied colony, '46 B-I (*Eciton burchelli*) passed through the several phases listed in Table 7.2.[4]

TABLE 7.2.
*Phases and cycles completed by colony '46 B-I* (Eciton burchelli)

| Cycles | Phase durations (in days) | | Days in the cycles |
|---|---|---|---|
| | Nomadic | Statary | |
| | | 21 [a] | |
| 1 | 10 | 21 [b] | 31 |
| 2 | 13 | 21 [a] | 34 |
| 3 | 12 | 22 [a] | 34 |

[a] All-worker brood.
[b] Sexual brood.

In this interval of about four months, the colony produced three all-worker broods (a) and one sexual brood (b). After an initial statary phase it passed through three complete cycles and had begun a new nomadic phase when the study ended. Colony behavior and brood condition, although somewhat more variable, corresponded essentially as described for *Eciton hamatum*.

From these and later results we conclude that colonies of these *Eciton* species normally continue their nomadic-statary cycles through the year

---

[4] The first nomadic phase of this series, in which a sexual brood in the larval stage was present, was shorter than the species average (Chapter 9). The following statary phase, which closed with a division of the colony (Chapter 10), I recorded as ending on the night most of the colony moved out of the statary bivouac (Schneirla, 1949, pp. 30ff.).

FIGURE 7.1.
Schema of the functional cycles characteristic of three doryline genera (group A). In the top row in each case (Cycle) phasic changes in colony behavior are indicated, with arrows standing for large daily raids and emigrations in the nomadic phase; dotted lines are for reduced raids and no emigrations in the statary phase. Concurrent conditions in brood coordination (*middle*) and queen's function (*below*) can be followed by the scale at top of diagram.

without a break (Schneirla and Brown, 1950),[5] with a consistent parallelism between brood condition and colony behavior.

In preceding chapters I have presented the criteria of colony behavior and condition, i.e., in bivouac formation, raiding, emigration, and brood condition, that we have used to distinguish the two phases of the functional cycle. Although these characteristics have been worked out mainly for surface-active species, they may distinguish the phases comparably in hypogaeic species of these genera, although, as in *Eciton* (Schneirla, 1949; Rettenmeyer, 1963), the latter may be more variable than the former in these respects. Table 7.3 gives results on phase durations for investigated species of four doryline genera.

We begin with a comparison of two *Eciton* species. In many investigated colonies of both *E. hamatum* and *E. burchelli,* nomadic phases are generally shorter than statary phases. In comparing doryline phase durations, one must bear in mind the characteristics of brood development discussed in Chapter 6 (see also Figs. 6.4 and 7.1). *E. hamatum* is less variable than *E. burchelli* in its nomadic phase durations, a difference that is paralleled by conditions of more precise synchronization and coordination in the broods of the former. Such characteristics should be examined in relation to species differences in colony size and other conditions that may affect processes centering on the brood, workers, and queen, and their interrelationships through the cycle (Fig. 8.6).

In a project carried out in Arizona in 1956 (Schneirla, 1958), I found equivalent criteria in colony behavior and condition for distinguishing the functional phases in *Neivamyrmex*. Ten colonies of *N. nigrescens,* when nomadic, were highly active in 51 of 69 nights observed, with vigorous raids, emigrations almost nightly, the queens contracted, and broods ranging from combinations of callows with embryos and microlarvae early in the phase to larger larvae alone and finally to prepupae. By contrast, in colonies judged to be statary, on 43 of 49 nights raids were weak or absent, there were no emigrations, and broods were in stages from the prepupal to mature pupal. While under study, three of the colonies observed in 1956 changed from the nomadic to the statary, and four changed from the statary to the nomadic condition. Each of the two longest studies involved a colony that passed through complete or nearly complete nomadic-statary cycles; also in Alabama in the summer of 1957 we followed a colony of this species through a complete 36-day cycle.

Evidence for *Neivamyrmex nigrescens* indicates that changes in brood

---

[5] E. Willis, in studying antbirds, followed the same colony of *Eciton burchelli* through a period of eight months, in which it completed seven nomadic-statary (N-S) cycles, as follows: (N)14–(S)20; 16–19; 12–21; 14–20; 18–19; 16–20; and 14–21 (see footnote 15).

TABLE 7.3.
Number of cases observed for respective durations of nomadic and statary phases in functional cycles of three group A genera and of nesting stops between emigrations in a colony of Dorylus (Anomma) wilverthi

| Species | Duration of phase (in days) | | | | | | | | | |
|---|---|---|---|---|---|---|---|---|---|---|
| | 0 | 5 | 10 | 15 | 20 | 25 | 30 | 35 | 40 | 45 |
| *Eciton hamatum* | | | | | | | | | | |
| Nomadic | | | | 9 | 13 | 6 | | | | |
| Statary | | | | | 2 | 5 | 20 | 1 | 1 | 1 |
| *Eciton burchelli* | | | | | | | | | | |
| Nomadic | | | 1 | 1 | 2 | 2 | 2 | 2 | | |
| Statary | | | | | 3 | 6 | 8 | 4 | | |
| *Neivamyrmex nigrescens* | | | | | | | | | | |
| Nomadic | | | | | 3 | 1 | 1 | 1 | | |
| Statary | | | | 5 | 4 | 3 | 1 | | | |
| *Aenictus laeviceps* | | | | | | | | | | |
| Nomadic | | | | | 2 | 6 | 3 | | | |
| Statary | | | | | | | 2 | 4 | 2 | |
| *Dorylus (Anomma) wilverthi* | | | | | | | | | | |
| Statary | | 1 | 1 2 | 1 1 1 | 1 | 2 1 1 1 | 2 1 2 | 1 1 1 | | |

*Source*: For group A genera: Schneirla (1949); Schneirla and Brown (1950); Schneirla (1958). For *D. (A.) wilverthi*: Raignier and Van Boven (1955).

coordination and queen's condition take place through the cycle corresponding to those found in *Eciton* (Fig. 7.1). In this species, nomadic and statary phases alternate in cycles through the active season, with both phases approximately 18 to 20 days long. This contrasts with other group A dorylines, in which the nomadic phase is shorter than the statary. Although cyclic functions are interrupted annually by winter dormancy in the species we have studied (Schneirla, 1963), tropical species of this genus probably continue through the year, as in *Eciton*.

For *Aenictus,* observations by others[6] suggested the presence of nomadic and statary conditions, and our systematic studies have confirmed this (Schneirla and Reyes, 1966, 1969). When colonies of *Ae. laeviceps* and *Ae, gracilis* have broods developing through the larval to the prepupal stage, they carry out nomadic phases with surface bivouacs, vigorous raids, and one or more emigrations daily; when they have broods in the pupal stage, they pass through statary phases with underground bivouacs, greatly reduced raiding, and no true emigrations.

In these surface-active species of *Aenictus,* we have found in a number of long-term studies, an invariable concurrence of colony behavior, brood development and coordination, and queen condition, clearly indicating functional cycles comparable with those in *Eciton* and *Neivamyrmex.*[7] It is interesting to note that although the nomadic phases of these *Aenictus* approximate 18 days in duration, or fairly close to results for *E. hamatum* and *N. nigrescens,* the statary phases are much longer than in these other dorylines (Table 7.3). This significant difference is discussed in Chapter 13.

Thus, colonies of all species of *Eciton, Neivamyrmex,* and *Aenictus* we have studied normally pass through alternating nomadic and statary phases, recurring so regularly in distinctive functional cycles as to justify my placing them together in group A to mark them off from other dorylines of quite different nomadic schedules. Basic to the cycles of all three group A genera are the developmental characteristics of the large, regular, well-coordinated broods described in Chapter 6.

The functional phases of *Eciton,* as a representative of group A, are so distinctive that a reliable description of them could be made after only about one month of investigation (Schneirla, 1933). The next step, deriving a theory to account for the operation of the cycle, was a much longer task (Schneirla, 1938, 1957a). In brood-stimulative theory, I

---

[6] These cases were observed by the late J. W. Chapman, missionary biologist of Silliman University (1964), in the Philippines, and by W. L. Brown, Jr., and E. O. Wilson (in Wilson, 1964), in China and in New Guinea, respectively.

[7] In other species of this genus, including hypogaeic species observed for short intervals (Wilson, 1964; Schneirla and Reyes, 1966, 1969), comparable correlations of colony behavior and brood condition suggest nomadic-statary patterns.

have postulated a set of causal concepts whereby changes in colony behavior and condition are derived from given properties of successive broods. It is thus critical for the cycle that brood-stimulative effects at a high level initiate and maintain the nomadic phase, but at a low level initiate and maintain the statary phase. There are two alternative hypotheses, one of which attributes control of the doryline cycle to periodic environmental events, the other to an innate rhythm in the queen (Chapter 8). Both of these ideas are unsupported.

There is of course much evidence demonstrating correlations between the rhythmic activities of various invertebrate and vertebrate animals and physical periodicities in their environments ( Harker, 1958; Cloudsley-Thompson, 1961). One classical example is the control of reproductive rhythms in marine annelids by lunar phases; another is the control of bird migration by seasonal changes in daylight. There seem to be no grounds, however, for the claim[8] that army ant cyclic functions are similarly controlled.

In long-term investigations of *Eciton* (Schneirla, 1949; Schneirla and Brown, 1950), *Neivamyrmex* (Schneirla, 1958, 1961), and *Aenictus* (Schneirla and Reyes, 1966, 1969), in which colonies of the same species operating in the same area at the same time were compared, there were no signs of the synchronization one would expect among them in their functional phases if the same extrinsic pacemaker were in control of all. See, for example, our detailed tables for *Eciton* (Schneirla, 1949, pp. 41 and 53; Schneirla and Brown, 1950, pp. 277 and 280). Rather, the correlations are between the behavior pattern of each colony and its internal condition: thus, the pacesetter lies within the colony.

Although the concept of an innate rhythm in the queen seems at first sight to account reasonably for the cycle (Sudd, 1967), there is no evidence for it. There are, on the contrary, circumstantial points opposing this concept (Chapter 8).

By contrast, brood-stimulative theory not only accounts logically for available evidence but readily suggests testable hypotheses. In Chapters 5 and 6, for each major turning point in the nomadic behavior of *Eciton* and other group A dorylines, I pointed to a correlated brood condition and suggested how that condition might serve as the cause. The same procedure recommends itself for the further discussion of known correlations between colony cyclic behavior and internal colony condition, summarized in Figures 7.1 and 8.6.

We begin with the proposal that in *Eciton* and *Neivamyrmex* each nomadic phase is initiated by the callow-arousal factor: a great rise in

---

[8] A report that nocturnal raids of a *Dorylus* species seem especially frequent on moonlight nights (Weber, 1943) is unaccompanied by evidence on colony cycles.

intracolony stimulation introduced by a pupal brood as it nears and reaches maturity. Chapter 6 presents evidence that nearly mature pupae arouse the workers increasingly, not only by mechanical effects from tarsal, antennal, and body-bending movements, but probably also by chemotactic effects (indicated by increasing pigmentation). The results of detailed observations suggest that workers remove the mature pupae from their cocoons in response to such stimuli. Workers are also attracted to the new callows, tugging at remnants of their pupal skins—which they consume— and licking body surfaces vigorously.[9] These stimulative effects expand rapidly in scope as a colony of *Eciton* passes its seventeenth or eighteenth statary day, and callows emerging in mounting numbers excite the adults more and more.

Colony arousal from this source is indicated in the last few statary days by daily increases in the frequency and vigor of raiding (Schneirla, 1949, Table 4, p. 27). The emergence of this brood brings the workers to so high a pitch of excitement that a maximal raid develops which, through behavioral steps discussed in Chapters 4 and 5, leads to the first emigration of the new nomadic phase. In *Eciton hamatum,* this peak as a rule comes on the very day all of the mature pupae emerge; in *E. burchelli,* with its larger colonies and wider brood range, it usually comes when not much more than two-thirds of the pupae have emerged (Schneirla and Brown, 1950; see footnote 2).

Physiological tests show that the excited activity of the workers beginning a nomadic phase has its basis in a heightened metabolic condition peculiar to this time. Samples of workers from the same continuously observed field colonies of *Neivamyrmex nigrescens,* taken at the start of a nomadic phase, when tested in the laboratory show a reliable increase in their records of oxygen consumption over those of samples from the same colonies in the statary phase. These records indicate a higher metabolic rate, which continues to increase in samples tested through the nomadic phase but which falls when the colony enters the statary phase.[10] This result would be expected from brood-stimulative theory, which postulates a physiological basis for brood-induced increases in worker activity in the nomadic phase (Schneirla, 1957a).

---

[9] In laboratory tests, groups of 3000 to 5000 workers of *Eciton hamatum* from a statary colony were kept unfed for a few days and then given a supply of mature (dark) pupae within cocoons. These they opened and consumed within a few hours; then they circled in column for twelve hours or more. Control groups given booty or an equal number of light cocoons yielded indefinite results. Substances obtained here through cannibalism may include those obtained in normal operations with mature pupae.

[10] These experiments were carried out by Howard Topoff (in prep.) as part of his work toward a doctorate in biology at the City University of New York. He also found correlated differences in the reactions of workers to light of different intensities in tests made during the two phases.

In Chapter 6, I discussed characteristics of brood synchronization and brood coordination in their relation to nomadic behavior. Results for the first third of the nomadic phase shed particular light on this relationship. Colonies of *Eciton,* after their first nomadic day, often lessen their activity in raiding and even have occasional days without emigration.[11] These features are especially prominent in daughter colonies after the division of a parent colony. Results with *Neivamyrmex nigrescens* (Schneirla, 1958, 1961) are comparable. In both of these genera, the indicated slackening of nomadic behavior may lie in decreased colony excitation when both the callow-arousal effect is waning and the stimulative effect of the young brood is still low.

By virtue of the coordination of these broods (Chapter 6) colonies of *Eciton* and *Neivamyrmex* have returned to their initial high level of raiding and emigration before the first week of the phase is ended. Observations and tests indicate that the stimulative role of the young brood has now replaced that of the callows as the primary factor maintaining colony nomadic behavior. Feeding operations, which underlie reciprocal-stimulative relationships, increase as the brood grows, then near maturity reach a peak correlated with heavy raids and emigrations.[12] Larvae of sexual broods, in this regard, exert an even greater stimulative effect that those of all-worker broods.

As studies of raiding show (Chapter 4), colonies are most active in the nomadic phase with larvae present and least active in the statary phase with the brood pupating. Our tabulated results for discovering colonies by means of their raids show that new cases in the nomadic phase are found much more often than in the statary phase (Schneirla and Brown, 1950, p. 319), as the nomadic forays are far better developed on the whole.

The end of the nomadic phase in group A dorylines corresponds to characteristic brood changes in the colonies, with interesting generic differences (Fig. 6.4). As described in Chapter 6, colonies of *Eciton* end the nomadic phase with cocoon spinning at larval maturity whereas colonies of *Neivamyrmex* and *Aenictus* remain nomadic until most of the brood is prepupal. With notable regularity, this phase ends in *Neiva-*

---

[11] In eight colonies of *Eciton burchelli* studied in one investigation (Schneirla and Brown, 1950), there were 10 cases of days without emigration in the nomadic phase, of which 6 fell between N-2 and N-6. In 10 colonies of *E. hamatum* studied in the same period there were 6 days without emigration in the nomadic phase, of which 5 fell in the interval N-2 to N-4. Of 4 similar instances in the 2 daughter colonies of '48 H-27 in the first 8 nomadic days after division, all fell in this same interval.

[12] In this interval, colonies of *Eciton hamatum* have notable success in attacking large colonies of the leaf-cutter ants, *Atta* spp.; likewise, colonies of *Neivamyrmex nigrescens* make heavy inroads against harvester ants, *Pogonomyrmex* spp. These attacks are most effective when the colony has a sexual brood in the larval stage.

*myrmex* when the brood is mainly mid-prepupal, but in *Aenictus* when it is in the *early* prepupal stage. For colonies of the last two genera, brood-stimulative theory postulates that excitatory effects (perhaps pheromonal) continue for a time even from inert prepupae. Thus, in these army ants, the end of this phase seems to be determined somewhat differently than in *Eciton*.

It is apparent that the described brood changes, although different between *Eciton* and the other two genera, nevertheless end the nomadic phase equivalently in the group A army ants as indicated by similarities in their phase durations (Table 7.3) and in their colony behavior (Chapters 3, 4, and 5). In the three genera the phase seems to end through the impact of massive brood changes effective within limited times, but with the striking difference that in the polymorphic *Eciton* and *Neivamyrmex* the main impact of this crucial brood effect is focused (as it were) through developmental convergence while in the monomorphic *Aenictus* it is reached through brood synchronization without the convergence (Fig. 6.4).

In *Eciton*, we have described how the level of colony excitement seems to fall abruptly on the day most of the larvae are enclosed. The change is notable in both *E. hamatum* and *E. burchelli* as colonies pass from dawn-starting, vigorous, and highly populous raids to slow-starting, smaller, and relatively sluggish forays of reduced pattern.[13] Investigating the changes in colony behavior and condition in these ants through the statary phase gives us interesting insights into their cyclic functions.

As observations of colony raiding show, this phase also is an interval of changing colony function. The forays, represented in Table 4.2 for *Eciton burchelli*, at first occur daily, then fall gradually to a low frequency prevalent through the intermediate part of the phase, but finally return to a high level. The results for *E. hamatum* are very similar (Schneirla, 1949, p. 27): detailed studies of the colonies suggest that such changes in behavior depend upon changes in brood condition through the phase.

As suggested in Chapter 6, the onset of the statary phase evidently centers on an abrupt fall in brood-stimulative effects. This change is transitional possibly because of two indicated factors (Schneirla, 1957a): One we may postulate as a holdover of the physiological excitation in the workers from the high level of the nomadic phase—$O_2$ consumption tests support this for *Neivamyrmex* (Topoff, in prep.)—the other, discussed for *Eciton* in Chapter 6, is a gradual decrease in stimulation from larvae

---

[13] Traffic counts summarized in Table 4.1 show that in the nomadic phase colonies of *Eciton hamatum* are much more active and with more workers engaged in the raids than in the statary phase. This comparison may minimize the difference because nomadic raids usually have their greatest exodus before midmorning and their greatest return after midafternoon, and our counts were made at other times.

that continue to spin. Consistent with the second point, *E. burchelli* reduces its raiding more slowly than *E. hamatum* in the first statary days presumably because its broods (of wider developmental range) require an appreciably longer time to complete spinning in their smallest members (Schneirla, 1949; Schneirla and Brown, 1950).

A gradual fall in colony excitation during early statary days is also indicated by the greater frequency of colony "shifts" (Chapter 5) during the first two days than in the following days, i.e., in the first statary days the colony reacts more readily when it is disturbed (e.g., by rain or by rupturing the bivouac). Prevalence of a low condition of colony excitation through the long intermediate part of this phase is indicated both by reductions in the frequency (Table 7.4) and in the duration and vigor

TABLE 7.4.
*Frequency of raids through the statary phase in a colony of* Eciton burchelli—*summary for four phases*

| Statary days | Total of days in four phases | Day-long raids | Afternoon raids only | No raid | Colony not visited | % days with raids |
|---|---|---|---|---|---|---|
| 1 to 3 | 12 | 11 | | | 1 | 100 |
| 4 to 6 | 12 | 8 | 1 | 4 | | 67 |
| 7 to 9 | 12 | 2 | 1 | 6 | 4 | 25 |
| 10 to 12 | 12 | 4 | | 5 | 3 | 44 |
| 13 to 15 | 12 | 2 | 1 | 4 | 2 | 33 |
| 16 to 18 | 12 | 5 | 1 | 3 | 4 | 62 |
| 19 to 21 | 12 | 12 | | | | 100 |

*Source:* Schneirla (1949).

(Chapter 4) of raids in this interval.[14] In both *Eciton* and *Neivamyrmex*, raids at this time are shorter and slower to begin, and raidless intervals are more common than at other times.

The indicated condition of physiological depression in this interval may be greatest in *Aenictus*, whose colonies seldom carry out surface raids through most of the statary phase (Table 4.2). Features of statary adjustment distinguishing this genus from others in group A are discussed in Chapter 13.

---

[14] Raidless days, common in the intermediate part of the phase, indicate that the colony is in a sluggish state. In *Eciton*, if a raid has not begun before midafternoon, none will take place at all, unless by external intervention. With statary colonies of *E. burchelli* that had not raided by 3:30 P.M., I found that machete slicing the bivouacs at intervals usually set off swarm raids that continued until nightfall.

In both *Eciton* and *Neivamyrmex*, raiding and other colony activities seem to begin a slow increase around midphase which continues thereafter. This change, as observations of the ants in artificial nests suggest, may center at least in part on reactions of the workers to the eggs and young brood. Near the end of the statary phase in these army ants, raids increase notably in frequency and vigor and, in *Aenictus laeviceps,* surface raids begin. Although, for reasons given in Chapter 6, this change seems attributable mainly to stimulative effects introduced through maturation of pupae, the young brood doubtless also plays a part.

Our discussion of excitatory agencies in the army ant cycle began with the callow-arousal factor as the probable initiator of the nomadic phase in *Eciton* and *Neivamyrmex*. The contrasting brood-excitatory patterns indicated for the initiation of nomadism and for cyclic colony function in *Anomma* are considered later in this chapter; for *Aenictus* in Chapter 13.

Both the regularity of functional cycles and the species-typical durations of the phases in group A dorylines seem based on properties of brood development necessary to brood coordination and therefore to continuing cyclic function in the colonies. These properties are: (1) the timing of brood stages as the chief factor controlling phase durations, (2) the condition of the brood or broods as influences on both the length of any phase and the current level of colony activity, and (3) changes in brood condition as a major influence on transitions from one phase to the next.

The close parallelism we have noted between phasic behavior and phasic condition of colonies in all dorylines studied leaves little doubt that serial brood production is essential to cyclic function in these ants (Schneirla and Brown, 1950; Schneirla, 1958; Schneirla and Reyes, 1969). Hypotheses for analyzing generic, species, and colony differences in this parallelism and for studying their causal basis may be readily worked out from brood-stimulative theory in terms of the above three properties. That group A colonies approximate species norms in the durations of their successive phases indicates their ability to establish bivouacs that serve as efficient brood incubators (Chapter 3) and to maintain a schedule assuring a standard supply of booty (Chapters 4 and 5). The intimacy of relationships between colony function, brood timing, and phase durations is illustrated by the shorter durations of nomadic phases in colonies with sexual broods than in those with all-worker broods (Chapter 9).

As Table 7.3 shows, phase durations recorded from colonies of a given species operating in the same natural area exhibit (statistical) central tendencies typical of the species but also show an appreciable range of variations. Repetitions of the studies in different years (Schneirla, 1949;

Schneirla and Brown, 1950) and by different observers[15] yield similar results in these respects. Preceding chapters have discussed features of bivouac-control, raiding, and other colony behavior that allow a species average to be approximated. The results also permit understanding deviations from the average, differences in seasonal and annual averages, and the like. Thus, we have observed variations in successive nomadic phases of the same colony evidently related to differences in temperature or in food supply sufficiently great to noticeably shorten or lengthen a phase, by their effects on larval development.[16]

Properties (1), (2), and (3) listed above as controlling phase durations in group A dorylines are those known empirically to influence the coordination of broods and the processes of reciprocal stimulation between broods, workers, and queen (Figs. 6.4 and 7.1). Important to this question are studies of generic and species differences in intervals between the maturation of successive broods. This value usually falls between 35 and 38 days in *Eciton hamatum,* between 31 and 37 days in *E. burchelli,* and between 36 and 40 days in *Neivamyrmex nigrescens,* but it is usually close to 46 days in *Aenictus laeviceps* (Table 7.3).

---

[15] Studies of phase durations in *Eciton burchelli* were carried out independently on Barro Colorado Island in our investigations (Schneirla, 1945, 1949) and in the studies of E. Willis on antbirds (1967). A summary of the results follows.

*Duration of functional phases in* Eciton burchelli

| | Nomadic | | | Statary | | |
| --- | --- | --- | --- | --- | --- | --- |
| | | Durations (in days) | | | Durations (in days) | |
| Observer | No. of cases | Range | Average | No. of cases | Range | Average |
| Schneirla | 10 | 11–17 | 13.4 | 18 | 19–23 | 20.7 |
| Willis | 42 | 12–19 | 14.7 | 51 | 19–24 | 20.1 |

Dr. Willis kindly loaned me his field data for a tabulation of these phase durations. Criteria of phase beginning and ending were much the same; important differences are the larger number of cases in the Willis series, also that they are more widely distributed through the year than ours.

[16] I collected a large part of a field study colony of *Eciton burchelli* (Schneirla and Brown, 1950, pp. 289–290) with an estimated one-third of its mature larval brood (then spinning) and transported it by air to New York. From March 20, 1948, the date of collection, until April 20, these ants and their brood were kept at low temperatures most of the time. This condition apparently slowed development as the pupae did not mature until around April 30. With only a minor part of this brood emerged, the workers consumed the many thousands remaining in cocoons, then began a circling in column which continued for more than twenty-four hours. This case may illustrate an unduly long statary phase, followed by a short interval of nomadic behavior.

Factors influencing the interbrood interval in each species—and therefore brood coordination and phase durations—are: genetic properties and the colony situation normal for development; temperature, humidity, food supply, and other external conditions; trophic and stimulative factors in brood-worker-queen relationships; and agents influencing egg laying by the queen. I have given reasons for centering our theory on the role of brood as paramount in doryline cyclic functions. The problem is discussed further in Chapter 8 in relation to the queen.

In a different approach to doryline cycles, we now compare the colony schedules of emigration in group A genera with those of group B, represented by the driver ants of the subgenus *Anomma* in Africa. From their investigations in the Congo, Raignier and Van Boven (1955) report that colonies of *Dorylus* (*Anomma*) *wilverthi* carried out lengthy emigrations at variable intervals. These movements never occur in series, as in a group A nomadic phase, but always as single changes of bivouac separated by stops which can vary from a few days to around two months. At first sight this behavior, summarized for numerous colonies in Table 7.5, seems too irregular to admit any comparison with *Eciton* and others of group A.

TABLE 7.5.
*Inter-emigration stops recorded for 27 colonies of* Dorylus (Anomma) wilverthi

| Durations of stops (in days) | Number of stops |
|---|---|
| 6 to 15 | 7 |
| 16 to 25 | 8 |
| 26 to 35 | 5 |
| 36 to 45 | 2 |
| 46 to 55 | 0 |
| 56 to 65 | 3 |

*Source:* Raignier and Van Boven (1955), p. 233.

The driver ant movements are cyclic, however; the investigators emphasize that most of the colony stops range between 16 and 35 days and average between 20 and 25 days in duration. Furthermore, cyclic changes in the colonies are indicated by increases in raiding, heaping of earth over the bivouacs, and other indications of distinct colony arousal preceding an emigration (Chapter 5), with these activities decreasing in the days following an emigration (Cohic, 1948; Raignier and Van Boven, 1955).

The results for *Anomma* reveal a clear parallelism between brood condition and changes in colony behavior, marked particularly by the "emergence of a great population of mature pupae near the time of each new exodus . . ." (Raignier and Van Boven, 1955, p. 236). Significantly, the most frequent interval between emigrations in these driver ants approximates the 20 days (or, perhaps better, 20 to 25 days) the investigators estimate as the time required for the development of an all-worker brood. In other words, each emigration seems to be aroused mainly by effects from the pupal maturation of an all-worker brood developed from eggs laid around the time of the preceding emigration (Fig. 6.4).

From these results, Raignier and Van Boven (1955) recognized that nomadic behavior in *Anomma* is "of the same nature" as I had reported (Schneirla, 1938) for *Eciton*. As they say, it is related to the reproductive cycle, and each emigration is caused by the brusque liberation in the colony of hundreds of thousands of young workers which by their "réserve énergétique" and great activity excite the colony to a change of bivouac (Raignier and Van Boven, 1955, p. 241).

The separate emigrations of driver ant colonies, because of the conditions under which they are usually set off, seem mainly attributable to callow-arousal effects and thus functionally equivalent to the initiation of nomadic phases in *Eciton* (Schneirla, 1957a) and *Neivamyrmex* (Schneirla, 1958). There are, of course, differences: Whereas callow arousal always initiates nomadic phases in *Eciton* and *Neivamyrmex* and usually sets off the single emigrations of *Anomma,* larval excitation is sometimes the main cause in *Anomma* (Raignier and Van Boven, 1955, p. 236) and always in *Aenictus* (Schneirla and Reyes, 1969). Conversely, the stops of colonies of *Anomma* between emigrations, although variable in their durations, seem attributable to prevalent brood effects at low intensities and thus equivalent to the statary phases of group A army ants (Schneirla, 1957b).

I conclude that the nomadic-statary cycles of group A dorylines are fairly precise and well timed because colony functions are dominated by regular changes in the stimulative effects of well-coordinated broods. By contrast, nomadism in *Dorylus* is much more variable because of the less regular coordination of successive broods, including the frequent intervention of male broods differently effective in group A (Chapter 9).

Results mentioned in Chapters 4 and 5 for various dorylines, pointing to correlations between variations in colony behavior and changes in brood development, suggest that cyclic nomadism in some form is widespread in the subfamily. As mentioned, nomadism takes on important differences among the genera, related especially to the degree of surface adaptation. The genera of group A, for example, which are most precise and complexly organized in colony cyclic functions, all contain surface-adjusted species. Brood synchronization in early stages and the degree of coordination between broods are both much more pronounced in *Eciton* and exert a correspondingly greater degree of regularity and precision on its cyclic functions than in the hypogaeic *Dorylus*.

Gradations of nomadism may also exist within each genus, the most surface-adjusted species of each having the most advanced and best organized cyclic nomadic function, and the most hypogaeic species the least. For example, in *Neivamyrmex,* as mentioned, we find indications that the range of size types in early brood stages is narrower in *N. nigrescens* than in the more hypogaeic *N. carolinensis,* with the predictable consequence that *N. nigrescens* has more precise brood coordination functions and more regular cyclic behavior than *N. carolinensis*. As another example, in *Dorylus* the colonies of *D. (Anomma) wilverthi* emigrate far more frequently and perhaps more regularly than those of the more hypogaeic *D. (A.) nigricans*. This the investigators (Raignier and Van Boven, 1955) attribute to differences in the bivouacs (Chapter 3), making for a slower development of the brood (and, we may suggest, a lower degree of brood coordination) in the latter.

For *Nomanyrmex* and *Labidus* of the New World, a few recorded observations of emigrations show they have brood synchronization and probably brood coordination, thus indicating cyclic behavior patterns (Luederwaldt, 1918; Weber, 1941)—see Chapter 6. All the circumstances suggest that great colonies, when a pupal brood matures (Chapter 5), come to a state of high excitation, bivouac near the surface and raid maximally, and finally emigrate. Hence these army ants, like *Dorylus,* may carry out discrete emigrations separated by nesting stops of variable durations, and could well be related functionally to group B.

I have called the narrowing of brood range through differential growth "developmental convergence." This condition accounts for greater summations of stimuli from the large broods, thereby influencing colony behavior more strongly and more precisely than would otherwise be possible. This feature, which arises through intrabrood synchronization, clearly must contribute strategically to interbrood coordination and thus to the timing of cyclic behavior characteristic of the group.

From the evidence and discussion of Chapters 3 through 7, I conclude that brood-stimulative theory explains the varying intensity of raids as well as the emigration schedules that typify cyclic behavior in all army ants. In the dorylines, brood-stimulative effects initiate changes in worker's response to booty and to other environmental conditions as well as to brood, queen, and other workers within the colony. To mention one example of changing environmental responses (discussed in Chapters 3 and 5), when the level of brood excitation rises sufficiently, doryline colonies undergo such ecological adjustments that colonies of epigaeic species increase their tolerance of open surface conditions, and colonies of hypogaeic species equivalently become more tolerant of conditions nearer the surface.

A relatively unexplored but potentially rich area of study lies in the many genera of ponerine ants known to carry out predatory forays. The simplest pattern may involve "a limited and aberrant form of nomadism" of the type Wilson (1958a) observed in the amblyoponine ant *Myopone castanea* of New Guinea, in which workers evidently carried their larvae to spots where they had killed beetle grubs too large to be transported to the nest. Wilson suggests that in a more advanced stage group predation may often be combined with booty retrieving and emigration. The circumstances often indicate brood-excited colony behavior. In one case, for example, he observed a strong column of about 2000 workers of *Leptogenys purpurea* in New Guinea that was clearly an emigration as around 1000 cocoons along with some "mature to nearly mature larvae" were being transported, and another case in which a colony of this species raided in two divergent columns from a bivouac that contained a brood of "mature to nearly mature larvae and cocoons." Colonies of these legionary ants may not emigrate very frequently or very regularly as I have often seen their columns and their raids in the forest and have often found their surface bivouacs without having observed one movement that was clearly an emigration.

To my knowledge, the first investigator to report a clear case of cyclic nomadism in a nondoryline ant is Robert W. Taylor (pers. comm.), research scientist in the Division of Entomology, C.S.I.R.O., Canberra, Australia, who found—in an undescribed species in the ponerine genus *Onychomyrmex*—that the "ants emigrated nightly, using temporary bivouacs during the day, and that this behavior is correlated with the presence of larvae [alone] in the brood." A nomadic phase is thus indicated, and a statary condition prevails at other times of year. These ants contrast in two ways to the dorylines: Not all species of *Onychomyrmex* thus far studied in the area exhibit this brood-correlated cyclic behavior; and no species seems to show more than one cycle each year.

Although types of group predation and other features reminiscent of the dorylines have been reported in various species of *Simopelta* (Gotwald and Brown, 1966), *Cerapachys* (Wilson, 1958b), and numerous other ponerine genera (Wheeler, 1936; Wilson, 1958a), the colony functions of these ants have not yet been investigated in detail, and thus we do not know how far the similarities may be carried. The colonies of such ponerines are comparatively small, ranging from a few hundred workers as in *Cerapachys* spp. (Wilson, 1958b) to one or two thousand as in *Simopelta oculata* (Gotwald and Brown, 1966). In other words, colonies of even the most populous legionary ponerines seem to be only about one-fiftieth as large as those of *Aenictus,* which has the smallest colonies of any doryline yet studied. From this difference alone, the needed continuous and systematic studies of colony functions may place these ponerines distinctly below the dorylines in the organization of raids, emigrations, and cyclic behavior.

For the present, therefore, we term these ponerines "legionary ants" on the resemblance of their functional patterns to that of the dorylines or army ants. Even so, this resemblance is intriguing and undoubtedly significant for study as it includes in many cases the presence of progressive group forays with emigrations (Wilson, 1958a), dichthadiigyne queens and synchronized broods (Gotwald and Brown, 1966), and—in at least the one case mentioned (R. W. Taylor, pers. comm.)—cyclic colony behavior. Legionary ponerines and army ants are compared in Chapter 13 where we also contrast the organization of colony behavior in two doryline patterns: that of the relatively simple *Aenictus* and that of the complex and advanced *Eciton.* The evolutionary implications of these comparisons are considered in Chapter 14.

These comparisons related to colony function are carried forward in Chapter 8, which centers on the queen. Although arguments for discussing so important a member of the colony much earlier in the book were understandably strong, the development of a brood-stimulative theory for colony function in Chapters 6 and 7 must have revealed my chief reasons for resisting this temptation. Nonetheless, doryline queens and problems associated with them represent a major part of our subject, as Chapters 8 through 11 show.

# 8

# The Queen

The doryline queen is the sole reproductive agent of her colony. Without her not only would the colony lose its unity and integrity as a social group, but it would perish. These creatures are so secluded and so self-protected that for a long time they were rare insects in collections.[1] They are so different from others in their colonies that their classification as members of the same group was long delayed. From the appearance of the queen (Fig. 8.1) we are immediately struck by the long evolutionary advancement she must have undergone from ergatoid or worker-type females, the probable source of the first doryline dichthadiigynes.

Army ant queens are very probably raised from fertilized eggs equivalent to those that ordinarily produce workers. Their extraordinary form and function, however, arise from exceptionally rich stimulative and nutritive conditions in the colony that set them apart (Schneirla and Brown, 1952). In group A dorylines, the ordinary broods contain workers only; the exceptional broods contain many males and a few young queens (Chapter 9).

The queen (Fig. 8.2), like the workers, is wingless throughout life; like them, whether or not she has eyes depends on the genus. There the resemblance ends. She is much larger and stouter than the workers and so much stronger that in an emigration she often carries numbers of them on her back. The queen of *Eciton burchelli,* even when fully contracted, is approximately 21 mm long as against body lengths of 3 to 13 mm in the workers. Even the first time I saw a doryline queen, when I opened the bivouac of a colony of *E. burchelli* and was fortunate enough to find her within, workers balled tightly around her, but there was no mistaking her. It was not only the queen's larger size but also her

---

[1] The first doryline queen to be found, which was described (André, 1885) as *Pseudodichthadia incerta,* proved to be a physogastric queen of *Labidus coecus* (Weber, 1941; Borgmeier, 1955). The first queens of *Eciton* to be found were described by Wheeler (1921, 1925). Thereafter, discoveries of doryline queens in widely different groups increased in frequency (Bruch, 1934; Borgmeier, 1955; Raignier and Van Boven, 1955).

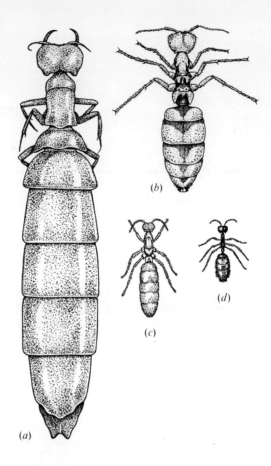

FIGURE 8.1.
Queens of four genera of dorylines, drawn in the contracted condition. Body lengths are: (a) *Dorylus (Anomma) wilverthi*, 52 mm; (b) *Eciton burchelli*, 21 mm; (c) *Neivamyrmex nigrescens*, 12.5 mm; (d) *Aenictus gracilis*, 8 mm. Estimates of their batches of eggs are 1–2,000,000, 225,000, 50,000, and 30,000, respectively.

glistening armor (groomed and polished to a high sheen by the workers), her unique form, and still other characteristics to be mentioned. For so bulky an individual, she moved surprisingly fast and was so promptly covered over by workers each time I exposed her that capturing this first queen of mine was exceedingly difficult.

In structure the queen differs radically from the workers. Her larger head contains a proportionately large brain and much glandular tissue,

FIGURE 8.2.
Functional queens of *Eciton burchelli* photographed in the laboratory (*a*) in the contracted state and (*b*) in the physogastric and egg-laying condition.

both essential to her complex behavior and reproductive functions. In contrast, the head of the major worker contains a relatively tiny brain though a much larger proportion of jaw muscle tissue than in either the queen or the other workers. In most army ants the slender, sickle-shaped mandibles of the queen differ greatly from those of the workers, which in *Eciton,* for example, are long and tonglike for the majors and relatively stubby for other workers. Likewise, the queen's eyes are larger, more convex, and more sensitive than the minute eyes of the workers, which in some species of *Neivamyrmex* and *Labidus* are even lacking. Queens and workers are both eyeless, however, in *Dorylus* and *Aenictus.*

Instead of the one or two beadlike segments that join the worker's thorax and abdomen, the queen has stout coupling segments, which in *Eciton* are marked dorsally by two great hornlike structures whose function is still undetermined.[2] The queen's legs are always stout. In *Eciton,* in which she often runs long surface emigrations, they are also long; in hypogaeic army ants, they are shorter. The queen's abdomen, even when contracted, is far bulkier than that of any worker. In *E. hamatum* its typical dimensions are 9 mm long and 4 mm wide as against 2.5 mm long and wide in the major worker. Periodically, the abdomen of the queen swells with fatty tissues and ova as she enters the gravid state, then contracts again, whereas that of the worker remains minute throughout life, in keeping with its austere diet and no egg laying.[3]

The queen stays in the bivouac at all times except when she runs the emigration. Accordingly, her form, through natural selection, reflects in detail the colony situation typical of the species. Queens of surface-adapted *Eciton* species, for example, have more rotund bodies and long legs in comparison with the elongated, slender bodies and stubby legs common in queens of subterranean-adapted species (Fig. 8.1).[4] Doryline queens, in their evolution toward a sequestered bivouac life and great reproductive capacity, also specialized physiologically. Wheeler (1928), pointed out two especially important adaptations of the dichthadiigyne: a great respiratory system permitting her to live in stuffy recesses among

---

[2] But almost certainly, the epinotal and petiolar horns prominent in *Eciton* queens (Wheeler, 1921) and the homologous structures in other genera (Borgmeier, 1955) have a protective function. Also, my observations of mating in *Eciton* (Schneirla, 1949) indicate that they may also serve as a hold-fast for the male during the time he is inserted (Chapter 11).

[3] Some at least of the largest workers in *Eciton hamatum* (Whelden, 1963), *Neivamyrmex nigrescens* (Holliday, 1904), and other dorylines have a few ovarioles capable of producing mature eggs. These doubtless would be unfertilized eggs, probably nonviable and consumed later by the colony and brood. No observations of egg laying by doryline workers have been reported.

[4] In *Neivamyrmex,* the queen of the surface-active *N. nigrescens,* when physogastric, has an ovoid although somewhat elongated gaster with closed tip (Schneirla, 1958); by contrast, the queen of the more hypogaeic *N. carolinensis,* when physogastric, has a greatly elongated gaster with gaping tip.

close masses of workers, and an extraordinary ability to amass energy reserves in fatty tissues. Such assets are all the more impressive when we see how ably she can shift from a condition of action and small body size, when the colony is nomadic, to a highly vegetative and reproductive condition when it is statary (Schneirla, 1944a).

The origin of the doryline queen is subject to speculation. Emery's suggestion (1920) that these great specialized fertile creatures evolved from an ergatoid (workerlike, fertile) female seems hard to believe, so different are existing workers from their queens. But this hypothesis gains support if we recall that the largest workers of some species have traces of a few ovarioles or egg tubes which may at times even produce eggs (see footnote 3). Also, in some of the existing small-colony ponerines that are, however, legionary in behavior, one or more ergatoid females have been found that evidently serve as reproductives (Wheeler, 1936; Wilson, 1958a). Another clue to the queen's origin may lie in the brief but intensive process of colony division, begun with a number of virgin queens as contenders but ended with just one or two survivors centered in daughter colonies (Chapters 9 and 10). The problem of origin is discussed in Chapter 14.

We find monogyny prevalent in the colonies of all known dorylines.[5] The queen in these ants is the only reproductive female in the colony, a key individual distinguished in many ways from the fertile queens of other social insects. Her output of eggs is prolific and in many dorylines is periodically scheduled. Her secretory potencies, moreover, both contribute to colony organization and are uniquely essential to the survival of the colony. She is irreplaceable, for without her the colony must join another of its species or perish.

Army ant queens lay an astounding number of eggs. Even the smallest queen represented in Figure 8.1, that of *Aenictus gracilis,* produces an estimated 30,000 eggs in each batch at roughly forty-six-day intervals. From this we go to the queen of *Eciton burchelli* which produces a great mass of around a quarter million eggs at approximately thirty-six-day intervals, and then the great queen of *Anomma wilverthi* with her one or two million at variable intervals. The three smallest queens shown in Figure 8.1 lay their large egg batches at regular intervals, then rest from egg laying in between. Queens of group A dorylines follow an impressively strict on-off schedule in egg laying, a feature lacking in queens of driver ants (Chapter 6).

Quantity typifies them all. Even the little queen of *Aenictus gracilis*

---

[5] In many colonies of *Eciton, Neivamyrmex,* and *Aenictus* that we have studied, the monogynic condition, i.e., only one reproductive female in each colony, has prevailed. There is no evidence that doryline colonies are ever polygynic except for limited times when a sexual brood is produced (Chapters 9, 10, and 11). But see footnote 15, Chapter 10.

produces an estimated 240,000 eggs per year. The queen of *Eciton burchelli* produces about that many in a single batch laid within about ten days, at about thirty-six-day intervals, which gives her an estimated 2,400,000 eggs annually. This number, however, the queen of the driver ant *Anomma wilverthi* [6] can top with only one laying. Interesting enough, by comparing Figure 8.1 and Table 6.1, we note a general correspondence between the body size of queens and magnitude of broods and of colony populations. Reproductive magnitude, therefore, and the part it plays in colony conditions may bear very significantly on species adaptation (Chapter 14).

All doryline workers are strongly attracted to their queens. This is vividly demonstrated by emigrations of colonies in group A species (Reichensperger, 1934; Schneirla, 1938) when the queen passes with a great entourage of excited workers massing around and behind her. If one attempts to snatch up the queen, she is likely to slip away; then back on the ground she is at once lost to sight within masses of workers. Observations and tests show that the queen's normal attraction for workers has its basis in secretions she alone produces (see footnote 9). When she is returned after a test removal from her colony, her scent can arouse (and orient toward her) workers moving in column several centimeters away (Schneirla et al., 1966). Her odor, scented acutely by workers, not only provides a basis for centralizing colony bivouacs but, as tests show, is indispensable to normal colony unity.

The queen is able to carry out her colony functions well for at least a year or two. Our studies, in which all functioning colonies of various *Eciton* species were found each with a single functioning queen, leave little doubt of this (Schneirla, 1949). In a systematic survey, we marked queens of two *Eciton* species distinctly, then returned them to their colonies (Schneirla and Brown, 1950). Because the study was done on Barro Colorado Island, we were able to make many recoveries as the ants could not leave. Of thirty-two colonies of *E. hamatum* with marked queens, we found several again after several months and some after a year or two. The best record was that of queen '48 H-15, *E. hamatum* (Fig. 8.3), marked on December 22, 1947, next observed in February, 1952, and finally taken up as she was energetically running in an emigration of her colony on May 25, 1952. Both she and her colony, which had a larval all-worker brood, were then in good condition. This queen then had functioned in her colony for at least four-and-a-half years during which time she may well have produced about fifty all-worker broods (or about 4,000,000 workers) and one or more sexual broods. The longest record

---

[6] The total annual output of eggs from one of these queens may exceed even that of termite queens (Emerson, 1939a), generally considered the most prolific among social insects.

FIGURE 8.3.
Colony queen of *Eciton hamatum*, marked on December 22, 1947, and returned to her colony on Barro Colorado Island, then rediscovered and studied further during the dry season of 1952. At her recapture in June of 1952, she had functioned in her colony for a minimum of 4½ years.

thus far for *E. burchelli* is that of a queen I marked in 1952 that was recovered in 1956 (Rettenmeyer, 1963). Three years may be a conservative estimate for the average life span of one of these army ant queens in her colony.

Although most queens of *Eciton* probably end by being superseded in colony divisions (Chapter 10), some of them doubtless perish through disease, parasites, or accident. I found colony '38-H (*E. hamatum*), for example, with an abnormally small worker population and in a poor condition, as indicated by a horde of golden-yellow mites on both queen and larvae.[7] Even so, this colony might have survived since its queen's general behavior seemed normal, as also was her reproductive condition as indicated by histological study (Hagan, 1954). The case of colony '48 H-O (*E. hamatum*), however, was a very different matter (Schneirla and Brown, 1950). Rambling trails more than 1000 m long were followed by very thin files of workers. Searching disclosed no colony center, but

---

[7] On the discovery of this colony ('38-H, *Eciton hamatum*), mites covered all parts of the queen's body including the head (Schneirla, 1945). Although many escaped when she was picked up with tweezers, forty-six of them were counted in the collection vial.

finally at one of the few trail junctions I found within a tight little ball of workers the headless body of a dead queen of the species. This was doubtless the former colony queen, evidently dead for many days, her body being almost completely dismembered with the gaster a mere dry shell. But many workers still clung to the remnant, holding on as I moved it about with tweezers and climbing back on after having been shaken off.

In nature, army ant queens are never absent from their colonies. Day and night the queen is securely sheltered in the bivouac within a cluster of workers, except on the emigration when she is closely accompanied by workers. Biological and behavioral relationships between queen and colony are mutual and close. Most ant queens, with reasonably good care, survive well in captivity; those of *Formica,* for example, last for considerable time even apart from their workers and brood. A captive doryline queen, however, can live only a few days away from her colony, in *Eciton* only about a week, at best. The circumstances indicate that beyond the food and almost incessant stimulative grooming the queen receives from the workers (Fig. 8.4), she also gets secretory and other biochemical agents essential to her well-being and reproductive functions and indeed to her very life.

This process goes two ways, for the queen is also indispensable to the function of her colony. Colonies of group A species, deprived of their queens, deteriorate steadily and face extinction unless they can fuse with other colonies of their species. The dorylines ordinarily cannot replace a lost colony queen as can, for example, a flourishing colony of honeybees with its queen-cell process (Butler, 1962). In these ants, a lost queen can be replaced only through the development of a sexual brood from eggs laid by the parent queen herself (Chapter 9).

Let us illustrate from studies of *Eciton* the intimacy and intricacy of the queen's relationships with her colony. Early in each nomadic phase she is usually found somewhere in the upper center of the bivouac, above the brood, enclosed in a tight capsule of workers. In *E. hamatum* and *E. burchelli* the workers thus staying with the queen are the darkest and evidently among the oldest. So strongly and persistently are they attracted to her that they seem to constitute a "queen's guard." Observations show that they function both as guards, attacking predators that may molest her in the bivouac or on the emigration, and as keepers, restricting her movements. This group varies through the cycle both in functions and hence in which of its members are nearest the queen. At egg-laying time, for example, it is the minor workers that surround her closely (Schneirla, 1949, p. 55). Workers of the queen's guard not only feed the queen but above all through social grooming must integrate queen

FIGURE 8.4.
A colony queen of *Eciton hamatum* with workers in a laboratory nest, much reduced in physogastry after having laid most of her current batch of eggs.

and colony by transmitting bilaterally between them a variety of stimuli and chemical products including pheromones.

As a rule, the queen of *Eciton* emerges from the bivouac only after dusk or at night when the colony emigrates.[8] Then she moves steadily along in the column, over leaf-strewn ground, along vines and tops of logs, wherever the route leads. Usually she runs at or near the head of a great wedge-shaped entourage of excited workers (Fig. 5.6), at times even carrying or dragging some of them. Arriving at the new bivouac, she disappears inside at once and often can be seen mounting through strands toward the top. The queen's security in the emigration is insured by her worker guard, by her sensitivity in following the trail, and by her strong and quick responses to disturbance. In the nomadic phase, when the queen is called upon for long nightly emigration treks, she runs with

---

[8] In laboratory tests with strong directed light, queens of *Eciton* turn quickly away from the source each time the light is turned on them. By contrast, the workers do not turn directly from light although light disturbs them and usually causes them to cluster (with brood) in available dark places.

agility as her figure is then in its most petite condition with her gaster fully or nearly contracted. General behavior in the emigration is much the same with the queens of *Neivamyrmex* and *Aenictus* species we have studied.

The workers keep close protection over their queen at all times and in the bivouac can fight off almost any predator except man with their biting and stinging. In an emigration the queen generally sets out after half or more of her colony has gone so that ample worker forces precede and follow her; her protection usually includes also numerous excited major workers that provide a good buffer against predators. At such disturbances as rain or gusts of wind against a leaf or vine bridge, she stops and is quickly hidden beneath a cluster of workers from which she must actually burrow her way to resume her trek when all is again calm. So rapid and effective are the devices of contact-odor communication in army ants that it must be a rare event for an emigrating colony to lose its queen even from such catastrophes as falling trees or flash floods.

During a heavy rain one night in Panama I found a queen of *Eciton burchelli* marooned with a few hundred workers on a rock in a stream bed where their line of emigration had been cut by the waters. But once the stream fell low enough, within two hours the ends of the column rejoined from the banks into a procession across the mud, rocks, and vines in the gully. Then, once on the bank, queen and retinue moved off and soon disappeared into the forest toward the new bivouac.

To understand better the strong attraction that an army ant queen exerts upon the workers of her colony, we have made tests in the laboratory that have shown that the workers are extremely attracted to her odor (Schneirla et al., 1966).[9] Furthermore, when she is returned to the colony after the experimental removal, the workers gather excitedly around her and reaccept her (although foreign queens would be killed). Her presence in a laboratory nest stocked with workers of her colony greatly raises the behavioral tone of the assemblage. It is misleading, however, to say, as Wheeler (1921) suggested, that removing the queen

---

[9] Preparing for the odor test, I place the queen for ten minutes on a new paper disc, then admit workers of her colony into an arena in which this disc is placed beside a fresh one (Schneirla et al., 1966). In dim light, workers of *Eciton, Neivamyrmex,* and *Aenictus* virtually always cluster promptly on the (presumably) queen-saturated disc but not on the fresh alternative. Workers are usually attracted particularly to the anterior part of the gaster of their queen (Schneirla, 1949; Rettenmeyer, 1963). Hagan (1954) found unicellular glands with individual ductules leading close to or into the cuticular exoskeleton in two segmentally arranged masses near the vulva—possible sources of secretions attractive to workers. Watkins and Cole (1966) have shown that workers of three *Neivamyrmex* species are more attracted to the odors of their respective colony queens than to those of other queens of their species and are more attracted to the odors of queens of their own species than to those of other species.

from the midst of her scattered bivouac causes a "perceptible apathy" among the workers. In such cases the workers, after having been scattered, first pass through an interval of excitement that is long or short depending on the degree of disturbance, then quiet down. This occurs whether or not the queen is taken out, hence, constitutes a recovery from a period of intense stimulation and shock rather than from the queen's absence.

As queen-removal tests show, however, a prolonged loss of the queen does in fact affect the colony in specific ways. When a queen of *Eciton* taken from the nocturnal emigration is held in the laboratory overnight, very definite changes take place in her colony, apparent the next morning, or about fourteen to sixteen hours later. Within the first few hours of the raid, a file of workers is seen filtering back over the previous night's emigration line while at the same time the raid expands in other directions. Ordinarily, the back trail is followed only during the first hours of a new raid.[10] But under these exceptional conditions, the back-tracking goes on, drawing in increasing numbers of ants and extending in scope until it has entered the trails of earlier days. This response to a prolonged removal of the queen is in fact so predictable with *Eciton, Neivamyrmex,* and *Aenictus* that we have used it to discover the route by which a new-found nomadic colony reached its current location.

Preliminary results indicate[11] that the members of the colony evidently most disturbed by the queen's absence are darkly pigmented workers. These workers, I suggest, are among the oldest in the colony and the ones most closely affiliated with the queen. In all probability they are the ones most disturbed by the absence of her odor, on which they have regularly centered, and therefore the first to scatter excitedly and leave the bivouac in numbers. Once outside, they enter an abandoned route of emigration not merely because it is an available route but predictably because it is the one trail from the bivouac that bears a heavy brood odor. These pioneers, in running back and forth, attract increasing numbers of nestmates, much as they would be drawn into a new extension of raiding.

---

[10] In the first stages of a nomadic phase raid, workers of *Eciton, Neivamyrmex,* and *Aenictus* often enter all available trails from the bivouac but usually soon withdraw into new trails, tapping richer sources of booty (Chapter 4). Use of the back trail (i.e., the previous emigration trail) after loss of the colony queen involves behavior quite different from that normal to raiding.

[11] From studies with *Eciton hamatum* the following test may be suggested and results predicted. The queen in her bolus of workers is removed from the bivouac at dusk before she has joined the emigration. About half of these workers are marked distinctively on the thorax with fast-drying lacquer and returned to the colony—with the queen held captive in the laboratory. The next morning workers so marked are likely to be more numerous in the early phases of back-tracking than would be expected from their proportion in the colony.

Conversely, within one hour after the queen has been returned to the bivouac, the back-trailing column as a rule begins to withdraw.[12]

Loss of the queen, as results show, basically impairs colony function and unity. Normally, workers from two different colonies of the same species never mix when their raids meet. The two masses collide, with some combat, but continue their separate raids (Schneirla, 1949). Clearly the disturbance is caused by differences in the colony odors, for in laboratory tests workers readily accept the odor of nestmates (on paper discs) but are agitated by, and recoil from, odors of other colonies of their species. One day or longer after the colony queen has been taken away, however, differences in the reactions of workers to their own colony odor and to foreign colony odors have begun to be unreliable.[13] Such results indicate the queen's essence must be the critical component distinguishing the odor of one colony from that of other colonies of the species.

Thus, from the time the queen is lost, colony reactions change to such an extent that when in back-tracking workers of the queenless colony meet raiding or emigrating groups of a normal (queen-right) colony, they join with the others, often with few signs of conflict. Usually, within an hour or two after workers of the two colonies have met on the trails, the process of union has begun. By stages, ants of the queenless colony move from their bivouac into the columns of the other, at length carrying their brood with them, and finally merging completely. Significantly, the brood of the queenless colony is then consumed within a day or two whereas that of the normal colony survives (Schneirla and Brown, 1950). Beyond this evidently protective influence of her odor, the queen also fills a very important role in her relations to the brood (Chapter 14).

For investigated species of *Eciton, Neivamyrmex,* and *Aenictus,* results indicate that the queen's chemical furnishes the critical component that distinguishes the odor pattern of her colony from that of others. Observations and tests indicate the following. (1) The workers of normal colonies of these ants all bear the distinctive colony scent, based on pheromones of the queen, which are passed from the source through the colony by frequent licking and grooming of the queen by workers, of

---

[12] When the queen is set down on the back trail near the bivouac, the process can be observed in detail. First, workers rush up from both directions and cluster over her. Later—usually after a few minutes—she moves off toward the bivouac and enters it with her retinue. Thereafter, evacuation of the back trail continues steadily until this trail is empty.

[13] Colony affiliation tests with army ant workers are therefore unreliable if the workers tested have been away from the colony for more than a few hours. In specific tests, workers from the colony of *Eciton, Neivamyrmex,* or *Aenictus* held in the laboratory for three days usually are then accepted about as readily by another colony of their species as by their own. Control workers, taken from the first colony and tested within one hour, are accepted by the first colony but attacked by the second.

workers by one another, and of the brood by workers. (2) This scent, especially because of its queen component, is accepted by nestmates but is disturbing to members of other colonies. (3) The queen odor is constantly reinforced through the processes of reciprocal stimulation, but only a few hours of nonrenewal can so weaken a worker's "colony-odor coat" that she may be accepted by members of foreign colonies. These conclusions are supported not only by results of queen-removal tests but also by major deviations in the responses of workers to some members of their own colony and to their own parent queen when a sexual brood develops in the colony (Chapter 9).

The facts, interestingly enough, indicate both that workers carry a queen-based odor that must be regularly renewed and that they are habituated (i.e., conditioned) to this odor through normal stimulative exchanges without which they would accept members of other colonies. These two factors seem best developed in members of the queen's guard group, which fight with (and are fought by) workers of a colony even when their nestmates are merging with it.

The army ant queen, centered in the bivouac much of the time, thus gives her colony a distinctive scent that integrates it and therefore aids it greatly in forming and resettling well-patterned bivouacs. Within only a few days after losing its queen, a nomadic colony begins to deviate from its normal bivouacs (Schneirla, 1949). Colonies of surface-adapted *Eciton* species lapse first into ragged, separated clusters and then into carpetlike masses over the ground, with less and less centering evident. Colony '48 H-O, carrying its dead queen, had reached an extreme stage of dissolution and seemed close to extinction as a colony. Contributions of the queen to the organization of her colony are discussed in Chapters 10 and 11.

Looking again at the queen as the colony egg producer, we can appreciate how unique she stands in this respect among insects. Not only does she lay an immense quantity of eggs throughout a long life, but this is combined with an often precisely timed schedule. Even the lowest output known for army ants, that of the *Aenictus* queen, ranks high for an ant—though it is only about one-fiftieth that of the driver ant queen (Table 6.1). But still more impressive is this combination in queens of group A genera; here we find large-scale egg laying very precisely repeated and timed in on-off intervals, a unique phenomenon among social insects.

It is not easy to understand how in the history of doryline study, cyclic function in colonies of these ants was so long overlooked. For a long time, doryline queens were rare, and finding one was a notable discovery (Wheeler, 1921, 1925). But as interest increased, they were taken more often (Borgmeier, 1955). At length, a fifty-year census disclosed that among scores of queens recorded from colonies representing eighteen species of *Neivamyrmex,* about half had been captured in the physogastric condition and half in the contracted condition (Schneirla, 1958). This interesting circumstance, considered in relation to the state of the brood in the colonies, might well have suggested that these two conditions arise in the same queen at different times and perhaps alternately. But instead, leading authors (Wheeler, 1925; Bruch, 1934) continued to describe contracted queens as "virgin, not yet gravid," and the belief seemed to prevail that once a queen became gravid, she must remain so for the rest of her functional life.

In 1936, on Barro Colorado Island, I was able to investigate the matter with the same colony queen (Schneirla, 1938, 1945). Having followed a colony of *Eciton hamatum* from the nomadic phase into the statary phase, I waited the (empirically) predicted nine days, then cut open the hollow log containing the bivouac. In the upper center of the ant mass and above the pupal brood I found the queen, fully physogastric and apparently just starting to lay eggs. It was difficult to believe that this was the same female I had seen (and marked) one night just two weeks before as she was running the colony emigration in a fully contracted condition. She was the first gravid queen of *Eciton* to be reported and the first doryline queen to be traced from the contracted to the physogastric condition. These two conditions are shown in Figure 8.2 for *E. burchelli.*

It is now well known that in at least three doryline genera the single colony queens go on functioning regularly once they have begun. By removing queens from the bivouacs of their colonies at intervals for observation and measurement—and returning them promptly to avoid interferences—we learn what changes in condition normally take place. Figure 8.5 illustrates summarized data on the gaster lengths of queens of *Eciton hamatum* and *E. burchelli* so studied (Schneirla, 1957a). We note, for example, in Figure 8.5*b,* that queen No. 14 (*E. burchelli*) on N-5 measured 10.7 mm in gaster length, but on S-12 (when she was physogastric) she measured 21.5 mm. Such data, supplemented by histological studies of queens preserved at corresponding times (Schneirla, 1949; Schneirla and Brown, 1950; Hagan, 1954), clearly indicate reversible changes in the reproductive condition of the queen as her colony passes through a functional cycle.

These changes we may summarize as follows. Through most of the nomadic phase the queen is fully contracted and maintains her gaster at its smallest volume of the cycle. But by the last two or three nomadic days her abdomen begins to swell so that white stripes appear and widen as the intersegmental membranes are stretched more and more. The swelling increases rapidly through about the first week of the statary phase by which time the queen has become fully gravid and has begun her period of egg laying. This interval, clearly indicated in Figure 8.5, lasts about seven days in *Eciton hamatum* and ten days in *E. burchelli*. Then the physogastry subsides, but the queen may not become fully contracted in the remaining statary days (evidence on this point is inconclusive). As our results show, however (Fig. 8.5), queens of these two *Eciton* species commonly have somewhat distended gasters in the first days of the nomadic phase before returning once more to full contraction.[14]

Our results for these two species of *Eciton* (Schneirla, 1949; Schneirla and Brown, 1950) clearly indicate that these changes from the contracted to the physogastric condition and the reverse recur regularly, from cycle to cycle, through the lifetime of the colony queen. The queen of colony '48 H-15 (Fig. 8.3), for example, in the period of nearly five years we had her on record, must have alternated about fifty times from physogastry and egg laying to contraction and nonlaying. Findings for investigated species of *Neivamyrmex* (Schneirla, 1958) and *Aenictus* (Schneirla, 1965) leave no doubt that these doryline queens also function comparably throughout the long lives they pass in their colonies. Except for interruptions during winter dormancy in nontropical habitats (Schneirla, 1963), long periods in the colonies with many repetitions of massive egg laying would thus seem common among doryline queens.

There is an impressive similarity in the queen's egg-laying schedules among group A genera (Fig. 7.1). Generally, in all species of *Eciton, Neivamyrmex,* and *Aenictus* studied, the first traces of physogastry appear within two or three days before the nomadic phase ends. In these dorylines, the queen is gravid and begins to lay eggs about one week after the start of the statary phase, completing the series in about one week midway in the phase. In *E. hamatum,* the layings begin at intervals of about thirty-six days; in *E. burchelli* intervals between layings are shorter and seem to be more variable in length; and in *Aenictus* they are about forty-six days long.

From the results for *Eciton hamatum,* presented in Figure 8.6, we see that, on the one hand, regular correspondence prevails throughout the functional cycle between colony behavior and brood condition (Schneirla, 1949; Schneirla and Brown, 1950) and, on the other, changes in the queen's reproductive condition. Anatomical and histological studies of many queens of this species preserved at respectively different times in the functional cycles of their colonies were carried out by Hagan (1954).

---

[14] Apparent exceptions are of two types. The first, very old queens, may be illustrated by the superseded parent queen of colony '52 B-I, which had an elongated, flabby gaster and was unusually dark (Chapter 10). The second exception, very young queens, is illustrated by queen No. 27 (Fig. 8.5a) taken on N-8 with her gaster somewhat enlarged, probably through overfeeding (cf. colony '52 B-I, Chapter 11).

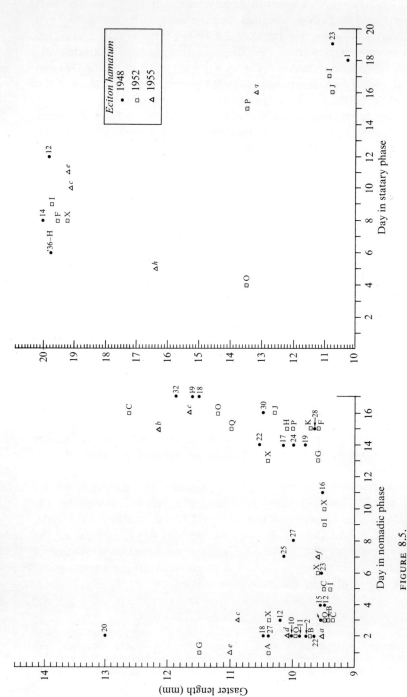

FIGURE 8.5.
Gaster length of queens of *Eciton hamatum* and *Eciton burchelli*, as a crude indication of reproductive condition, measured at different times in the functional cycles of their various colonies. *Eciton hamatum*: Gaster lengths of 9.4, 10.0, and 10.2 mm correspond to body weights of 0.122 g, 0.132 g, and 0.135 g, respectively. 1948 colonies are

given in Arabic numerals (one exception is '36-H); 1952, in capital letters; and 1955, in lower case italic letters. Some queens are measured more than once. *Eciton burchelli:* Gaster lengths of 10.2 and 10.5 mm correspond to body weights of 0.143 g and 0.152 g, respectively. 1948 colonies are given in Arabic numerals; 1952, in Roman numerals; and 1955, in italic Roman numerals.

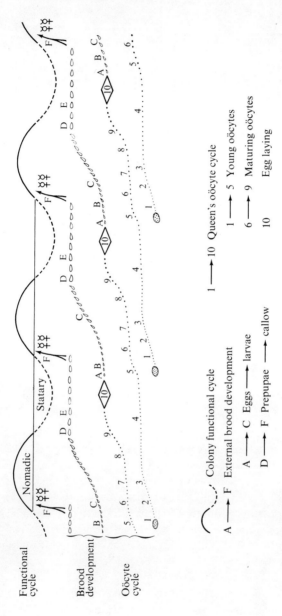

**FIGURE 8.6.**
Schema to represent corresponding conditions in the behavior and function (*top*) of a colony of *Eciton*, the coordinated development of its broods (10 A to F) and the queen's reproductive processes (1 to 10, *below*), the last based on Hagan's research (1954).

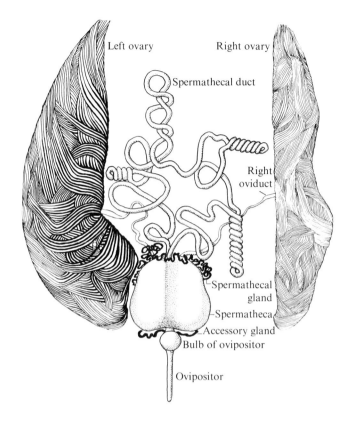

FIGURE 8.7.
Anatomy of reproductive apparatus of the queen of *Eciton hamatum,* with the right ovary and its ovarioles shown in part. (Hagan, 1954.)

This pattern holds comparably through the cycle in species of *Eciton, Neivamyrmex,* and *Aenictus* that were studied.

Clearly, in order for the group A cycle to occur, immense broods must be delivered separately and at well-spaced intervals by the queen; she seems adequately equipped to do this. The reproductive equipment of the queen of *Eciton hamatum,* as drawn by Hagan (Fig. 8.7), resembles that of the honeybee queen in having a wide median ovary with oviducts branching from it like the arms of a horseshoe. Along each oviduct the ovarioles form a profuse series, differing in number and in length of tubules according to the species. Investigators counted, in the queen of *Neivamyrmex nigrescens,* around 500 ovarioles on each side (Holliday, 1904); in the queen of *E. hamatum,* more than 2300 on each side (Hagan, 1954); and in the queen of *E. burchelli,* at least twice as many tubules, which were even longer than in *E. hamatum* (Whelden, 1963). The number of eggs produced must be related to the number of ovarioles, but actual counts of oöcytes in the gravid queens of any species are well below estimates for the total number of eggs produced in single batches by queens of that species. The answer may be that one or more divisions of the oöcytes occur as they descend in the tubules. In any case, doryline queens clearly have the resources essential to produce eggs in the quantities typical of the species.

The smooth coordination of reproductive functions in the army ant queen with the functional cycle of her colony is remarkable. The capacity to produce great batches of eggs in an on-off fashion that distinguishes the queen of group A species is essential for the occurrence of regular and well-marked alternating nomadic and statary phases. Wheeler (1928) called attention to the importance of an exceptionally developed respiratory system that enables the queen to carry out prodigious reproductive labors in the recesses of a bivouac. Such features are all the more remarkable considering what is now known about the queen's relation to the colony cycle, which involves her making periodical and very radical changes. From the nomadic to the statary phase, at intervals of a few weeks, her metabolic level must change from low to high in oxygen consumption, fat reserves, and many other respects.

If the occurrence of physogastry in an insect is already a complex biological problem (Mergelsberg, 1934), the reappearance of a physogastric, maximally gravid condition at regular intervals makes it all the more complex. For dorylines, research on these questions lies in the future. But, fortunately, significant research has been carried out on the females of roaches (Wigglesworth, 1965) and of certain other solitary insects (Scharrer and Scharrer, 1963) capable of producing series of oöcytes at intervals. The products of certain neurosecretory tissues associated with the brain, it has been shown—and particularly the corpora allata—are essential to the reproductive functions of these insects.[15] Because these tissues must be well represented in doryline queens, it is important to note that their cyclic functions in the females of certain other insects may be influenced both by internal (Engelmann and Lüscher, 1956) and by external (de Wilde and Stegwee, 1958) agents.

Although we have seen that the durations of colony functional phases are influenced by numerous conditions (e.g., colony nutrition and temperature), it is clear that whatever agents control the timing of periodic egg production by the queen must be in chief control of timing the cycle. There are three hypotheses to account for the scheduling of egg

---

[15] The evidence has been summarized and discussed by Topoff (in prep., b) in relation to factors controlling phasic variations in doryline nomadic-statary cycles.

laying in the army ant queen: (1) periodic environmental stimuli, (2) a pacemaker intrinsic to the queen, and (3) a primary control through the functions of colony and brood.

Hypothesis (1), although applicable to cycles in the reproduction and general activities of many other animals (Harker, 1958; Cloudsley-Thompson, 1961), receives no support from evidence available on army ants (Chapter 7). On the contrary, it is opposed by a lack of correspondence between the cycles of numerous colonies studied in parallel and any external periodicity, and by a similar lack of parallelism between successive cycles in the same colony and external periodicities (Schneirla, 1949; Schneirla and Brown, 1950).

Hypothesis (2) has been stated by Sudd (1967) as follows: "The queen's reproductive rhythm may be innate, but it is reinforced by the increase in the amount of food the queen gets after pupation has removed the tens of thousands of hungry larval mouths (Schneirla, 1957)." (P. 89.) Although the assumption of such an innate rhythm in the queen —in primary control of the colony cycle—might seem reasonable by analogy with certain other insects (Harker, 1958; Wigglesworth, 1965), I have never favored this idea for army ants as it lacks support in their case. On the other hand, evidence has mounted for a theory (Schneirla, 1938, 1945, 1957a) postulating very different factors as the necessary controllers of cyclic processes in colony and queen.

We may characterize hypothesis (3), which proposes that cycles of egg laying in the queen are timed dominantly by intracolony processes external to her, as the "colony-situation-feedback" concept. In preceding chapters I have supported the conclusion that brood-stimulative processes initiate phasic changes in the colony cycle. We thus view brood stimulation as the dominant factor that regulates increases in stimulative, trophic, pheromonal, and other conditions that (e.g., by acting on the corpora allata) account for physogastry and egg laying or for contraction in the queen.

Let us now discuss the queen's role as it changes through the colony cycle to find how changes in the functional condition of the colony might initiate changes in her reproductive processes at critical times; also, how functions of the queen might influence the course of events in the colony as feedback effects.

In the group A species studied, queen and colony are so well coordinated that in the nomadic phase, when the colony is emigrating nightly over long, difficult routes, the queen is contracted and well able to stand the exposure and exertion. The plates of her gaster are then pulled tightly together into a solid, armored capsule protecting her vital reproductive organs, and powerful legs carry her lightened body easily and safely over the routes. Until nearly the end of this phase, the queen seems to be confined most of the time within her cluster of guard workers, evidently receiving minimal social stimulation and food while the larval brood grows. Also, colony pheromones, perhaps from both workers and larvae, may hold her ovulation processes to a low level through most of the phase. That *something* does this is suggested clearly by Hagan's results, summarized in Figure 8.6.

As the larval brood nears maturity, however, and is more widely distributed in the bivouac, social stimulation, which I conclude (Schneirla, 1957a), arises through intensified worker responses to the larvae—now larger and more widely distributed in the bivouac—may have the effect of releasing the queen from her previous restraints (Fig. 8.6). From evidence indicating that the queen's position in the bivouac changes radically in the latter part of the phase[16] and that signs of early physogastry appear at the time (Fig. 8.5), together with results on worker-queen relationships discussed in Chapters 10 and 11, I infer the following. Conditions of increased worker arousal and of food surplus, incident to pupal maturation, now promote a crescendo of social stimulation from the workers focused on the queen, which soon initiates a variety of changes in her.

Before long, the queen, in her turn, evidently becomes increasingly active and responsive and enters a new physiological condition that rapidly increases her feeding frequently to gorging at food heaps. The entire process seems attributable to her excitation by workers at a time when ample food is present—spared by the brood, so to speak. In chapter 6 we considered an important developmental transformation in the maturing larva whereby the brood feeds decreasingly, reducing its competition for food with queen and workers, but at the same time exciting the colony increasingly, energizing great raids, and insuring large quantities of booty. I suggest that this change is the basic factor that starts the queen toward physogastry. In *Eciton, Neivamyrmex,* and *Aenictus,* her physogastry begins in the last days of the nomadic phase, then progresses steadily.

Observations on laboratory nests show that reciprocal stimulation now becomes complex and may pass through distinct (hormonally changing) stages as the queen lays down abdominal fat and accelerates in ovulation. The process is intensive and mutual: With this change, the queen increasingly stimulates the workers through her behavior and (doubtless also) her chemical products. On the last emigration of the phase, the queen's enlarged abdomen with its now widened intersegmental bands

---

[16] Through most of the nomadic phase the queen of surface-bivouacking *Eciton* usually can be found in the upper center of the bivouac, enclosed in a capsule of guard workers. In the final days, however, her position is much more variable as she evidently roams the bivouac a large part of the time.

shows us how her condition is changing. This is a time of very important transition in the colony cycle.

I suggest, furthermore, that as the broods of group A colonies mature (Chapter 6), the larvae similarly introduce many behavioral, stimulative, trophic, and secretory changes into the colony that bear directly on bringing the queen to full physogastry. Histological studies (Hagan, 1954) of queens preserved at this time, represented in Figure 8.6, show that each queen not only rapidly matures a great batch of ova but also has a spurt in the early production of a new generation of oöcytes or potential eggs. The best interpretation of these facts, I believe, is that the queen responds to a brood-initiated change in the colony situation by entering physogastry, then continues toward the fully gravid state—provided that the transition of the colony from the nomadic to the statary condition proceeds normally.

There are several results that favor this interpretation as against the hypothesis of a control intrinsic to the queen. For one, when a sexual brood begins its larval maturation several days earlier than is usual with all-worker broods, the queen begins physogastry at a correspondingly earlier time. Also, in a colony of *Neivarmyrmex nigrescens* that is in the process of emerging from winter dormancy (Schneirla, 1963), when the queen enters her first physogastry, she shows clear signs of responding to a rising stimulative and trophic condition in the colony. As the obverse of this event, when a colony of this species nears the end of its active season in autumn (Schneirla, 1958), the queen remains contracted at the time of nomadic-statary transition evidently because the colony's stimulative and trophic properties are low and do not arouse her to a gravid condition. Correspondingly, in preliminary studies with underfed colonies of *Eciton* in laboratory nests, I have found no indications of physogastry at any time in the queens. The exceptional manner in which physogastry appears in the callow queens of daughter colonies after division (Chapter 11) may be interpreted, I believe, as a response to more intensive stimulative and trophic properties than hold for parent queens at the corresponding time in the cycle.

Previous chapters have shown that as a statary phase begins in group A colonies, existing circumstances advance the queen in physogastry. Although the brood is not feeding, colony raids continue regularly (Table 4.2), insuring her ample food. Laboratory observations have indicated the critical role of the workers in the continuation of physogastry during this interval. Workers are intensely attracted to the gravid queen, constantly moving around her, busily engaged in grooming and feeding her. In these ways, feedback effects emanating from the queen are major factors bringing her to full physogastry.

Egg laying in queens of *Eciton, Neivamyrmex,* and *Aenictus* begins at

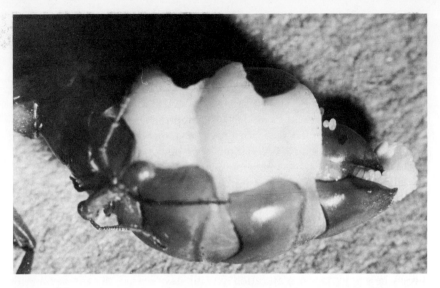

FIGURE 8.8.
Close-up of gaster of physogastric queen of *Eciton hamatum* at the moment a new stream of eggs begins to emerge. Scale: length of egg 0.3 mm.

about the sixth statary day, then rises to a peak within a few days. At this time physogastric army ant queens are truly in a delicate condition: Exposure to bright light, dryness, heat, or shaking will reduce their egg laying. Intense light exerts severe traumatic effects which may be even fatal. Although gravid queens studied in the laboratory can move about, even carrying workers with them, running in an emigration at this time might lead to dangerous or even fatal accidents (e.g., through falls or abdominal punctures). Fortunately, the queen's reactions are so well intermeshed with events in the cycle that when she is in such a vulnerable state, she remains safe and well cared for within the dark recesses of a well-sheltered bivouac.

The delivery of a great batch of eggs by an army ant queen, illustrated by observations on *Eciton hamatum* (Schneirla, 1938, 1944a), is a virtually continuous operation, lasting about one week. The eggs emerge in long bursts, each starting with a shivering of the gaster. The shaking soon expands into an oscillation of legs and body as peristaltic contractions, one after another, begin in anterior segments and sweep over the entire abdomen. Before long the eggs emerge from the cloaca in a wide thick ribbon that slides over her shelflike hypopygium, much like toothpaste squeezed from a tube (Fig. 8.8). The workers, in licking her body, are especially busy at the abdomen and its intersegmental membranes. They avidly lick up a milky fluid that at times emerges from the queen's vulva after sessions of egg laying. These operations are closely related to licking and packeting the eggs as they emerge.

Egg laying (e.g., in *Eciton hamatum,* Schneirla, 1944a) is not a continuous operation but goes on for intervals of four or five minutes at a time, separated by nonlaying intervals of roughly the same length of time, during which the queen remains active. In one two-hour observation of a queen of this species in a laboratory nest, on the second or third day that she has been laying eggs, there were fifteen intervals of quiescence— in each of which from 200 to 400 eggs were laid—and fourteen of nonlaying. Shortly before each interval of laying ends, there is a brief play of antennae; the queen then stirs, and once the eggs have stopped emerging, she immediately begins to move about. In the laboratory, during such intervals, the queen runs about with workers or stands among them and feeds; in the bivouac, she probably remains in place, groomed and fed by workers that cover her. The queen of colony '46 B-IV, *Eciton burchelli,* shown in Figure 8.2, when captured in the fully gravid condition, was centered in a large bolus of minor workers that, as laboratory observations showed, must have been engaged in grooming and feeding her and in handling the eggs (Schneirla, 1949). Under natural conditions, it is probable that the on-off periods constitute a secondary rhythm within the overall act of egg laying, the phases of which may quicken in their tempo as egg laying rises to a peak, then slow down as it tapers off.

Figure 7.1 summarizes evidence showing correspondences between colony, brood condition, and reproductive changes in the queens of *Eciton, Neivamyrmex,* and *Aenictus.* I hold that trophic and stimulative effects *initiated by the brood* influence the queen's reproductive functions in ways that are crucial to the ongoing events of the colony cycle. In other words, excitatory conditions *external* to the queen, but within the colony situation, control the queen's egg-laying cyclic processes, which, on the other hand, run parallel to colony events. Significantly, Hagan (1954) found indications of notable accelerations in the ovulation processes of the queen at each of the two times in the cycle when the colony reaches a peak in its trophic and stimulative functions. Feedback effects from colony to queen underlying these changes are indicated by arrows in Figure 8.6.

A comparable but briefer feedback effect seems to arise in the late statary phase and to continue into the early nomadic phase. After the queen has laid her eggs, she rests in the statary bivouac and changes toward the contracted condition. She may not then reach full contraction, however, because of new developments. These concern an increasing stimulative and trophic resurgence in the colony, evidently based on effects introduced through the maturation of the pupal brood (Chapter 6). Because these changes in the colony situation occur at a time when the brood is not feeding, the queen may have a new interval of increased social stimulation and heavy feeding. This effect, although it seems to last only a few days, aids her recovery from the physiological drains of the gravid period. Also, as Hagan's results (1954) indicate, it may cause two significant reproductive changes within her: first, a spurt in the development of the oncoming young generation of oöcytes, and, second, the initiation in her germarium of a new generation of oöcytes (Fig. 8.6). It is impressive to discover that the queen not only has two successive oöcyte generations under development at the same time but that these broods may concurrently receive colony-stimulated developmental impulsions.

But within a day or two after the next nomadic phase has begun, the queen returns to full contraction (Fig. 8.5) and internally (Fig. 8.6) undergoes a leveling off in the progress of both lots of oöcytes developing within her. This may be due to the restriction put on the queen's feeding and grooming by the heavy competition from the newly emerged callows. Then, with the growing larval brood continuing this restriction, the queen next passes through a period of limited movement (and feeding) that hold her reproductive processes to a low rate through most of the nomadic phase. In such ways, I suggest, the queen's reproductive functions throughout the cycle are basically controlled by the brood in that critical changes in her function seem to arise through brood-initiated changes in the colony.

Queens of group A species evidently can stop egg production altogether for a considerable interval (while oöcyte growth continues slowly), then resume it promptly on appropriate stimulation from the colony. Most impressive, the feedback effects from one interval of physogastry (dependent on developmental timing in the brood) evidently operate to delay another such episode until the queen has recuperated and is well able to undertake a new one. This next gravid interval comes about on appropriate stimulation from the colony. In simplified communications-theory terms, one might say that in effect the colony "signals" the queen to advance each new generation first slowly, then more rapidly, and finally to complete and deliver it.

Evidence is lacking for a direct timing of the colony cycle by functions intrinsic to the queen. The queen enters into a variety of relationships with the colony, however, that may influence the cycle. Some of these relations may be suggested by evidence for ants (Brian and Hibble, 1963, 1964) and other social insects (Weaver, 1966), pointing to how pheromones from the queen influence the colony and how those from workers or brood influence the queen. For example, substances of the queen may inhibit reproductive functions in the workers, as in honeybees (Butler and Fairey, 1963) and certain ants (Carr, 1962). In the dorylines, for example, pheromones from workers and brood may subdue reproductive processes of the queen through most of the nomadic phase, or substances from the callows (added to colony effects already mentioned) may stimulate the queen's reproduction near the end of this phase.

Also, the queen may directly affect the cycle in limited ways. To illustrate, a group A queen, after maturing a mass of eggs in one series, may undergo a refractory phase, i.e., a period within which she cannot repeat the process, that could set a minimal limit on the time from one gravid period to the next. From our results for group A species, however, the range of interphysogastric intervals even in group A queens is too great to suggest that this factor can influence timing of the cycle except within wide limits.[17]

Clearly, important influences of the queen on colony function center on the type of eggs she produces in any given batch, i.e., those of a potential all-worker or sexual brood. This difference, introduced in Chapter 6 and to be discussed further in Chapter 9, bears, among other things, on the duration of the current nomadic phase and on the way the colony behaves at the time this brood reaches pupal maturity.

Finally, there are reasons to believe that the queen's egg-laying rate influences the distribution of types of workers in the colony. In Chapter 2 we saw that in polymorphic dorylines there are species-typical frequencies of worker size types not only among the adults (Figs. 2.2 and 2.3) but also in the brood; furthermore, we noted that adult and brood size-type distributions are likely to be very similar. One factor accounting for this, and I suggest possibly the principal one, is the queen's rate of egg laying, which, from our empirical results, seems to start slowly and gradually to pick up more and more speed until it reaches its peak late in the series. Then, rather abruptly, it falls off. Such differences in rate of egg laying might affect yolk content of the eggs, worker responses in feeding and handling brood at later stages, and still other developmental conditions influencing the distribution of polymorphic size types in the mature brood (Chapter 6). For reasons to be discussed in Chapter 13, however, this apparent correlation seems feeble in the monomorphic *Aenictus*.

In contrast to the relatively precise reproductive schedule holding in group A queens, queens of group B (*Dorylus, Anomma,* and perhaps *Labidus* and *Nomamyrmex*) function in much less regular ways although

---

[17] Interphysogastric intervals (not the same as interbrood intervals) are calculated from the beginning of one physogastric period, roughly two days before the end of a nomadic phase, to the beginning of the next such period, two days before the end of the next nomadic phase. Lengths of nomadic and statary phases recorded for *Eciton hamatum* may be noted for all-worker brood series in Table 7.1. A comparison of the lengths of the nomadic phase for all-worker and sexual-brood series is shown in Figure 9.3.

they have a much greater egg output. Species in group B genera evidently have colony and reproductive patterns that are based on a brood coordination that is relatively variable and weakly organized. Although in group B queen and colony presumably interact by their reciprocal feedback effects equivalent to those in group A, they apparently are much more variable than in group A.

Accordingly, in Figure 8.6 we present the relationships between queen's function and colony condition within the terms of our brood-stimulative and colony-feedback theory. Raignier and Van Boven (1955) conclude that queens of *Anomma wilverthi* become fully gravid and lay eggs at peak, with a period of excitation and emigration then stimulated by the callow-arousal effect. It is even possible that egg delivery in these queens is continuous—presumably maintained at a high level for many days when the colony has settled in place after an emigration, then falling off somewhat, and rising again as a new brood nears pupal maturation. Their results also suggest that even when brood-excitation effects may be too weak or diffuse (Fig. 6.4c) to arouse a colony emigration, it may nonetheless bring about a new peak of egg delivery in the queen, taking place about halfway through a long (e.g., fifty-day) colony stop. The maturation of this new brood in its turn some twenty-five days later can then lead to an emigration. By interpreting their results in this way, we can account not only for episodes of physogastry in the queen but for the emigrations coming most frequently around twenty-five days as well as those around fifty days in this species. The much longer intervals between emigrations in *Dorylus* (*Anomma*) *nigricans* (Raignier and Van Boven, 1955) are probably representative of those in more hypogaeic taxa of this genus in which ecological conditions impose slow growth and looser synchronization (Chapters 3 and 6) on the broods and more variable and extended patterns of brood coordination on the colony (Chapter 7). The adaptive significance of the consequent feedback effects on colony queens is discussed in Chapter 14.

Available evidence thus favors the idea that the reproductive functions of doryline queens are finely intermeshed with the operations of their colonies as these operations themselves, through their feedback effects, control the queen's functions. The relations involved seem always to be reciprocal, with physiological and reproductive conditions in queens of group A species closely related to stimulative-trophic colony conditions, and the worker population for its part responding intimately to effects emanating from brood and from queen at the time. In these interrelationships, from our findings, the brood is the dominant agency.

Such a pattern of subgroup interrelationships unquestionably attains its keener precision through the regular on-off reproductive functions of the group A queen. Conceivably, a state of internal colony affairs sufficiently precise to account for the well-defined nomadic-statary cycles of group A species can arise only when colony and brood populations (and queen's output) are still relatively small. But when these agencies assume much larger proportions, as in group B species, the overall functional picture becomes diffuse and variable. The possible reasons are considered further in relation to colony communicative processes in Chapter 12 and to colony adaptive patterns in Chapter 14.

# 9

# Males and Young Queens

Most ants produce their sexual individuals after the colonies have grown for a few years. Then, usually in the spring or summer—at a time characteristic of the species and when food is plentiful (Brian, 1965)— males and young queens appear in the broods. Honeybees produce males in times of plenty (Plateaux-Quénu, 1961; Butler, 1962) and develop new queens in response to changes in hive conditions brought on by the loss or overaging of the colony queen (Simpson, 1958). The production of males and new queens by army ants, a long-unsolved problem, did not seem to fit the pattern known for other social insects. Reports of winged males in large numbers issuing from nests of *Dorylus* species in Africa (Mayr, 1886; Brauns, 1901) intimated some kind of mass production process while records on species of *Neivamyrmex* in South America (Gallardo, 1915; Bruch, 1923) suggested that the males appear in large, distinct broods. Evidence on male flights, although indicating seasonal trends (Smith, 1942; Borgmeier, 1955), involved confusing differences among cases having to do with species, locality, and climate.[1]

There were no reports on young queens, however, until Wheeler (1921) in British Guiana found two callow queens among hundreds of male cocoons in a bivouac of *Eciton burchelli*. As we now know (Schneirla, 1949; Schneirla and Brown, 1950), these broods in many army ants contain both reproductive sexes, i.e., both males and potential queens.

For reliable information on the timing of sexual broods, direct research on the colonies producing them has proved to be more reliable than reports on males captured at insect light traps (Schneirla, 1948). From our studies on species of surface-adapted *Eciton* in Central America, we have been able to discern a basic pattern of sexual-brood production

---

[1] Although males of *Neivamyrmex* species have been reported (Smith, 1942) for various parts of the southern United States as commonly flying from September to early November, in southeastern Arizona males of numerous species are captured at times from mid-July to late August (Schneirla, 1961).

(Schneirla and Brown, 1952). Among 141 colonies of four species studied in times of rain, we found 189 broods, but only 2 of these were sexual broods. In three dry-season projects in Central America (Schneirla, 1948; Schneirla and Brown, 1950, 1952), we studied in all 109 colonies of five *Eciton* species, finding a total of 171 broods of which 21 were sexual broods. Clearly, colonies of *Eciton* produce all-worker broods at all times of the year, but only a minority of the colonies bring forth sexual broods, and evidently under exceptional conditions.

What these conditions may be is suggested by a review of our results for the dry seasons of 1946 and 1948 on Barro Colorado Island (Schneirla, 1949; Schneirla and Brown, 1952) and particularly by a comparison of sexual-brood records for the first and second halves of the dry season in these two years. In 1946 we found six colonies of *Eciton hamatum* and *E. burchelli* with sexual broods that had been initiated in the first half of the dry season (i.e., December 15 to February 15) and just two in the second half; in 1948 all four of the sexual broods discovered in colonies of these two species were initiated in the first half of the season (Schneirla and Brown, 1952, p. 20). This suggests that exposure of colony queens to dry-season conditions at the time of egg laying, and particularly early in the season, is a major factor in the production of sexual broods.[2]

This assumption clarified other results pertaining to species ecological differences. In the dry season of 1946, sexual broods definitely began earlier in *Eciton hamatum* than in *E. burchelli,* but in 1948 our findings for the two species were much the same in this respect. This difference becomes understandable when we take into account how the typical modes of bivouacking of the two species may affect the degree of exposure of their colonies to conditions of drought or humidity.

Late in the nomadic and early in the statary phase, when these conditions are presumably determinative for the type of brood produced, *Eciton hamatum* typically bivouacs on or beneath the surface while *E. burchelli* often bivouacs in elevated places (Chapter 3). In the first part of the 1946 dry season, surface conditions were moist, but the air was dry most of the time; hence colonies of *E. burchelli* were then more often exposed to dry conditions than those of *E. hamatum.* But in 1948, from the beginning of the dry season, colonies of the two species generally met with

---

[2] I found my first sexual brood in a colony of *Eciton hamatum* in southern Mexico (Schneirla, 1947) in a period of changeable weather and in highly varied terrain in which many colonies with all-worker broods were also discovered. I studied a colony of *E. hamatum* for more than four months in the 1946 dry season in Panama; during this time it produced five all-worker broods; but in the same period a colony of *E. burchelli* had six broods, of which one was a sexual brood (Schneirla, 1949).

similar ecological conditions as both the forest floor and the air were then generally dry (Schneirla and Brown, 1952, pp. 19–20).

From this standpoint, colony differences within the same species are clarified. Barro Colorado Island is an ecologically diversified area, where in the dry season colonies bivouacking on plateaus and in open forest are exposed to drought but those bivouacking in ravines or humid forest elsewhere are less exposed. Study of our records in terms of these differences suggests that type of bivouac site is an important factor in sexual-brood production. For example, in the same season, colony '46 B-I occupied an elevated and wind-exposed statary position when the queen laid the eggs of what proved to be a sexual brood whereas colony '46 H-B occupied only humid statary sites and produced only all-worker broods (Schneirla, 1949).

These findings gave rise to the "dry-impact" hypothesis by which colonies of *Eciton* located in an area that uniformly exposes them to drought will usually produce sexual broods. A test of this kind was planned for the Tuira River area in Darien, Panama, a region where dry weather ordinarily begins in late December and persists for about four months (Schneirla and Brown, 1952, pp. 313–316). In the relatively level forested terrain of these lowlands, most surface-operating colonies of *Eciton* should be exposed to dry conditions on entering their first or second statary phases after the season began. Allowing (empirically) about fifty days for sexual brood development from the time the queen begins to be physogastric[3] and assuming that sexual broods were initiated after the start of dry weather, most of the colonies should have broods by mid-February. Actually, of eight colonies of *E. hamatum* and *E. burchelli* found in Darien in a ten-day survey carried out after February 15, six had sexual broods that must have been initiated after drought began in December. Clearly, the hypothesis was supported.

Results on Barro Colorado Island also supported the hypothesis. In 1946 when the rains continued moderately until early January, the records indicated that 7 of 8 colonies found with sexual broods must have initiated them after January 15. But in 1948 when the rains stopped in mid-December, by the same token 5 of the 10 sexual broods found in that season must have been initiated before January 15 (Schneirla and

---

[3] On an empirical basis (Schneirla, 1948), it seemed that the production of a sexual brood (rather than an all-worker brood) might be determined when the queen was entering physogastry. Thus, for *Eciton hamatum,* a sexual brood found as mature larvae could have been initiated about thirty-two days before, and such a brood found as mature pupae could have been initiated about fifty days before (Schneirla, 1949). This formula predicted the conditions under which colony '48 H-12 produced a sexual brood (also the general results of our Darien survey, Schneirla and Brown, 1950).

Brown, 1952, Table 5).[4] Altogether, of the 18 colonies found there with sexual broods between December and May of these two years, 16 must have initiated them before February 15. Early dry-season production of these broods thus seems predominant.

But apparent exceptions have come to light, namely, four sexual broods of *Eciton* that were produced in rainy weather. Three of these involved surface-adapted *Eciton:* (1) a colony of *E. burchelli* with callow queens and mature male pupae, found in British Guiana on July 21, 1920 (Wheeler, 1921); (2) a colony of *E. burchelli,* found with mature sexual larvae in Trinidad on July 10, 1950 (Schneirla and Brown, 1952); and (3) a colony of *E. hamatum* found on Barro Colorado Island with the remnant of an alate male brood on August 4, 1949. Investigation into these cases (Schneirla and Brown, 1952, pp. 20–23) disclosed that the first three colonies had all passed through earlier rainless periods that could well have coincided with the maturation and egg-laying periods of the respective colony queens.[5] The fourth case, a colony of *E. vagans* found on Barro Colorado Island with a mature pupal sexual brood on July 5, 1956 (Rettenmeyer, 1963), has yet to be examined in this respect.

From the evidence, I suggest that colonies of *Eciton* in areas of distinct seasonal change commonly initiate their sexual broods in response to the first adequate impact of dry weather—coming after times of rain—when a queen is maturing and laying eggs. I propose that in areas of variable terrain, variable annual weather, or both of these, queens lay the eggs of such broods when sequences of dry days coincide with this same reproductive interval. This idea and its significance are discussed later in this chapter.

The dry-impact hypothesis also applies to the sexual broods of other group A dorylines. We have evidence to this effect for *Neivamyrmex* species in the canyon areas of southeastern Arizona. Annually, after a post-winter resurgence in April and May, colonies of *N. nigrescens* produce all-worker broods regularly and are then nomadic (Schneirla,

---

[4] On December 21, 1947, we found colony '48 H-12 (*Eciton hamatum*) in the early part of a statary phase with its queen laying the eggs that produced a sexual brood (Schneirla and Brown, 1950, p. 283). Rainfall had been very light from December 15 on, as indicated by a generally dry and cracked condition of the forest floor.

[5] In the one case of these three in which rainfall records were unobtainable, that of the mature sexual brood of *Eciton burchelli* found in British Guiana on July 21, 1920 (Schneirla and Brown, 1952), I asked the late Dr. William Beebe about rainfall in the area during his entire stay that year. Without knowing my purpose, he answered that the entire period from February through July had been rainy, except for the month of June and particularly its first half, in which he had never needed his poncho in the forest. I had estimated earlier that this brood might have been initiated early in June.

1963). But there follows, usually after early May, a distinct dry interval which lasts through June. Significantly, in several annual studies (Schneirla, 1961), we have found colonies of this species (and of *N. opacithorax*) with sexual broods at times from mid-July through August. Also, light-trap collections (Forbes et al., in prep.) have yielded alate males of six hypogaeic species of *Neivamyrmex* at times from early July to early September. The results suggest that seasonal ecological conditions related to weather, and specifically to the annual late-spring drought, underlie the production of sexual broods in all of these species.[6]

From results with surface-adapted species of *Aenictus* studied on Negros Island in the southern Philippines, the hypothesis seems also applicable to that genus. Confirming Chapman's reports (1964), we have found several colonies of both *Ae. laeviceps* and *Ae. gracilis* with sexual broods at times ranging from mid-March to late May in five years of study.[7] This region has a well-marked annual dry season usually beginning early in February.

For investigated species of all three group A genera, therefore, results lead us to believe that the onset of dry weather somehow initiates sexual broods. This aspect of seasonal regularity in producing males and new queens holds numerous advantages for adaptive colony functions (Chapter 14). The driver ants may diverge from this pattern, as Raignier (1959), investigating *Dorylus (Anomma) wilverthi* in the Congo, found colonies with sexual broods in all months of the year. Such broods, in fact, seemed to be most frequent in the rainy months as fifty-five in a total of seventy-five sexual broods were discovered then. This apparent difference from group A may arise through variable annual climatic conditions or may represent a generic difference in reproductive patterns.[8]

All species of group A dorylines studied are very regular in the timing, size, and composition of their sexual broods. In *Eciton, Neivamyrmex,* and *Aenictus,* they are species-typical in size and contain distinct populations, nearly all males with a few young queens as we first found them

---

[6] So distinctly hypogaeic are all of these species that their males and workers have not been taken together and so are still unrelated taxonomically (Borgmeier, 1955). In the one case for which a synonymous relationship has been confirmed, the male named *Neivamyrmex harrisi* and the workers named *N. wheeleri* (Smith, 1942; Watkins, 1968), the workers are seen very infrequently on the surface.

[7] As an example, the first week of February, 1961, was entirely dry although light rains preceded and followed (Schneirla and Reyes, 1966). After early March we began to find colonies of these species with sexual broods (at first in the larval stage), and these records continued through April.

[8] Raignier (1959) suggests that sexual broods are produced in *Anomma* as a result of periodic depletions of sperms in the queens, then renewed through rematings. Although this hypothesis may be eventually supported for *Dorylus*, it seems inapplicable to *Eciton* (Whelden, 1963), *Neivamyrmex,* and *Aenictus,* in which all functional queens studied histologically are found with ample supplies of sperms (R. M. Whelden, pers. comm.).

in *Eciton* (Schneirla, 1949; Schneirla and Brown, 1950). Two nearly complete counts of males of sexual broods in *E. hamatum* totaled 1447 in the first and 1619 in the second (Schneirla and Brown, 1952). Counts for *E. burchelli* range from 2500 to 3500, much larger than for *E. hamatum* although smaller in proportion to the immense all-worker broods of *E. burchelli*. Colonies of *N. nigrescens* have sexual broods containing from 1000 to 1200 males, and those of *Ae. laeviceps* and *Ae. gracilis* as a rule are about this same size.[9] From our results, therefore, the populations of males in sexual broods number only a fraction, around 2 to 3%, of all-worker broods of these army ants. Clearly, whatever conditions limit the populations of these broods must operate rigidly and invariably.

Limitations on young queens are far stricter. In all our cases the number of these individuals never exceeded six. In this study (Schneirla and Brown, 1952, p. 11), by examining the largest specimens from various brood stages, we found in *Eciton hamatum* two queens among a large sample taken in an early larval stage and just one in a sample taken at an intermediate stage—all others were males. In the same study, among the only twenty specimens in a mature larval brood, fourteen were queen larvae.[10] In the same way, we took the first larval queens of *E. burchelli* on record by searching the sexual-brood area of a bivouac, finding at the base of this cluster the only two enclosed larvae (both queens) in a mass of otherwise unenclosed larvae—all males (Schneirla and Brown, 1950, p. 306). By keeping a close watch on the emigrations of colonies with sexual broods nearing larval maturity, we have found young queens both in *E. hamatum* and in *Neivamyrmex nigrescens*. As mature larvae, the potential queens are the first to spin cocoons and are very different in appearance from the males (Fig. 9.1). In a continuous study of *N. nigrescens* (Schneirla, 1961), I found a callow queen beneath a rock—that lay above the probable sexual-brood area of the statary bivouac—active in a group of workers near clusters of workers that held still-enclosed male pupae. A. Y. Reyes (Schneirla and Reyes, in ms.), in digging up a statary colony of *Aenictus gracilis,* found five lightly pigmented queen pupae among hundreds of still-unpigmented male pupae.

---

[9] Professor Alfredo Reyes of Silliman University, my collaborator, made two nearly complete counts of males for *Aenictus gracilis* as they escaped from the bivouacs on successive evenings. Those of two colonies totaled around 1200 and that of a third, 1074. Males may be somewhat more numerous in *Aenictus* in proportion to colony size than in other group A dorylines.

[10] All fourteen queen larvae in this brood, but only six of the males, had started to spin their cocoons. Randomly selected male larvae, however, when plunged into scalding water (as in silkworm practice) extruded wads of silk. Clearly, they were about ready to spin as a subsequent histological study showed (R. M. Whelden, pers. comm.).

FIGURE 9.1.
Specimens of mature larvae from the sexual brood of colony '48 B-XVII (*Eciton burchelli*). (*a*) cocoon (length 24.3 mm) containing mature queen larva; (*b*) mature queen larva removed from an identical cocoon; (*c*) and (*d*) two male larvae still naked though ready to spin.

From these results, young queens seem to be always present in the sexual broods of these army ants—though in very small numbers—and always more advanced in development than the males.

Group A sexual broods are distinctive not only in their small and dimorphic (queen-male) populations but also in the size distributions of the males. In one study we measured body lengths in large samples of male larvae from the sexual broods of three colonies of *Eciton hamatum* (Schneirla and Brown, 1952). The distributions obtained are shown in Figure 9.2. A sample from the brood of colony '48 DH-1, preserved at an early larval stage, gave a unimodal distribution very close to a theoretical normal frequency curve. Another sample of males, taken at an intermediate larval stage from the brood of colony '48 DH-4, had a very similar distribution. A curious partial deviation from this result turned up however in the sample of nearly mature larvae from colony '48 DH-3, which did not fit the normal frequency pattern quite as well as the others did. One difference was that the larger larvae in this sample were slightly more numerous than the smaller larvae; another was that this brood contained about fifty male larvae which were much smaller than the rest.

Male populations in sexual broods of *Neivamyrmex nigrescens*, like those described for *Eciton,* are also well-synchronized groups close to the normal frequency pattern of distribution (Schneirla, 1961). Those from sexual broods of *Aenictus* we have studied are similar, except for having

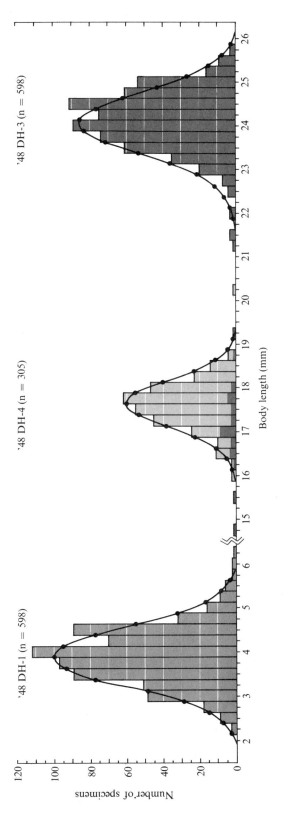

FIGURE 9.2.
Distribution of individuals in large samples of the sexual broods of three colonies of *Eciton hamatum* in the early ('48 DH-1), intermediate ('48 DH-4), and late ('48 DH-3) stages of larval growth. With a few young-queen specimens excluded, these curves fit a normal frequency distribution at the 0.01, the 0.1, and the 0.05 levels of acceptance, respectively. Note the "dwarf" cases of the mature sample overlapped with the intermediate sample.

sharper peaks and being moderately skewed toward the largest extreme.[11]

From the characteristics of group A sexual broods, with their few precocious females and, as pointed out later in this chapter, their many symmetrically distributed males, we may derive significant clues as to how these exceptional broods are produced. Also, as in *Aenictus,* these results may give us a basis for predicting the time of maturation of the males and hence of their departures from the colonies (Chapter 11).

The sexual broods of driver ants differ from those of group A genera in both number and timing. In fifteen sexual broods of *Dorylus (Anomma) wilverthi,* Raignier (1959) found the numbers of males ranging from around 650 to around 3400, with the smallest two containing only 200 and 400 individuals. The brood numbers of this species not only varied more than those of group A species, but they also seemed to be far smaller in proportion to the worker populations than those of *Eciton, Neivamyrmex,* and *Aenictus.* The young queens that could be found in colonies of this driver ant were, much as in group A species, few in number and always ahead of the males in development. The sexual broods of this *Dorylus* species also seemed to lack any distinct seasonal timing and to occur more often in any one colony than in group A army ants. These differences, to be discussed in Chapters 10, 11, and 14, may be significant for different patterns of colony function and species adaptation in group A and group B dorylines.

We also find a sharp contrast in developmental patterns between doryline sexual and all-worker broods. The greater size and different appearance of the potential queens and males at corresponding stages is illustrated in Figures 6.3 and 9.1. The very different rates and attainments of growth in these two types of brood are represented for *Eciton* in Figure 9.3. Particularly notable is the marked initial acceleration in size of the sexual larvae in contrast with the much slower start of all-worker larvae. We find comparable differences between the sexual and the all-worker broods of *Neivamyrmex* (Schneirla, 1961) and of *Aenictus* (Schneirla and Reyes, in ms.). In all three of these genera, individuals in sexual broods reach larval maturity about one week earlier than all-worker broods (Fig. 9.3). Also, at this stage they have a body volume nearly twenty times greater than the largest worker larvae at maturity.

One important outcome of the striking difference in growth rate is that colony nomadic phases are considerably shorter with sexual broods than with all-worker broods. In *Eciton hamatum,* for example, nomadic phases

---

[11] For example, a brood of 1057 male prepupae of *Aenictus gracilis* studied in our laboratory had the following distribution of body lengths: 7.5 to 8 mm, 67 (specimens); 8 to 8.5 mm, 774; 8.5 to 9 mm, 172; 9.0 to 9.5 mm, 44. June Kaiser carried out this study as a Lincoln Ellsworth fellow of the American Museum of Natural History.

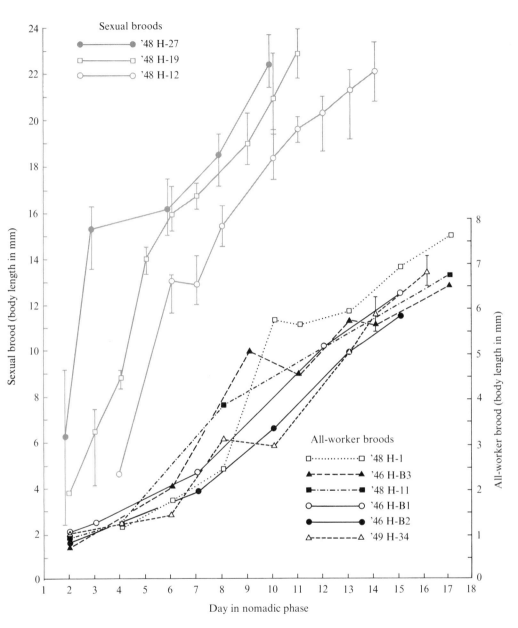

FIGURE 9.3.
Larval growth of two brood types represented by series of samples from colonies of *Eciton hamatum* in the nomadic phase. *Upper:* three colonies with sexual broods. *Lower:* six colonies with all-worker broods. Curves represent average of body lengths, with range indicated in some cases.

of colonies with developing sexual larvae ranged from 10 to 13 days whereas those of colonies with all-worker larvae ranged from 15 to 17 days. Our shortest record for *E. hamatum* is only 8 days, our longest 13 days; the corresponding records for *E. burchelli* are 9 and 12 days (Schneirla and Brown, 1950, Fig. 6). Our field records indicate that the length of nomadic phases when sexual broods are present fluctuates widely because of colony situation and condition. One important factor affecting the colony in these respects is local food supply as a sexual brood in the larval stage consumes more food (despite its smaller numbers) than an all-worker brood.

For the durations of the statary phase, however, we find no such differences brought on by the sexual broods (Schneirla and Brown, 1950, Fig. 7). This becomes understandable when we recall that pupation, which both types of brood are undergoing in this phase, is a time of nonfeeding.

Our results for the other two group A genera are comparable with those for *Eciton* in these respects. In *Neivamyrmex nigrescens,* for example, nomadic phases are generally 18 to 20 days long with all-worker broods but only 8 to 11 days with sexual broods present. One colony ('59 N-III, Fig. 10.6) in fact settled into its statary bivouac after just 6 nomadic days without further emigrations although great raids continued on the three nights following the last emigration (Schneirla, 1961). The nomadic phase is shortened comparably in colonies of *Aenictus* when sexual broods are present.

In both *Eciton* (Schneirla and Brown, 1952) and *Neivamyrmex nigrescens* (Schneirla, 1961), the nomadic phase ends when the sexual larvae become mature and spin their cocoons. In *Eciton,* the stimulative effect of the brood on the colony is equivalent to that with all-worker broods. In *N. nigrescens,* however, the all-worker larvae spin no cocoons at maturity, and therefore brood function may somewhat differ in its relation to the colony at this time. In *Aenictus,* in which neither the all-worker larvae nor the sexual larvae spin cocoons at maturity, a somewhat different pattern is indicated, perhaps similar to that suggested in Chapter 6 for the all-worker broods of *Neivamyrmex.*

For group A dorylines, however, we may say that, with either a sexual brood or an all-worker brood present in the colony, the functional cycles continue on much the same causal basis, except for the durations of the nomadic phase. (But, as noted in Chapter 7, and to be discussed later, this is not the case with *Dorylus.*) Thus, in the processes of reciprocal stimulation and feedback based upon brood effects (Chapter 6) and implicating workers and queen (Chapters 7 and 8), both types of brood are coordinated equivalently with the broods preceding and following them in the series.

On the other hand, sexual larvae of group A army ants exert notably more intense stimulative effects than all-worker broods, as indicated by much more vigorous and extensive raids. Other differences, as, for example, less frequent emigrations in the larval stage (Chapter 5) and signs of colony fission at pupal maturity (Chapter 10), point to stimulative and pheromonal effects from the sexual brood which are distinctive of that brood and more potent than in all-worker broods.

To understand what role these broods play in the colony, we must go back to investigate the conditions under which they are produced. Considerations of both polymorphism (Weaver, 1966) and sexual dimorphism (Rothenbuhler, 1967) are involved. The question of polymorphism—here a matter of accounting for the appearance of young queens (rather than workers) in sexual broods—we considered in Chapters 6 and 8 and will pursue further later in this chapter.

The principle of haplodiploidy may apply hypothetically to the army ants in the simpler form of Dzierzon's rule (Chapter 6) to the effect that females develop from fertilized eggs (i.e., presumably with doubled sets of chromosomes), but males develop from unfertilized eggs (i.e., presumably containing single sets of chromosomes). The production of males through the female's laying unfertilized eggs seems to characterize the Hymenoptera (Whiting, 1945; Flanders, 1965), including doryline ants (Schneirla and Brown, 1952).

As mentioned earlier in this chapter, the sexual broods of group A army ants, nearly all males with a very few precocious queens, arise through specific (dry) seasonal conditions. Our findings (Schneirla and Brown, 1952) indicate that in these broods: (1) the number of viable eggs (i.e., capable of developing) is relatively small; (2) only a few of the eggs are fertilized, the rest unfertilized; and (3) the survivors, because of their relatively small numbers, receive food, social stimulation, and (doubtless also) pheromones in amounts and forms never effective in all-worker broods. The young queens, for example, although presumably produced from fertilized eggs genetically equivalent to those in all-worker broods, have trophogenic and other advantages that permit them, within a shorter developmental time, greatly to exceed the developmental threshold of workers (Fig. 9.3). From our results, these conditions come about for group A colonies when a period of dry weather coincides with the interval of physogastry and egg laying in the colony queen.

As a result, the queen lays a batch of eggs which in outcome differs in several respects from any all-worker series. Although at first sight the effects of desiccation seemed to offer a reasonable hypothesis (Schneirla and Brown, 1950), the mechanism may be much more complex. It is well known that the metabolic processes of insects are delicately responsive to changes both in environmental humidity and temperature

(Uvarov, 1931; Ludwig, 1945) and that the physiology of ovarian tissues is affected by even slight changes in individual metabolism (Ezikov, 1922; Wigglesworth, 1965). Our observations reveal that when doryline queens are physogastric, they are acutely sensitive to drying and to sudden increases in light intensity and in temperature, which can even be lethal (Schneirla, 1944a). Food, grooming by workers, and pheromones may also be implicated. It is quite possible that in the surface-adjusted army ants, which presumably evolved under tropical conditions, the effects of a combination of such factors rather than any one, introduced at the propitious time in the queen's reproductive cycle, may lead to a sexual brood.

To suggest how the queen is thereby induced to change her egg series from the kind that normally produces an all-worker brood, let us review the one case in which a queen (of *Eciton*) known to have produced all-worker broods was observed while laying eggs from which a sexual brood developed (Schneirla and Brown, 1950). We had studied the colony ('48 H-12, *E. hamatum*) through November and early December of 1947, during which it produced two all-worker broods and passed through two functional cycles. Early in the study we marked the queen, then contracted, so that even though the continuous study later stopped, the colony could be reidentified when found again.

On December 21, or about ten days after dry weather had set in, the colony was rediscovered in a statary bivouac, clustered near the ceiling of a large open log-shell in a hillside clearing exposed to the south (see footnote 4). The ground nearby was cracked, and the colony at this site must have been reached daily for long intervals by dry winds that swept uphill across the clearing.

We found the queen on December 22, physogastric and still laying eggs when she was removed from the cluster of minor workers that enclosed her. The graduated development of the brood mass below her, which contained eggs, embryos, and early larvae, indicated that egg laying had been proceeding for several days. This brood mass (which was both covered and permeated by tiny workers) was not noticeably different in size from that usual for potential all-worker broods of the species. In the interval of three days to December 24, in all nearly 1400 specimens were taken from it, or a little more than the total number common for sexual broods in *Eciton hamatum*.

Most surprising of all, when the brood mass was broken open again on the third day of this interval, we noted the presence of a great number of "curious egglike objects" within. They were especially numerous in its outer layer, just under the outer sheath of minor workers, and also along the walls of several of the galleries that these workers had evidently created through the interior. Examined at the laboratory, these proved to be empty chorions, i.e., the shells of eggs from which the contents had probably been removed by workers. Cytological studies of eggs from this brood, moreover, indicate that many of them would not be capable of developing.[12] This study continued during the eleven-day nomadic phase in which the sexual brood passed through the larval stage (Fig. 9.1) and through subsequent events of colony division[13] to be discussed in Chapter 10.

From these results and others (Schneirla and Brown, 1952) I offer the following hypotheses for sexual-brood production. First, because the army ant queen reaches an especially high metabolic level when she is physogastric (Schneirla, 1944a), such abrupt environmental changes as those described can then radically affect her reproductive processes and can readily block the fertilization of eggs that are matured and laid at the time. It is this inhibition of fertilization[14] that sets the pattern of the unique doryline sexual brood. Its physiological basis, though unknown, may resemble in certain respects that of male production studied in other insects (Schneirla and Brown, 1952; Wigglesworth, 1965; Rothenbuhler, 1967).

Second, I suggest that the eggs are laid in great quantities approximating those of a potential all-worker series but under the conditions are greatly reduced in number and otherwise changed. This series, I propose, consists of three types of eggs which are, in order of their laying: (1) fertilized eggs, laid in one short initial burst; (2) unfertilized but viable eggs, laid in a longer but time-limited series; and (3) unfertilized, nonviable eggs, laid in a long terminal series—which may exceptionally include short series of viable eggs.[15]

Because the young queens in these sexual broods are closely similar to one another, are very few, and all precede the males in development, we can assume that series (1) of fertilized eggs ends soon after laying

---

[12] Dr. Roy Whelden (pers. comm.), from cytological studies of intact (Bouin-fixed) eggs from this brood, stated that the contents of many of them had a scattered appearance, suggesting that these eggs would never have developed. Such eggs, reported for other ants (Brian, 1965) in which they evidently serve the colony as food, are often larger and softer than the viable eggs.

[13] Attempts to duplicate this study in the dry seasons of 1948 and 1952 on Barro Colorado Island, with numerous colonies followed from the time of egg laying to larval maturation, were unsuccessful. For such investigations, areas of relatively uniform ecological conditions are clearly more promising.

[14] Processes that control fertilization, so far as they are known for other insects (Flanders, 1946; Wigglesworth, 1965; Rothenbuhler, 1967) are complex and diversified. For the dorylines, such hypotheses as a vitiation of sperms through physiological changes (e.g., dehydration) in the queen seem preferable to assumptions of specific valvular actions blocking the passage of sperms.

[15] Thus the "dwarf" males we found in the sexual brood of colony '48 DH-3 (Fig. 9.2) may have developed from a burst of viable eggs laid among otherwise nonviable eggs in the terminal series.

begins, when the mechanism inhibiting fertilization comes into operation (see footnote 14). The young arising from these eggs all become queens because, in the "opulent" (Flanders, 1962) colony situation peculiar to these broods, the earliest individuals are maximally fed and stimulated from the beginning of development. Furthermore, the first food that these exceptional members receive as microlarvae may be a special diet,[16] consisting of series (3) eggs and perhaps others as well. These potential queens are enormously accelerated in growth over workers in ordinary broods, reaching a developmental state of exceptional size and pattern presumably because they are fed to the limit and are maximally groomed at all stages.

Because the males develop as a distinct, graduated group with a normal frequency pattern behind the queens, I assume that egg series (2) follows (1) and is fairly short. The graduated population of closely similar individuals and the normal frequency pattern common in this brood (Fig. 9.2) indicate that in developing, its members compete for food and stimulation in a substantially chance situation, i.e., one that allots them equal and freely varying developmental resources (Schneirla, 1948). Heavy feeding (i.e., "overfeeding") must inevitably take place among these potential queens and males simply because of the much smaller numbers competing for food in a sexual brood than in an all-worker brood. Heavier feeding is also assured by the superior attractiveness of male and female larvae to the workers, owing, doubtless to a great extent, to their unique secretory properties. Any reduction of the developing females and males by worker cannibalism must operate within strict limits because group A sexual broods consistently are found to have species-typical numbers. For example, male populations average around 1500 in *Eciton hamatum* and around 3000 in *E. burchelli*.

Part (3) of this series may well be eggs of the nonviable kind known as "alimentary eggs," which investigators have found to be a common part of the diet in the brood rearing of *Atta* (Autuori, 1956), *Leptothorax* (Le Masne, 1953), and numerous other ants. In group A army ants such eggs, presumably soft and readily squeezed open by workers, may serve as a specialized food of the queen larvae and of the males also. Overfeeding of the potential queens and male larvae, as in honeybees (Haydak, 1943; Butler, 1962), contributes to their growth from the start as even on the first nomadic day the sexual larvae have reached a body size equaling that of the largest worker larvae of the species at maturity (Fig. 9.3). Observations in field and laboratory indicate that all sexual larvae are continually groomed and plied with food and undoubtedly

---

[16] A resemblance to the well-known case of royal jelly in the rearing of potential queen larvae in honeybees (Haydak, 1943; Grout, 1949; Butler, 1962) is thereby indicated.

gorge themselves throughout the stage. Because with sexual broods raiding and booty return reach a higher level at corresponding times in the nomadic phase than with all-worker broods, the less than 2% as many individuals in a sexual brood may consume more food than does an *entire* all-worker brood. Among about 1500 males in a sexual brood of *Eciton hamatum,* for example, each may receive in all more than fifty times as much food as any worker larva within an estimated population of 80,000. In these two types of brood, therefore, development differs greatly in its rate and pattern, and their effects on colony function and behavior also differ in important ways.

When a sexual brood is present, group A dorylines not only increase their raiding to a maximum but also substantially reduce the frequency of their emigrations. Colonies of *Eciton burchelli* may not emigrate on as many as half of the nights, and colonies of *E. hamatum* also often miss emigrations—as they seldom do when all-worker larvae are present (Schneirla and Brown, 1950, 1952). Much the same holds for colonies of *Neivamyrmex nigrescens* (Schneirla, 1961) and of *Aenictus laeviceps* and *Ae. gracilis* with sexual broods (Schneirla and Reyes, 1966). The deterring effect, whatever it is, seems to be extreme in the driver ants as Raignier (1959) found that colonies of *Dorylus (Anomma) wilverthi* do not emigrate at all when males are present.

The main reason for the inhibition of emigration by a sexual brood cannot be the difficulty of carrying the large male larvae, as Brauns (1901) suggested for *Dorylus.* In any case, the workers of all group A species studied can transport these larvae even when they are mature and in their bulkiest state (Fig. 9.4*a*). The answer, as nest observations suggest, may stem from what we call a "stimulative fixation" of workers on the highly attractive sexual larvae.[17] At all stages the actions, secretions, and odors of the sexual larvae are clearly much more potent as attractants to workers than the equivalent in all-worker larvae. We observe in our laboratory nests that the workers cluster more closely and in greater numbers around the male larvae, spend much more time in grooming and shifting them about, and feed them more frequently. Getting samples of a sexual brood from the colony bivouac is usually far more difficult than with an all-worker brood, particularly in advanced stages when the workers hold the sexual larvae tenaciously. As a consequence, after raiding falls off at dusk, the workers are fixed much more on the sexual larvae and so at times are less apt to carry the brood from the bivouac than with an all-worker brood.

Once an emigration begins, however, the complexity and high pitch of

---

[17] Comparably, Raignier (1959) has described how male larvae of *Dorylus (Anomma) wilverthi* larger than about 10 mm in length make "violent peristaltic movements" with the anterior end that greatly activate the workers.

(a)

(b)

FIGURE 9.4.
Emigration columns of *Eciton burchelli* transporting sexual **larvae**. (*a*) Male larvae (*left*) being carried ahead of the queen's retinue. The queen, stopped by the disturbing effect of a photo-flood light, is somewhere under the mass of workers at the right. (*b*) Emigration route at a time of column stoppage (*upper right*). Several major workers have formed a "rosette" disturbance group over a queen larva; three submajor workers carry a male larva in tandem (*at the left*).

activity in transporting the sexual larvae emphasize how strongly this brood energizes the colony. Even before the brood has grown very large, little groups of workers run beside and after individual larvae carried in the emigration. The raids, large from the first day, increase steadily in scope and vigor. This excitation reaches a peak, illustrated in the accelerating magnitude of "ant roadways" and other mass behavior associated with raiding and emigrating (Chapter 12).

Another and highly significant difference between the effects of these broods concerns the far more effective competition exerted by a sexual brood with the colony queen (Schneirla, 1956a). In the course of time, as workers become more and more strongly attracted to the sexual brood, great numbers of them form close affiliations with this brood rather than with the parent queen. As Chapter 10 will show, the more the sexual larvae develop, the more this attachment spreads among the workers, setting up a nascent conflict of subpopulations that provides a basis for an eventual overt division of the colony. By the time the sexual larvae are mature, this exceptional process has advanced so far that if the adult queen is then removed from the colony for even a few hours, she is unlikely to be reaccepted. Her treatment under these conditions contrasts radically with the prompt and lively reacceptance of the parent queen— even after she has been kept away for a day or two—when an all-worker brood is present.

In these changes, indicated within the worker population as the sexual larvae grow, the young queens play an increasingly prominent role. As larval maturation approaches (if not before), the queen larvae are in fact established individually around the base of the bivouac, each within a little cluster of satellite workers. Significantly, the adjacent groups are separated by appreciable distances, which observations suggest are adjusted through disturbances and even conflicts that arise when any one group moves too close to another. In the last emigrations of the nomadic phase, as observed in *Eciton,* the queen larvae are as a rule transferred near the end of the movement, each carried by submajors running in tandem amid a small crowd of workers. It is not in the movement itself, when these queen groups often pass one another without friction, but in resettling at the new bivouac that combat is often observed among the worker groups as they adjust their positions. Such conflicts, governed in frequency by bivouac dimensions, may at times lead to fatalities among potential queens, hence, may be considered an important aspect of queen competition.

The sexual broods of group A army ants, from our results, are discrete populations, each of which appears in a relatively precise, species-typical coordination with the all-worker broods that precede and follow it. The described characteristics of population size and bisexual patterning indicate that regular and stable processes of colony function underlie the production of these broods. These processes may be rather different in the driver ants, whose sexual broods, as Raignier (1959) found for *Dorylus* (*Anomma*) *wilverthi,* seem to appear among colonies in any area, at any time of the year, and perhaps more than once a year for any one colony (see footnote 8).

Basic differences may exist between the driver ants and the group A dorylines in the biological basis of the sexual broods. There seem to be greater variations in the population range of these broods in the driver ants than in group A dorylines, as well as in their relation to colony biology and behavior. In the production and timing of these broods and in their pattern of relationships to colony function, *Aenictus* resembles *Eciton* and *Neivamyrmex* far more closely than it resembles those driver ants that have been studied. The resemblance of *Aenictus* to *Eciton* and *Neivamyrmex,* as we will find in Chapter 10, extends to the role of the sexual broods in colony division.

Colonies of all three group A genera are alike in the shorter durations of their nomadic phases with sexual broods than with all-worker broods present. They are substantially alike in that they also initiate and maintain these phases through brood stimulation as, in the nomadic phase, populations of sexual brood pass through the larval stage in all three. They differ, however, in the relationship of the mature pupal males to the last part of the statary phase, much as when all-worker broods are present. The relevance of these differences to patterns of colony division will be discussed in Chapter 10.

Because of the bearing of sexual-brood production on colony division, the apparently equivalent seasonal conditioning of these broods in *Eciton, Neivamyrmex,* and *Aenictus* should not be underestimated. Two clearly adaptive aspects of this relationship are discussed in Chapter 14. This type of environmental adaptation is not indicated in the driver ants. We find as described in Chapter 11 that colonies of these three group A army ants differ significantly in the ways their mature males leave the colonies.

Our results point to processes of competition among potential queens that operate during brood development (Pardi, 1951).[18] For the last stages, the hypothesis of contesting groups of queen-affiliated workers in

---

[18] Similar conditions have been observed in other social insects. In the common nests often started by overwintering females of the wasp *Polistes gallicus* in the spring (Pardi, 1948, 1951), one individual becomes "dominant" over the others and exhibits typical queen behavior; the others, meanwhile, behave more and more like workers, and their ovaries undergo a reduction in size. If the ascendant female is removed from the nest, interindividual relationships change; then one of the subordinate females can become "dominant" and the functional queen of the series. Comparable changes have been observed in certain bumblebees when no functional queen is present (Free, 1955).

limited bivouac space suggests how the number of mature queens is reduced to a species-typical and always low number prior to colony division. Then in the processes of colony division for which the production of a sexual brood provides the prerequisites, the young queens enter a last stage of competition within the colony, and the males leave the colony.

# 10

# Colony Division

Most ants form new colonies through what is known as the "claustral pattern" (Wheeler, 1933). The young queens, impregnated in a mating flight from the parent colony, lose their wings on reaching the ground and seal themselves in isolated chambers, where each starts a new colony. Such a pattern is reminiscent of the female solitary wasp which lays her eggs and may even feed the larvae in a cell she herself dug after having mated (Roubaud, 1910; Wheeler, 1928).

One of the exceptions to this pattern among ants is the doryline mode: the fission of parent colonies. It was a favorite hypothesis of W. M. Wheeler (1923, 1928) that army ants must produce new colonies by splitting old ones in a manner analogous to bees in swarming. But although he himself once had such a case nearing its climax (Wheeler, 1921) before his eyes, he neither elaborated on this hypothesis nor supported it with evidence.

As shown in Chapter 9, in two lengthy dry-season studies in Panama (Schneirla, 1949; Schneirla and Brown, 1950), we found that exceptional processes introduced by a sexual brood provide the basis for new colonies. By raising broods combining males and young queens, the colonies introduce the functional and behavioral prerequisites for colony division. Raignier's findings (1959) indicate a similar basis for colony division in driver ants.

For group A, results discussed in Chapter 9 indicate that although the queen lays her eggs at her usual time in the statary phase, this is a special batch through which unusual processes arise. The first-laid eggs are probably fertilized and through a specialized diet and overfeeding produce a few young queens; the next-laid eggs are presumably unfertilized and through overfeeding of the larvae produce males; the last-laid eggs in the series may be nonviable and may furnish the specialized food that starts the survivors on their course of development. This brood is unique in that from early in its larval stage it strongly attracts the workers, drawing them to affiliate with it, and in effect competing with the parent queen (Schneirla, 1948). As the larval stage of the sexual brood advances, the

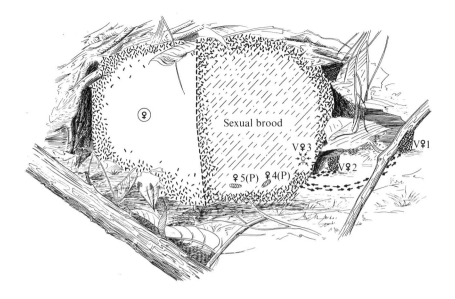

FIGURE 10.1.
Schema of conditions in the bivouac of a colony of *Eciton hamatum* with a mature sexual brood just before overt colony division. *Right:* sexual-brood pole of colony with current positions of five virgin (V) queens indicated from right to left in the empirically determined order of their maturity, i.e., queens 1, 2, and 3 have emerged whereas 4 and 5 are still pupae (P) in cocoons. Position of the functional queen is indicated on the left (brood-free) side of the bivouac.

fission that this brood introduces into the colony increases although it remains latent until the brood is mature.

From observations on how the larval sexual broods are placed in the bivouacs of *Eciton hamatum* and *E. burchelli* (Schneirla and Brown, 1950, 1952), we found that these broods are gradually moved to a part of the colony bivouac away from and opposite the parent queen. This placement, moreover, through the activities of brood-affiliated workers, becomes more definite with each new resettling of the bivouac after an emigration. A review of some study cases will illustrate the point.

To test the concept of an increasing polarized placement of the sexual brood, in the Darien project (Schneirla and Brown, 1950) we examined in detail all colonies of *Eciton* with such broods. Even early in the larval stage the young were already massed in a cluster of strands near one side of center in the bivouac. The parent queen and her worker cluster in turn were found to the other side distinctly separated from this brood (e.g., colony '48 DH-1 and Fig. 10.1). Significantly, the queen with her cluster is not over the brood as she is in colonies with all-worker larvae.

An unusually good opportunity to test sexual-brood versus parent-queen *distance* at larval maturity came with the finding of colony '48 DH-3 (Schneirla and Brown, 1950, p. 315). At 10:15 A.M. on February 21 I located this colony through its great raid. The bivouac was a large half-cylinder massed well beneath the overhang of a giant log, all within a low grotto formed by a thick mass of vines. At first the area was cool; I cleared away the vines, however, and increased the temperature by using sheets of tinfoil to reflect the sunlight against the massed ants. With the colony thereby lethargized, I used large forceps and a trowel to take the cluster apart strand by strand, working alternately at the two ends, separating workers and brood in large jars. The operation, which lasted from 11:00 A.M. to 4:00 P.M., was successful, especially because the workers held so tightly to the large male larvae that they could be pulled out with them in strands.

The parent queen, then slightly physogastric, was posted within a bolus of workers near the extreme left pole of the bivouac in the part of the bivouac area (about one-third) that was brood-free. The mature larval brood, most of it suspended in strands of workers, was massed to the right of the queen's area. Around the base of the bivouac in the brood section in both front and rear, small clusters of workers were located at intervals, each containing a larva engaged in spinning. Most of the twenty larvae which had already spun envelopes were at the base; fourteen of these turned out to be queen larvae; otherwise there were 1325 male larvae still unenclosed.

This colony gave us an overall picture: The sexual brood is sequestered laterally from the parent queen in the bivouac, with the young queens mainly in separate worker groups spaced around the bivouac base. It clarified an earlier case of ours (Schneirla and Brown, 1950, pp. 293ff.) on Barro Colorado Island.

Our study of another colony, '48 H-19, dated from December 28, 1947, when it had a brood of nearly mature all-worker larvae through the period of a sexual brood. We took the queen (found in a cluster above the brood in the bivouac) to the laboratory for marking and returned her to the colony the next morning. She was received in a "normal manner," without any signs of worker disturbance. The back-tracking column, then under way, stopped within an hour.

During a statary phase that began on December 31, with the colony bivouacked inaccessibly in a hollow log, this queen laid the eggs of what proved to be a sexual brood. The colony began a nomadic phase on January 20. Next day we found the queen contracted and in good condition; she was reaccepted promptly by the workers when we returned her after having kept her out of the colony for three hours. On January 29, after nine nomadic days with fast-growing sexual larvae and great raids,

*10 / Colony Division* 221

the colony was bivouacked beneath a large log-shell, and its raid was small. Around the rear base of the bivouac we saw about ten small clusters of workers, all connected by wide, short columns to the bivouac. Next afternoon, January 30, we probed into the bivouac to capture the parent queen and found her (partially physogastric) in a brood-free area near the top, on the end opposite the spinning clusters.

We kept the queen in the laboratory overnight; by the next morning, however, the colony had shifted into a new position in the hollow log so that we could not return her as intended to the exact place where she was removed. Also, vigorous back-tracking was then under way, with the ants carrying away their bulky male larvae—many already partially enclosed—in a wide column. The case now being abnormal, we held the queen until early afternoon of the next day, February 1, when we found the entire colony on the march with its large brood of sexual larvae now all enclosed. Early in the afternoon we set down the queen (with about twenty workers from her cluster) close to the column a few meters from its end, which had then pulled away from the bivouac log. Seven hours later, that evening, she was at the same spot, in good condition and active; "however, each time she shifts, some of the workers grasp at her legs and body, holding her in place. Two dozen or more dead and maimed workers lie around her." Under the conditions, we put her back in the same air vial for another trial elsewhere. At 11:00 P.M., three hours later, we set her down near the head of the column, in advance of the cocoon-carrying section. The result was the same: At dawn the queen, still active and in good condition, was in the midst of a small group of workers holding her firmly in place, close to where she had been released.

The results showed clearly that from the time the queen was readily reaccepted on January 21, great numbers of the workers in her colony had changed their reactions to her. Now, ten days later, they responded to her not by killing her—as they would a foreign colony queen—but with the disturbed, tense reactions of a group-restraint process we term "sealing off." [1]

With these results, we were alert to find another colony with sexual brood to study without artificial intervention as a normal case. Colony '48 H-27 (Schneirla, 1956a), which we discovered in a statary bivouac within a hollow log, emigrated on the night of February 10 with a brood of just-emerged callow workers and, as we found in their new bivouac the next morning, a brood of young sexual larvae. The queen, removed for marking at the laboratory, was readily reaccepted by the workers when

---

[1] The behavior of workers in a sealing-off group differs from that in a cluster formed normally about a parent queen in that it involves a noticeably greater tension which increases the tightness of the enclosure and hems the queen closely in.

FIGURE 10.2.
Bivouac of a colony of *Eciton burchelli* ('66 B-I), showing the sexual-brood pole at the left where cocoons are being opened.

we returned her three hours later. This was the last and only time we interfered with her during the study.

During the next week, raids were all large and the larvae grew rapidly (Fig. 9.4). The colony showed great excitement, forming sections of ant roadway in the emigrations, which increased in magnitude as the larvae grew. When taking a brood sample on February 17, the last day of the short nomadic phase, we found the mature sexual larvae massed on one side of the bivouac in an arrangement similar to that described for colony '48 DH-3 (Fig. 10.2).

That evening the colony settled in the hollow base of an old, partly buried log, which became its statary bivouac site. During the evening and on the next day, there was much milling around of workers within the cavity. During the night, similar activities continued behind the bivouac deeper inside the log: Wide columns tugged unenclosed larvae to heaps of wood detritus and returned partially enclosed larvae to the bivouac. This evidently ended on the night of February 20.

After the fourth statary day (February 21) with the larvae enclosed, extrabivouac activities fell sharply with small raids the rule and on some days no raids at all. Within the bivouac, however, interesting events were going on, visible both through a narrow chink in the wall of the log and at

FIGURE 10.3.
Queen cluster of the type formed about queens (including "sealed-off" young queens) in colony division.

angles through the opening at the end.[2] The sexual brood, as we discovered in taking a daily sample of two cocoons, was stationed on the side of the bivouac near the center of the log, the parent queen evidently on the brood-free side next to the wall of the log. Signs of intrabivouac friction among the workers increased, particularly at the boundary of the brood zone and the brood-free zone. Here, after February 23, gentle probing with long tweezers often brought forth empty pupa cases. Doubtless, enclosed pupae in the area of apparent tension were exposed to cannibalism by workers as we found only intact cocoons elsewhere.[3]

Colony excitement increased notably after March 5, as marked by the raids, from which columns of laden workers returned later and later in the night. Meanwhile, we noted centers of apparent disturbance among the workers in places near the base of the bivouac. From one of these places, on the night of March 5, there erupted a short column of workers and, after a local turmoil, a pale golden-yellow queen (callow No. 1) appeared there. She pushed her way out in a large worker group, moving steadily in the column, stopping at length in a little hollow under a chip at 25 cm from the bivouac where a cluster enclosed her at once (Fig. 10.3). Next morning (March 6) this cluster was in the same place; it had now grown

---

[2] The observer used a small flashlight with red bulb, introduced into the log at the end of a rod, for detailed examination of the bivouac, excluding outside light with a photographer's hood.

[3] A previous study (Schneirla and Brown, 1950, colony '48 H-12), which involved probing the interior of the statary bivouac each day to take brood samples, brought clear signs (e.g., empty cocoons) of brood cannibalism by workers at the border of the brood-free zone.

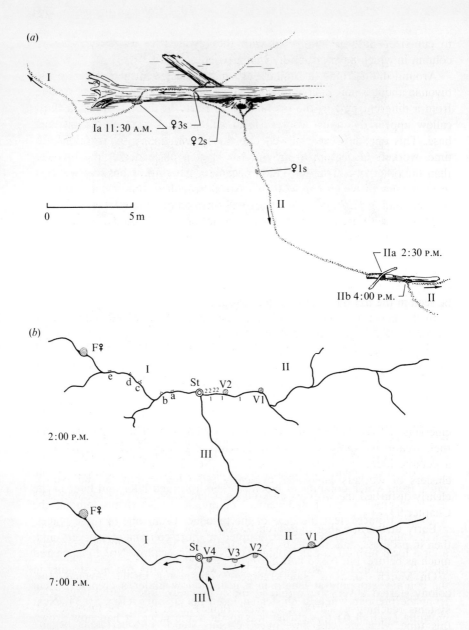

FIGURE 10.4.
Sketch of principal raiding systems and queen movements of a colony of *Eciton hamatum* on the day of actual division. (*a*) After five hours of raiding on three principal trail systems, the functional queen is at Ia, and three young queens are at 1s, 2s, and 3s. Later, the functional queen moves out on trail I, and one of the young queens (probably 1s) becomes dominant on line II, at IIa and IIb. (*b*) *Top:* situation of the same colony to show successive positions of parent queen (a, b, c, d, e, and finally to F, *at left*), principal stopping points of a secondary young queen (V2), and the position of a cluster formed around the leading young queen (V1,

to cup-size and was connected with the bivouac by a steady two-way column in which pieces of booty were carried outward.

Around midmorning, a similar but smaller cluster formed closer to the bivouac about a new callow queen (No. 2) that emerged with her retinue from a different part of the bivouac base. Early in the afternoon, a third callow appeared among workers from still another part of the bivouac base. This one, however, moved out slowly with many interruptions. In time, workers also grouped around No. 3 at a place nearer the bivouac than the other two. This formation became a flat ring of bustling workers in the center of which callow queen No. 3 remained almost constantly in motion (and in view), weaving back and forth among the workers.

These three queen clusters, formed near the mouth of the log hollow, stayed in place for two days (i.e., into March 8): that of queen No. 1 growing slowly and sustaining its "food-line" to the bivouac; that of No. 2 always smaller but also with a column and food; and that of No. 3 constantly a ring of active workers around her, with hardly ever any booty to be seen in the liaison column to the bivouac.

In the meantime another callow queen (No. 4) appeared, first poking out and withdrawing her head at intervals at a place in the bivouac wall where we had noted local turbulences among the workers. She stayed near this place for hours, appearing and disappearing, weaving from side to side among workers which stopped her whenever she began to make headway.

Now, by the nineteenth statary day, with still no reliable signs of male emergence, four callow queens were in circulation. These queens, from their treatment by workers (as in the case of colony '48 H-12), fell into a definite hierarchy in which No. 1 and No. 2 remained in their stable clusters of workers and were evidently fed whereas No. 3 and No. 4 clearly disturbed the workers and did not seem to be fed (see footnote 18, Chapter 9).

Early in the morning of March 9, with male emergence from cocoons then in progress, callow queens No. 1 and No. 2 were still in their clusters much as before, but No. 3 and No. 4 had disappeared.

On March 10, with male emergence continuing in the bivouac, the colony started a very busy raid at dawn that developed three large trail systems reaching at 10:00 A.M. more than 125 m from the bivouac. At this time what we call the "queen-exodus stage" of operations began. These events, which went on as the raid continued, are represented in Figure 10.4. Shortly after 10:00 A.M. the parent-queen appeared, ac-

*at right*). 1 1 1: successive principal stoppages and worker clustering points incident to the outward passage of virgin queen 1. 2 2 2 2: principal stoppages of virgin queen 2. St: statary bivouac. *Bottom:* With the colony emigration underway divergently on trails I and II toward cluster F (*left*) of the parent queen and toward V1 (*right*) of the leading young queen, positions V2, V3, and V4 mark sealing-off clusters formed about secondary young queens.

companied by a group of more than 1000 excited workers. She passed directly into trail I (Fig. 10.4a) that started from the brood-free side of the bivouac and soon was moving steadily outward on this route at the head of a 2 m long mass of workers. For a time, this event looked like a queen's passage in emigration; however, after 6 m the route was blocked by returning raiders, whereupon a large cluster of workers formed about her beside the trail. Some of the raiders seemed excited and disturbed when near her; these were shunted off, however, by her large retinue. Within thirty minutes she was again moving outward, having burrowed her way from under the cluster, which soon dissolved into a long group once more sweeping after her. In this way, by stages of progress and of reclustering (a to e, Fig. 10.4b), the parent queen disappeared and her workers moved outward on trail I.

Around 11:00 A.M., or about one hour after the parent queen had started, callow queen No. 1 began to move from her four-day-old extra-bivouac cluster, with many workers around her, into the adjacent trail II (Fig. 10.4a). At first progress was slow, both because the young queen hesitated frequently and because workers in her group had occasional conflicts—now and then verging on open combat—with raiders on the trail. A close watch on this callow queen at regular intervals assured us that she held the lead constantly on trail II and that she was the same queen present in the cluster formed around 2:00 P.M., some 15 m from the statary bivouac (Fig. 10.4b, V2). Under way again later, this queen and her group moved outward more rapidly than before and by dusk had clustered at a place (V1) about 25 m from the statary bivouac.

Also by dusk, the parent queen, who with her group had remained in the outermost position on trail I, was at point F. From the start, therefore, neither the parent queen nor callow No. 1, her daughter, had ever fallen from the lead on her respective trail.

Meanwhile, throughout that same day, we kept track of the others. Within ninety minutes after callow queen No. 1 had started on trail II, three other callow queens at intervals entered this same trail behind her. In contrast to callow queen No. 1, these three had noticeable difficulty in making their way outward on this trail. Not only were their worker groups smaller and less constant in running with them than the group of No. 1, but interference from other workers on the trail was in general stronger and in time blocked their outward progress altogether. From midafternoon, as callow queen No. 1 pushed outward by degrees, the other callow queens seemed to spend their time isolated in small clusters beside the column of raiders or running back and forth at points nearer to the statary bivouac. By late afternoon two of these callows were enclosed within tight clusters not more than 10 m from the statary bivouac; the third and fourth could not be found.

During all these events, none of the callow queens was seen on either trail I or trail III, both of which continued in use by heavy raiding columns. They might have, however, entered these trails briefly.

This state of affairs ends the queen-exodus stage of colony division, which is often more complicated in *Eciton hamatum* and is always more complicated in *E. burchelli*. Let us briefly analyze the movements of the queens in '48 H-27. Very probably the parent queen simply entered the route nearest her part of the bivouac and callow queen No. 1 also took the trail nearest her cluster. Each of these queens may thereby have entered an avenue on which large numbers if not a majority of their respective affiliated-worker groups were present as raiders. Callow queen No. 1 may also have been attracted to trail II by its odor (as may the parent queen to trail I). As for the other callow queens, if they ever entered trail I, they might well have been sealed off at once by workers. Possibly they were more readily attracted to trail II by odor; probably they kept off trail III (or were excluded from it) both on the basis of odor and worker reactions. (Frequent disturbance reactions from workers on a main raiding route may exclude all queens from it.) In the studied *Eciton*, therefore, queen-exodus operations normally are restricted to two routes from the bivouac although in more complex cases (as in colony '46 B-I) still other routes may come into use briefly.

The hypothesis of odor, as the basis for responses to trails in the queen-exodus stage is supported by events in worker and queen behavior as the leading queens move outward on their respective routes in the course of time. Usually, in *Eciton*, the parent queen is the first to appear and the first to become established in a terminal cluster. Once a queen is so established, her cluster seems to furnish a nucleus of increasing strength for worker reactions on that route, both stabilizing her in the leading position and excluding other queens from the route.

In surface-adapted *Eciton*, as a rule, the queen-exodus stage takes place in a more or less distinctive period of radial movement by the queens and their worker groups, concurrent with raiding. The final stage, that of the "overt division," is illustrated in the most concise form by postdusk events in '48 H-27 (Fig. 10.4b). In this colony, with the two leading queens in their respective clusters and the secondary queens sealed off, overt division began even before dusk, much in the manner of a normal emigration. In this exceptional situation, however, with two divergent lines of exodus established, the main part of the worker population—still in the bivouac with the males and the young brood—seemingly must split. This in fact did happen to colony '48 H-27; late in the afternoon, columns of booty-laden workers began to move divergently from the statary bivouac site toward queen-clusters F and V1, to which branch-raiding columns also returned from the periphery. Here, unlike the queen-exodus stage, there is no indication that odor governs trail responses as traffic from the bivouac seems to move into one or the other of the two columns quite by chance.

In case '48 H-27, after dusk, as mixed columns of unladen and booty-laden workers left the statary bivouac divergently on trails I and II, workers returning from the raid on trail-system III divided between them. After a time, also, workers began to emerge from the bivouac carrying the young brood in packets.[4] Presently the alate males also, as they came from the bivouac, first distinctly apart and at length in a procession, seemed to enter one or the other of the columns by chance. Their movement reached its peak near 9:45 P.M. and ended after 11:00 P.M. Further events show that both the males and the young brood (carried by workers) were divided between the daughter colonies roughly in the proportion 4:3, with the parent queen's colony drawing the larger portion of each.

Colony '48 H-27 completed its division around 2:00 A.M. on March 11, or about twenty hours after the first dawn raiding on March 10 had opened the queen-exodus stage. Thereafter, as is usual, the daughter colonies were connected by a thin bidirectional column over the intervening trails and through the deserted statary bivouac site. This column, as is usual in the simplest cases, disappeared late on the day after the division.

Returning now to the queens, we see that during the queen-exodus stage, while worker reactions to the leading queens (i.e., parent queen and leading callow) take on more and more the nature of clustering or following, reactions to the secondary queens become increasingly agitated (i.e., "disturbed") and marked in time by more prompt sealing-off operations. Thus, in colony '48 H-27, as callow queen No. 1 and her group pushed on trail II, worker sealing-off reactions to the other callows increased so that by late afternoon these queens could not pass beyond 10 m from the statary bivouac. This persisted even during the overt division, with the secondary queens blocked from advancing on *any* trail. Just once, as far as we saw, one of these queens broke loose for a brief run with the column on trail II but then quickly received a sealing-off treatment that held her in place until the emigration ended.

Normally, this process leads to the final abandonment of the secondary queens. At 2:00 A.M. on March 11, in colony '48 H-27, just as the divergent emigrations ended, we noted three sealing-off clusters along trail II close to the statary bivouac. At dawn, however, they were gone; then at 8:00 A.M. we found two of the callow queens at the statary site itself, one with a few workers still around her, the other running around alone. By

---

[4] This all-worker brood, then in stages from embryos to early larvae (Chapter 6), was laid by the parent queen in the statary phase.

that time the liaison column by-passed the statary bivouac by 2 m and had thinned greatly; we therefore considered the colony division ended and picked up these callow queens for study—rather than leave them for predators.[5]

Another study, that of colony '48 H-12 (*Eciton hamatum*), corroborates and supplements case '48 H-27. On the eighteenth statary day of this colony with a pupating sexual brood, events began that were comparable with those described for the queen-exodus stage in '48 H-27. Because of our tests, however, colony '48 H-12 became an abnormal case. In the three days from January 27 to January 29, a close study of events indicated that four young queens emerged from their cocoons in the order shown in Figure 10.1. When on the morning of January 27 callow queen No. 1 emerged from the bivouac with a worker retinue and a trailing column, we gave this group only enough time to establish a cluster over the queen about 1 m from the bivouac, then removed her permanently. That evening, an exodus in column with male cocoons and young brood, leaving from the sexual-brood side of the bivouac, settled in a cluster in a hollow log-end 14 m downhill—we missed the queen, probably the parent queen. Then another callow queen (No. 2?) pushed out with workers, and a cluster formed about her. She was captured in her turn; but within two hours still another callow queen (No. 3) emerged from the bivouac and moved out with workers. Later in the evening she ran with workers to a cluster near the spot where No. 1 and her group were captured. Emergence of males was well underway the next day, February 28, and clusters formed along raiding lines to the east and west of the bivouac. In the evening of January 29, as the emergence of males ended, the parent queen and her group emigrated outward from the eastern cluster, but ants and callow queens from the other clusters withdrew toward the statary bivouac. In the early afternoon of January 30 at the virtually abandoned statary bivouac, we found two callow queens with scarcely a dozen workers and took them up.

With these complex events (Schneirla and Brown, 1950, pp. 286–287), the colony finally did not divide, doubtless because we removed two of the three leading young queens before they and their worker groups could get into the lines of radial emigration. Had these remained, a two-way division might well have taken place with or without the supersedure of the parent queen. The three secondary queens, out later with their groups on a line opposite hers, apparently held too weak an attraction for the

---

[5] Placed in the same air vial, these queens grew excited, nipped at each other's antennae and legs, and came to grips much as had two secondary queens from colony '48 H-12 in a similar test. In eventual histological studies, Dr. Roy Whelden (pers. comm.) found that all three of the abandoned callow queens from colony '48 H-27 had empty spermary receptacles, i.e., were unfertilized.

workers[6] as they abandoned two of these callows in midday and the third on the evening of January 30. Afterwards, all the workers moved away into the emigration of the parent queen, the center of the reunited colony (Chapter 11).

From our results, it is clear that normally the first young queens to emerge from their cocoons have the strongest worker groups about them, their advantage probably dating from their larval and pupal development (Chapter 9) and based on their specific (odorous) attractiveness. The superior attractiveness of the leading callow queens is shown by how readily the workers cluster about them and follow them in the queen-exodus stage, somewhat as in the retinue of a parent queen.

Clearly, a behavioral tension exists in the bivouac at the time of the young queens' emergence, indicated by their movement out of and away from the parent cluster and by the struggles of the secondary queens with workers. Disturbance in the bivouac probably comes to a head through the movement of the callow queens, which in turn strongly affect the parent queen, causing her—in *Eciton* and *Neivamyrmex*—usually to be the first to leave in the queen-exodus stage. The leading callows, from the time they quit their cocoons, exert a clear power to recruit workers. Recruitment of an entourage summates for No. 1 during the queen-exodus stage but (as our test with colony '48 H-12 shows) does not necessarily transfer to the other callow queens when the leading callows are taken out of the running. Cases thus far discussed show that sealing-off reactions to secondary queens may be somewhat variable ("ambivalent"?) in the early stage at the bivouac but can grow in strength in the queen-exodus stage to the point of forcing them back to the statary bivouac site. There, normally, they are finally abandoned but not killed.[7]

Division of colonies in *Eciton burchelli,* although by measures equivalent to those in *E. hamatum,* is more complex and thereby offers a convenient means for discussing how a species-typical pattern may vary. From our first studies (Schneirla, 1949; Schneirla and Brown, 1950) we have had evidence that fundamentally a bipolar colony organization also arises in colonies of *E. burchelli* through processes introduced by a sexual

---

[6] We had begun by removing during the colony's last emigration on the night of January 8 the first of four enclosed queen larvae carried near the end of the procession. Thus, from a total of six recorded young queens, we may have taken away three of those most attractive to workers. We recognized the parent queen both by her much darker pigmentation than the callows and by her distinguishing mark (Chapter 8). Callows were differentiated by means of a detailed sketch recording their changes in position in and near the bivouac.

[7] Although the killing of secondary callow queens seems to be common in the driver ant *Dorylus (Anomma) wilverthi,* I have seen it in *Eciton* only in one case (colony '52 H-J), in which the colony may have been affected by heat and dryness in the bivouac area.

brood. This is illustrated by the bipolar nature of the bivouac at the stage of pupal maturity shown in Figure 10.5.

Our clearest case of a normal colony division in *Eciton burchelli* was in colony '52 B-I on Barro Colorado Island (Schneirla, 1956a). This colony was bivouacked under the roots of a tall tree. As its sexual brood neared maturity, the colony pushed out strong raiding systems to the north and south. First to leave the bivouac in the queen-exodus stage was the parent queen, who, however, advanced only a few meters on the south trail before she was shunted off it. Throughout the afternoon she stayed near this trail on the bivouac tree itself within her large cluster of older (darkly pigmented) workers. Meanwhile, as the raid went on, one of the daughter queens with retinue moved out beyond this position and subsequently established a southern daughter colony (Chapter 11). Probably during midday, a second callow queen (who became the eventual queen of a northern daughter colony) moved into the north line. The three others that moved in behind her with smaller worker groups remained behind her throughout; they were sealed off repeatedly and by evening occupied small clusters near the statary bivouac. In a striking similarity to case '48 H-27 (*E. hamatum*) and others, none of the secondary callows went far into the southern route, if indeed they entered it at all.[8] In the evening, in '52 B-I, divergent emigrations of workers carried booty and young brood, with alate males running in both columns; thus the colony divided nearly equally in two. Here, as in '48 H-27, both the queen-exodus and overt-division stages of fission were completed in the actions of raiding and of emigration, respectively, of the same day. Except for the supersedure in the latter instance, the two cases were much alike.

At the completion of the division after midnight, when only a thin two-way column connected the new north and south bivouacs, I captured the parent queen and her large cluster of dark workers from its place of isolation, now just one meter higher on the tree, thus a little farther from the southern trail than in midafternoon. She turned out to be a real beldame, very darkly pigmented, with many ventral scars, and an elongated and sagging gaster. Worker reactions to her in laboratory tests left no doubt about the behavioral basis of her exclusion from the south line.[9] Clearly,

---

[8] My hypothesis is that workers, when coming upon the leading queen on this route, affiliate with her in increasing numbers. This leads promptly to responses of disturbance to secondary queens, excluding them nearly from the start. Note that in the case of colony '52 B-I, after the early sealing off of the parent queen, just one callow queen moved out on the southern route, with all others confined to the northern route.

[9] In laboratory tests on the same night, workers of groups taken at random from the northern and southern routes reacted to this queen with leg and antenna nipping and other signs of disturbance. But workers of a sample taken from this queen's cluster on the tree gathered over her showed no such behavior.

FIGURE 10.5.
Curtain-type bivouac formed by colony '46 B-I (*Eciton burchelli*) which is in process of division, having just matured a sexual brood. This mass, which exceeds 125 cm in width, contains at least three callow queens, many males, and a young all-worker brood. Note the septum in the center of the bivouac. (Photo courtesy of American Museum of Natural History, New York.)

the supersedure of a parent queen involves a high frequency of worker disturbance reactions. In the described case, the exclusion of the parent queen from the southern route, although similar to events in the sealing off of secondary callow queens, differed from these in the great size of the superseded queen's group and in the completeness with which it had been cut off from the trails.

Thus, although the parent queen in *Eciton* often leaves the bivouac first in the queen-exodus stage, headway is by no means assured for her unless she is attractive to numbers of workers on the raiding line she travels and, conversely, does not disturb too many of them. These conditions also were poorly met by the parent queen of colony '52 H-J (*E. hamatum*), the first to leave her bivouac in this stage (Schneirla, 1956a). She was so promptly and forcibly blocked off by disturbed workers in the raiding column that with her group she turned back into the bivouac log after only a few meters of slow outward progress and was seen no more (Chapter 11).

Supersedure of the parent queen opens the way to young queens, which, from our results (Schneirla and Brown, 1952), seem always present in the sexual broods of *Eciton*. The replacement may occur early, as in the case of colony '52 B-I, or may require hours, as happened in colony '55 B-IV. Here the parent queen appeared on the northern trail in the late morning but was shortly sealed off at the bivouac edge and disappeared from sight. Although at an early time that evening a callow queen with retinue worked her way out on the southern route to a stable clustering place, even by next morning there were no signs of further queen movement on the northern route, despite its continued use during the day in raiding. Then, late in the morning, a callow queen began to appear every few minutes near the exit of this trail at the bivouac base. She disappeared through midday but reappeared early in the afternoon, struggling as before within a group of workers plainly concentrated on her. This went on until after 5:00 P.M. when she burst from the bivouac; within thirty minutes she burrowed her way from one sealing-off cluster after another. Then, just before dusk, she moved out (now virtually unopposed) on the northern trail, soon heading a throng of workers that grew rapidly in numbers and excitement.[10] Once under way, this callow queen and her group moved steadily along to the place where she disappeared in a

---

[10] From this highly excited throng, which soon grew to about 16 cm wide and over 2 m long, arose a "sweet, musky scent" that I could easily smell while standing over the column. This odor, whatever its source among the ants (and queen), became perceptible only after this queen "broke loose."

rapidly formed cluster, the first base of her daughter colony ('55 B-IV-N, Chapter 11). Comparably, in colony '52 H-J (Chapter 11) a callow queen, after hours of struggling at the bivouac, gained ascendancy at dusk on the route from which the parent queen had been excluded.

We have studied other cases in which the parent queen must have been superseded but, possibly through insufficiencies in the callow queens, no division occurred. When the sexual brood of colony '48 B-XVII (Schneirla and Brown, 1950) was in the mature larval stage, we removed two cocoons containing queens from the lower margin of the bivouac and kept them out for study (Chapter 9). When most of the males emerged as callows twenty-five days later—abnormally slow—the colony carried out a single emigration in the evening, moving off undivided (with a young all-worker brood). In this operation and during the next five days the colony behaved normally; hence, a queen must have been present although we failed to see her. In this case, as in colony '48 H-12 (*Eciton hamatum*) and in colony '59 N-X (*Neivamyrmex nigrescens*), to be discussed below, by removing two of the young queens, we may have blocked a colony division although (in the latter case, if not both) others remained.

The intricacy of colony-division processes in *Eciton* is shown by the effects that artificial changes may produce under certain conditions. For example, when the queen-exodus stage of colony '52 B-V had evidently ended with two leading callow queens established in clusters, one on the eastern trail and the other on the western trail of raiding, I intervened with a test that reversed the latter completely. By removing from beside the base of the western trail a secondary young queen with a cluster of workers that had confined her for more than thirty minutes, I caused traffic—which at this time just before dusk had been smoothly moving outward on this trail—to reverse toward the previous statary bivouac high in the crown of the bivouac tree. This reversal extended outward until it reached the site of the west queen's cluster, about 13 m from the tree; it continued through the evening and into the night until not only raiders in outlying zones but all of the occupants of the (initially settled) west-queen cluster had moved toward and up the tree. The complex outcome is discussed in Chapter 11. This part of the results shows, however, that secondary queens may continue to play a necessary part in the colony division—a stabilizing effect, as results with both *Eciton* species indicate—even when they seem to be firmly sealed off in the rear and definitely out of the running.

Deviations may occur through numerous other causes and often through natural conditions. Thus, on the day of its completed male emergence and queen exodus, colony '52 H-J had only *one* strong raiding trail from the bivouac and seemed below the normal level of activity presumably because for days the area of its statary bivouac had been very dry and very hot in the daytime. The following unusual pattern of events occurred. First, in the morning, as described, the parent queen had been forced back into the bivouac after a short exodus, and thereafter, during most of the day, neither she nor any other queens emerged (on this, the one main trail). The presence of callow queens was doubtful as the day before I had seen workers haul two of them from their clusters close to the bivouac and kill them, the only two of this series then on record. Late in the afternoon, however, a third young queen appeared at the bivouac surface, weaving about within a group of workers. For more than three hours she put up a strong, continuous struggle against apparent restraint until finally, early in the evening, shortly after an emigration began, she burst into the column with a long queue of workers forming after her. As in the case of the callow queen '55 B-IV-N, although on a much smaller scale, this queen with entourage made a virtually uninterrupted run to a hollow log at 12 m where a cluster formed about her. She thus became the queen of the resulting undivided colony. Nonoptimal ecological conditions may have been responsible for the failure of this division since in addition to the (unusual) killing of queens, the population of alate males counted in the emigration was less than 800—exceptionally small for the species.

A comparable situation may have held for colony '52 B-IV (*Eciton burchelli*), a moderately large colony studied in the statary phase with sexual brood bivouacked within a hollow tree. From here, on the main day of male emergence, just one raiding trail developed on which, during the day, two callow queens managed to enter the large cluster formed beneath a log.[11] In the emigration of that night, one of these two callow queens was abandoned at the starting point; the other (with an entourage) moved to the new bivouac. In this case also a substandard raid was combined with a male brood (about 1000) undersized for the species. In the case of colony '66 B-I, a failure to divide was more difficult to understand as this colony, although somewhat undersized for its species, produced a normal-sized male brood (i.e., around 3000) and at least two callow queens. During the queen-exodus stage, which occurred mainly out of view under the broad log within which the colony had its statary bivouac, the parent queen was superseded. On the day of completed male emergence (Fig. 10.2) there was only one basal raiding trail, over which the colony emigrated that evening with just one callow queen. In this case,

---

[11] The parent queen of this colony was not seen at all and may have been superseded in events confined to the bivouac log.

the immediate cause of the failure to divide seemed to be the availability of just one avenue from the bivouac; why this happened, however, was not as evident as in colony '52 H-J.

One of the most complex cases of division on record was the exceptionally large colony, '46 B-I (*Eciton burchelli*), from which I reported the first description of a doryline colony division (Schneirla, 1949). This colony, on the night of March 28, its nineteenth statary day, began to move from its hollow tree statary site to form a great cluster beneath a nearby log (Fig. 10.5). In the evening, after very large raids that day, ants with alate males and young brood emerged from the right side of this curtain bivouac and emigrated 5 m to the northeast. There they formed a cylinder under a log; at the same time, ants from the left side of the large bivouac moved off similarly on the western raiding trail to cluster beneath a tree. On March 30, the twenty-first statary day (S-21), there were heavy raids to the northeast, west, and southwest, and in the evening the complex movement resumed: The section on the northeastern trail emigrated 25 m farther out to form its bivouac, north-2; that on the western trail continued to move into its apparent clustering place beneath the tree at 9 m; over the southwestern trail a third section emigrated to a clustering place at 6 m (south-1). Cocoons and young brood were carried outward and alate males ran in all three columns. Continuing during the night, these divergent movements emptied both the hollow (statary) tree and the curtain cluster nearby; meanwhile, a withdrawal from the western cluster had begun which was completed the next day. From dawn on that day, raids developed from bivouac north-2 to the north and from bivouac south-1 to the southeast, and in the evening the ants of these sections carried out massive divergent emigrations, forming the new bivouacs north-3 and south-2, respectively. The next day, after large raids, the sections moved to new bivouacs north-4 and south-3, now 125 m apart, and were connected in the evening by a rapidly thinning two-way column. A two-way division of the colony clearly had been completed although the northern daughter colony (which contained a callow queen) was only about one-third as large as the southern daughter colony (which contained the parent queen). The sequel is described in Chapter 11.

Colony '48 H-27, representing the simplest and the most clear-cut type of division in *Eciton,* with clearly differentiated queen-movement, queen-exodus, and overt-division stages, contrasts sharply with the case of colony '46 B-I, one of the most complex. In the latter, which took at least three days to run its course, operations in columns and clusters were so expansive that queen movements were hidden altogether. It would appear that the abortive exodus to the west centered on a leading callow that was sealed off and abandoned during the night of March 30. In this stage, which combined operations of queen exodus with divergent emigrations, the northern movement developed around a second callow queen (identified in bivouac north-4 on April 2). No doubt there were other callow queens beside these. The southern movement, which started from the statary site after the other two movements were under way, centered on the parent queen, as it proved.

*Neivamyrmex nigrescens,* because it nests underground and its surface activities are nocturnal, presents division processes much less accessible to study than those of surface-adapted *Eciton.* As a result, to investigate one normal case of colony division in these ants, I had to devote most of my time in the summer of 1959 to a single colony. This colony, '59 N-III, we found on July 11, just beginning a nomadic phase, with a sexual brood in the early larval stage. Nightly raids were heavy, booty plentiful, and the brood developed rapidly. After July 19, with the colony settled, raids were small or absent for some time (Schneirla, 1961).

A necessary adjunct of the study was the cordon or inspection lane, (Fig. 1.4) around the area of the underground bivouac site that is sketched in Figure 10.6. At first, in nightly patrols, little surface action was seen. Then, starting with a vigorous raid on the night of July 21, small raids issued on some of the nights thereafter to the west from holes $H_1$ and $H_3$ and to the east and south from beneath rocks $R_1$ and $R_2$. Even on nights without raiding, workers bustled around $R_2$, which was clearly a "hot spot." I rehearsed a way of overturning this rock and returning it to exactly the same place[12]; and so during raids on the nights of July 23 and 24, I saw "clusters of thousands of excited workers" in the bowl-shaped area underneath. On the night of August 5, a strong raid developed to the west from this spot, $R_2$, and on lifting the rock, I saw a great change. At the left side tight masses of workers were holding hundreds of large cocoons, which, from a sample, contained lightly pigmented male pupae. Other workers there were clustered around two or three thimble-size boluses of eggs. At the right side, in a bare space, a light orange-red callow queen circled in a definite path with workers after her. When I overturned the rock thirty minutes later for another look, nothing had changed. At 4:30 A.M., surface traffic carried booty from $R_2$ to entrances beneath rocks 4 m to the south. These events, which took place at the same time that separate night raids were going on to the east from $R_2$ and to the west from $H_1$ and $H_2$, indicated that the colony was sectioning underground. After August 6, an outward carrying of booty, which occurred frequently during nightly raids (both to the east from $R_2$ and to the west from $H_1$ and $H_2$), supported this idea. In my notebook, I labeled the area of $R_2$, "sexual-brood section?"

---

[12] Unless such a rock is reseated exactly in its original concavity after having been lifted, the microclimate beneath it is so impaired that the ants may discontinue its use.

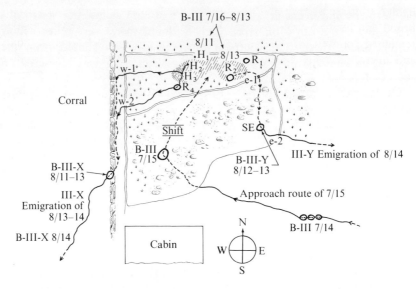

FIGURE 10.6.
Sketch of statary bivouac site of colony '59 N-III (*Neivamyrmex nigrescens*), indicating the line of approach on July 15 and the paths later taken by two daughter colonies (III-X and III-Y) in the division carried out on and after August 11. *Shaded area:* principal statary site of the colony underground; *double lines:* cordons or inspection lanes. $H_1$, $H_2$, and $H_3$: exits from the old queen's section of the bivouac; w-1 and w-2: trails used in the raiding and eventual emigration of section III-X; $R_1$ and $R_2$: rocks under which sexual-brood section was clustered; e-1 and e-2: trails used in the raiding and eventual emigration of daughter colony III-Y. Scale: lateral distance between cordons 10 m.

Finally, on the night of August 11, the overt division began with heavy raids to the east from $R_2$ and to the west from $H_2$. Then after 8:15 P.M., a broad column issued from $H_3$ carrying packets of young brood, stopping at a rock ($R_4$) a few meters to the southwest. At 9:35 P.M. the parent queen, distinguished by her dark coloration, ran with a heavy retinue from $H_3$ (over route w-1) to the site B-III-X and was followed by a column carrying young brood in packets, lasting about ninety minutes thereafter. This column was soon joined by another column moving from $R_4$ to the site B-III-X (over route w-2). After 11:00 P.M. there was a steady return of raiders with booty to B-III-X on the west and to $R_2$ on the east, indicating the actions of separate parts of the colony. Significantly, no males appeared in this operation.

On August 12 after dusk, a heavy raid to the south issued from $R_2$ and passed through the southeast (SE) to the east; on the other side, lighter traffic passed from $H_3$ and $R_4$ to B-III-X carrying both young brood and booty. A raid, but no emigration, issued from B-III-X, which remained connected with $H_3$ by a light, mixed column. From $R_2$, booty and microbrood (see footnote 4) were carried after 10:25 P.M. to the southeastern site under rocks at 5 m. No queen appeared.

On the next night, August 13, a large raid developed after dusk from $R_2$ to the southeast over routes e-1 and e-2; meanwhile, a steady column carried brood and booty to the southeast from $R_2$. Alate males began to run from $R_2$ to the southeast after 1:00 A.M., becoming more numerous after 2:30 A.M. Then as brood carriers increased to a thick colunm, a callow queen passed from $R_2$ to the southeast. This traffic continued nearly until dawn while heavy raiding traffic returned to the southeast.

Daughter colony III-Y, centered on the callow queen, had clearly separated.[13] The separation may have been well begun on August 5 when I saw the young queen and the male cocoons at $R_2$ (at which time I drew the shaded area in Fig. 10.6). The two features that emphasized the extent of the fission were the evacuation of the callow queen's section two nights after that of the parent queen and that this section contained all of the more than 1000 males, rather than the young brood's being divided nearly evenly between the daughter colonies. Doubtless the competitive processes of division had occurred underground prior to the emergence on that night of just one callow queen and her worker retinue.

Our cases show that more than one callow queen develops in the sexual broods of this species in advance of the males and that a competition must exist among them. From the last emigration of colony '59 N-X on the night of August 5, of three mature, already enclosed queen larvae in a brood of about 1000 mature but unenclosed male larvae, we took two of them for study. This intervention (as in '48 B-XVII, *Eciton burchelli*) may have blocked a division of the colony, for although after August 25 the colony staged nightly raids from opposite sides of the bivouac and so displayed a clear bidirectional tendency, it did not divide. Instead, on the night of August 31, the twenty-fifth statary day (S-25), the colony emigrated in a single column with just one callow queen, only 340 alate males, and a long train of workers carrying microbrood. The division process had miscarried, perhaps through our reduction of the young queens. Clearly, the parent queen had been superseded after delivering her eggs, and nearly two-thirds of the male brood had disappeared, perhaps through worker cannibalism in a disturbance of the colony on August 27, the twenty-first statary

---

[13] On August 14, by digging we traced out mammal burrows (probably) occupied by the parent queen's section beneath $H_3$ and a former nest of *Trachymyrmex* sp. (probably) occupied by the sexual-brood section beneath $R_2$ and found these spaces empty.

day (S-21)—corresponding to the time the parent queen's section had emigrated in colony '59 N-III, minus males![14]

Colony division in this species seems to be a two-way process, as in the *Eciton* studied. The difference may be that in *Neivamyrmex:* (1) only the terminal sectional emigrations occur aboveground; (2) the sectional emigrations are more separated in time than in *Eciton;* and (3) the males all emigrate with the sexual-brood part. In scores of cases observed in *Eciton,* we have found no deviations from a nearly equal division of the males between the daughter sections. From fragmentary evidence, other species of *Neivamyrmex* may be similar to *N. nigrescens* in their patterns of division.[15]

The pattern of colony division in *Aenictus,* as our preliminary studies indicate, is similar in general to that described for *Eciton* and *Neivamyrmex*. As the following observations show, however, the timing of male flights is strikingly different in relation to the overt division of the parent colony.

On April 27, colony '64 Ae. g.-1 (*Aenictus gracilis*) began a statary phase of 27 days in which it carried a sexual brood through pupal development. After many days without any surface raids, forays of mounting intensity began on May 19, the twenty-third statary day (S-23), and (late in the afternoon) alate males appeared on the surface. Observers saw 27 males fly from the bivouac on May 19 (S-23), 42 males on May 20 (S-24), 1110 on May 21 (S-25), and 187 on May 22 (S-26), ending the series. Thus, a total of 1367 males was counted leaving the bivouac on four successive days, most of them at times between midafternoon and dusk. Another colony of this species ('64 Ae. g.-2) had similar records of male-departure flights on successive days and at a corresponding time late in the statary phase. In this case, a total of 1261 males flew away between 4:20 and 5:10 P.M. on the second day of these flights, the peak of their exodus.

Thus, in both of these cases, the flight of the males from the parent colony was completed from the parent colony before it began the specific

---

[14] On August 27, the twenty-first statary day, at 3:30 P.M. the colony developed a heavy exodus to the east in full sunlight in a broad column that continued into a stage of raiding until after 2:00 A.M. This extraordinary emergence of the ants may have followed a mass cannibalism of male pupae in the bivouac.

[15] Two interesting cases have come to light for *Neivamyrmex carolinensis*. The first (M. R. Smith, pers. comm.) involved the collection in Sylvania, Georgia, on March 18, 1954, of ten callow queens from the same nest beneath a clump of weeds. The second (E. O. Wilson, pers. comm.) involved the finding of six callow queens of this species in Ravenal, South Carolina, on June 9, 1957, from the same nest in a rotted pine stump. In neither case were any males found. This circumstance, with the dates, might suggest that in this species the young queens spend the winter in the parental colony, with new colonies forming by a distinctive pattern early in the spring resurgence.

division, rather than occurring after division from the daughter colonies, as in *Eciton* and *Neivamyrmex*. In colony '64 Ae. g.-1, the overt division began early on the day after the last males had left and so promptly as to suggest that the exodus of the males may have played an important part in initiating the actual processes of fission.

After dawn on May 23 (S-27) the ants spread in great numbers from their underground bivouac on the north side of a bamboo thicket with columns presently rounding the margin carrying microlarvae in packets. Within an hour there began a complex series of operations on previous raiding trails. By midafternoon, in a series of steps, the colony had divided into three sections.

These steps, which seemed to combine actions of queen exodus and of emigration, were centered on three queens: A and B, callow queens (light reddish-brown), and C, the parent queen (black). Callow queen B with entourage moved out first, returned a short distance, then at 7:15 A.M. passed directly to a point 3.5 m on the northern trail where a cluster formed about her at 8:30 A.M., then to 8 m where a new cluster formed, then at 10:00 A.M. to 19 m where a still larger cluster formed. Emigration on this trail continued in column, and when the connection with the bivouac was broken at 10:00 A.M., the surface cluster of callow queen B at 19 m contained more than 50,000 workers[16] as well as large amounts of microlarvae and booty.

Events elsewhere were less clear. At 7:00 A.M. the parent queen C passed from the statary bivouac with a worker retinue which clustered at 3 m around the southeast side of the thicket. As the emigration continued, the queen's group moved onward with two clustering stops. This queen and her group may have rounded the dimly lit southern end of the thicket in midafternoon, clustering at a place near the southeast corner to which a heavy brood-laden column moved after 2:30 P.M. Later, after dusk, there were further moves by stages to the southeast, where this queen was identified during an emigration at 10:00 A.M. the next day. At that time, her section, then separate from section A, contained a large number of microlarvae and an estimated 15,000 workers.

A callow queen and her retinue moved from the statary bivouac at 7:10 A.M. on May 23, shortly after the parent queen C moved out. Callow queen A soon was enclosed in an elevated cluster, from which she moved with her retinue around the southwest side of the thicket shortly before 10:00 A.M. After two more short movements, she remained in a cluster on the southern edge of the thicket until late afternoon. At 4:10 P.M. she

---

[16] For these estimates of workers in sections A, B, and C, the volume of the workers in each preserved daughter colony was first measured; then the number contained in a measured fraction of the total was counted and the whole number approximated by multiplying.

and her group moved again, passing the cluster probably containing parent queen C, then veering away from the thicket about 10 m on the western trail into a cluster containing an estimated 20,000 workers with microlarvae. At dusk, therefore, all three of the queens in evidence had moved into distinct clusters. Further events are discussed in Chapter 11.

In this case, as in two others observed in *Aenictus gracilis,* the operations of queen exodus and of emigration were combined in the radial movements of subgroups centered on different queens. In colony '64 Ae. g.-1, a callow queen was the first to become separated in a daughter colony which, moreover, was at least twice the size of either of the other two sections. Thus there are distinct similarities to the patterns of *Eciton* and *Neivamyrmex* although that of *Aenictus* may be the most generalized of the three.

Doryline colony division, as first demonstrated from our studies on *Eciton* (Schneirla, 1949; Schneirla and Brown, 1950, 1952), derives from a condition of bipolarity introduced into the colony through counterattractions of large sections of workers to the sexual brood on the one hand, and to its parent queen on the other. As for the driver ants, Raignier (1959), by digging up many colonies of *Dorylus (Anomma) nigricans* and *D. (A.) wilverthi,* obtained evidence indicating a pattern of colony division comparable with that of *Eciton*—and of *Neivamyrmex* and *Aenictus* also —in this basic respect. From conditions in excavated nests, he derived a reasonable picture of the "normal process" as one in which the parent queen's section with most of the workers is stimulated to emigrate by the maturation of an all-worker brood, most of which it takes along. This exodus, which divides the colony, can happen when the males are at any stage from that of advanced larvae to that of callows.

The sexual-brood section left behind contains a variable population: a few callows queens, all of the males, and an indeterminate number of workers and of eggs (laid by the parent queen). Colony division in the driver ants thus seems to be a two-way process, as in group A but very different in detail. From Raignier's report, the bulk of the workers and young brood get into the parent queen's nest section while the males stay in the sexual-brood's section, complete their development, and fly away. There follows a pattern of colony founding based on the surviving young queen (Chapter 11).

Although continuous behavioral studies of colony division are lacking for the driver ants, Raignier's results suggest that, in effect, as sexual broods develop, a competitive process arises in their colonies whereby the parent queen finally heads an emigrating colony (or is superseded?) according to the potency of her attraction (e.g., odorous) for workers. As events in our cases (e.g., colony '52 B-I) indicate, supersedure involves a drastic limitation in the attractiveness of the parent queen for workers beyond the confines of her own group. Although a pattern emerges that is typical of each genus, with general similarities among the patterns of the three group A genera, plainly the course of events may vary from case to case according to circumstances. Variations may arise through colony size (e.g., '46 B-I), through the colony's current ecological situation (e.g., '52 B-I), and doubtless other factors.

The typical pattern of *Eciton,* which seems the most complex and precise of all, often runs its course with clearly differentiated stages of queen exodus and of overt division (by divergent emigrations) as in colony '48 H-27. At other times, factors enter that draw out the process and greatly complicate the relationships of these stages, as in '46 B-I. Normally, as in '48 H-27, the first divergent movement of leading queens prevails, despite indications of instability in their terminal nuclei (Chapter 11). At other times, some change or other in the situation (e.g., '48 H-12) introduces complicating variations involving the other queens. However secondary these queens may be, our results suggest that in group A divisions they play a necessary role in behavioral stabilization, thereby making important adaptive contributions (Chapter 14).

Although colony division in the three group A genera as well as in the driver ants is based on the rise of bipolar colony organization introduced through development of a sexual brood, there are important differences. In *Eciton,* for example, as well as in *Neivamyrmex,* the males go into the daughter colonies in the overt-division stage (dividing equally between the daughter colonies in the former but not in the latter). *Aenictus* and the driver ants seem alike but differ from these genera, however, in the nonparticipation of their males in daughter-colony emigrations. *Eciton* seems unique among the genera studied in the relatively clear separation of the group processes in its queen-exodus stage of queen competition from those of the overt-division stage in which the colony splits more or less equally. In *Aenictus,* however, these stages are not separated, but operations of queen-exodus overlap those of daughter-section emigration. The active queen-competition processes of *Neivamyrmex* all take place underground and possibly more slowly than in *Eciton* and *Aenictus* before the overt-division stage begins. In this respect, *Neivamyrmex* may be closer to the driver ants than to *Eciton* and *Aenictus.*

The pattern of events in the colony divisions of group A dorylines—from processes of brood coordination to a sequence of relationships between sexual-brood and parent-queen segments—seems relatively precise in contrast to the diffuse, variable processes indicated for driver ants. The differences may express an evolutionary background centered on hypogaeic adaptations and much larger colonies in the latter; they are, therefore, related to functional and particularly reproductive differences, as we pointed out in the comparison of cyclic patterns in groups A and B (Chapter 7). This comparison may continue to advantage in discussing the sequelae to colony division in the early adjustments of daughter colonies.

# 11

# The New Colonies

After the queen-exodus stage, an overt division of the colony takes place. This pattern, by which new colonies begin their independent existence, broadly represents a mode typical of the dorylines. It differs so widely from that of most other ants—except possibly for some of the legionary ponerines still unstudied in this respect—that it is best explained by limiting our discussion to the dorylines themselves (Schneirla, 1956a).

In group A ants the division of the parent colony seems complete when the two daughter colonies move apart and settle in well-separated places. Their behavioral independence from each other at this time, however, is tenuous because a slender two-way column still connects them. This liaison file, in which workers run back and forth, evidently not yet well affiliated with either of the new colonies, continues for many hours, then normally disappears during the day following the overt division.

Such was the course of events with colony '48 H-27 (Schneirla and Brown, 1950, p. 304). This colony (Chapter 10) separated into two sections, sharing the workers, young brood, and alate males of the original colony in the ratio 4:3, with the section of the parent queen taking the larger part. This queen, on which the new '48 H-27-E colony centered, was recognized at the time of division by her code mark and by her dark coloration and was readily distinguished from the light golden-orange callow queen of the west colony.

The division was normal and smooth-running with traffic on the trail between the two young colonies gradually lessening the morning after the divergent emigrations took place. An early disappearance of the liaison column, as in this case, we may take as an indication of greater stability of daughter sections than if this trail were used longer. In all cases, however, the new colonies continue to be somewhat unstable for a few days after the division, as their behavior indicates from tests to be described below.

Successive bivouac places, routes of emigration, and main raiding lines of the new colonies in this case are sketched in Figure 11.1. In their

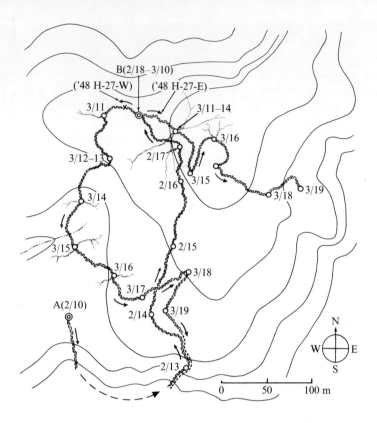

FIGURE 11.1.
Schema indicating line of approach (A, February 10–18) before division of colony '48 H-27 (*Eciton hamatum*) to statary bivouac site (B), where it matured a sexual brood, then divided. Subsequent divergent paths in successive emigrations of the two daughter colonies ('48 H-27-E and '48 H-27-W) after March 11 are indicated to the right and to the left. Note that on March 19, '48 H-27-W followed the old, February 13, trail of the parent colony.

first few days of independent activity after the division (B), both daughter colonies were more variable in their actions outside the bivouac and much less vigorous in their general behavior than is usual with established colonies of the species (Schneirla and Brown, 1950). Colony '48 H-27-E, after its first emigration (March 11), remained at its second site for three days during which it raided weakly and carried out no emigrations at all. Daughter colony '48 H-27-W was similarly low in activity. Thereafter, however, both colonies reached the level of daily raids and nightly emigrations typical of the species. After the fourth nomadic day (March 14), both queens were followed nightly in the emigrations by large worker retinues and seemed well established in their colonies. They were inspected from a distance, however, and not handled during this period, as we knew, especially from queen-removal tests to be described, that stable organization comes slowly to these new colonies.

On the afternoon of March 19, when the two daughter colonies were in the eighth day of their first nomadic phase and had moved widely apart, we judged them to be secure, independent units and etherized their bivouacs for study. Both queens, each enclosed in a bolus of workers located near the top of her bivouac above the (all-worker) brood, seemed to be in good condition and well established in their colonies. One important difference, however, evidently related to their age (to be considered later) was that the (adult) queen of the east colony was then fully contracted whereas the (callow) queen of the west section was somewhat physogastric. Each daughter colony, as it entered the nomadic phase, had a few hundred males and a large, young all-worker brood (i.e., roughly one-half of that from the parent queen's last laying). As later histological study (R. M. Whelden, pers. comm.) showed, the callow queen (west section) had been fertilized. Doubtless, had we not intervened, each of these daughter colonies would have completed its nomadic phase normally (Chapter 7) and would have continued in the cyclic function typical of the species.

In this case, the daughter colonies seemed to function about equally well after they separated. The success of daughter colonies issuing from the same parent colony in *Eciton* may be similar or different, however, according to what internal and what environmental factors influence them.

In these respects, colony '52 B-I, which completed its division with the parent queen superseded by one of the callow queens and thus had callow queens in each of the daughter colonies, seems more complex. Although these young queens were sisters from the same brood and seem very similar in physical appearance and behavior, the south queen (who superseded the parent) may have been somewhat superior to the other (e.g., in secretory and in reproductive function). At first the north colony seemed the more active of the two, as indicated by the greater vigor of its first raids and emigrations. But after the sixth day of the first nomadic phase, this difference was reversed! Eventually, the south colony completed this phase and matured its all-worker larval brood in sixteen days whereas it took the north colony eighteen days. Altogether the north colony needed about five more days than the south one to mature a callow brood and to end the ensuing statary phase.

For daughter colonies of about equal size, as these were, differences in performance may result from differences in queens (e.g., the south colony may have matured before the north one in the sexual brood) or in colony ecology. Environmental differences were present in the latter half of the first nomadic phase, when the south colony passed through an area of warm, moist forest on the south bank of a ravine in which it seemed to capture much more booty than the north colony could get in the generally cooler, drier forest on the north bank. Both physical differences between the queens and environmental differences thus may have favored the south colony.

There was less difference between the daughter colonies of '55 B-IV (Chapter 10), which began their first nomadic phase only a few hours apart and thereafter behaved similarly, e.g., each missed four emigrations in the first week. These colonies both ended this phase in thirteen days, with '55 B-IV-N in a hollow log where we captured it for study. The north colony, estimated as perhaps 25% smaller than the south one, contained a young queen in early physogastry, an all-worker brood of over 100,000 larvae, one dealate male, and many myrmecophiles.[1] In function and behavior these two colonies showed no reliable difference during their first nomadic phase, and doubtless both would have continued well.

Although daughter colonies of *Eciton* have unstable unitary organization just after dividing, results of the three normal cases described indicate that stability comes gradually, probably within the first week, as long as no accidents occur. The queen's role in this respect is important, as the following results show.

In colony '48 H-12 (Chapter 10) our removing two of the three leading callow queens in the queen-exodus stage did not prevent the colony from beginning its division with vigorous divergent movements to the southeast and to the west, involving the parent queen and three remaining callow queens, respectively (Schneirla and Brown, 1950, pp. 283–284). The two divergent movements, however, continued to be connected by a busy two-way column on which, early in the evening of January 30, after a divergent daytime raid, the west section began to withdraw to the center. In doing so, it dissolved a moderate-size bivouac cluster that had begun to form as well as at least one queen cluster nearby. During the night, what had now become the evacuation column of the west section passed back through the previously statary site (where the callow queens may have been abandoned) and joined the east section (based on the parent queen); then the reunited colony moved on (raiding and emigrating) as a unit. In this case, at least one of the secondary callow queens provided, to a certain extent, a basis for an independent daughter colony. But somehow as this section was forming its bivouac, a deficiency arose through which the potential daughter colony began to behave as though it were

---

[1] These samples and the contained myrmecophiles were studied by C. W. Rettenmeyer (1963), who worked with me in collecting both daughter sections of colony '55 B-IV.

## 11 / The New Colonies 249

queenless and proceeded to break down. Why this division miscarried I could not explain.

In *Eciton,* therefore, divergent movements based on young queens may begin well but then collapse, as was also illustrated by the initial westward movement in daughter colony '46 B-I (Chapter 10) after I had removed its young queen. Even as this section withdrew into the main colony, there continued both a divergent exodus to the north—which had begun about the same time as the west exodus—and an exodus to the south. On April 2, five days after the division had begun, the two daughter colonies developed independent raids. With the connecting column gone, I examined the bivouac of the north section, finding in it an estimated one-fourth of the original worker population, a few hundred alate males, masses of young all-worker brood, and a callow queen (Schneirla, 1949). Within four hours after I had removed this queen and her bolus of workers, ants of this colony were moving in column toward the south section. That evening a heavy emigration column, including the forces of a north section raid, traveled back through the statary bivouac site toward the south section. At 10:00 P.M., as a test, I set down the north callow queen, with about one hundred workers from her bolus, close to the emigration column then heading back toward the (then emigrating) south section. To insure that she would be presented to many workers from the north section, I placed her at a point near the statary bivouac tree. At intervals, workers from the column entered the open nest containing this callow queen and her worker group, only to withdraw. At midnight, as the emigration toward the south section continued, this queen and her workers were still in the open nest beside the column, the queen unharmed but also unabsorbed. By 2:00 A.M. she had lost more than half of her workers. At dawn, hours after the emigration had ended, I found the callow queen in the open nest, completely abandoned and in good condition. Meanwhile, the reunited colony, centered on the parent queen, was massed in a single bivouac more than 100 m away.

Such results indicate that a condition of instability persists in daughter colonies of *Eciton* even after the actual liaison column between them has been discontinued. A normal colony, after it has lost its queen, starts backtracking operations more slowly and with greater latency than does a new daughter colony under the same conditions. After the new colonies have passed their first few nomadic days, however, the difference is less apparent.

Beside these post-division studies, other investigation shows how the affiliation between the new queen and workers of a daughter colony increases in the course of time. One example we found in colony '52 B-V (Chapter 10). When on March 5 I removed the cluster of a secondary queen of this colony from beside the basal column, the west section began to withdraw. At this time the parent queen was apparently sealed off, and divergent movements to queen clusters to the east and west (about 40 m apart) were well under way. The general withdrawal of the west section continued, however, and during the evening became a steady emigration in which the entire section moved to the east. The next morning, March 6, the third nomadic day, after the east section had completed an emigration farther to the east (with the west section following in column), the colony was reunited in one long bivouac. From the extreme southwest end of this bivouac, I removed a callow queen in her bolus. I returned the latter at once, however, assuming that she was the queen initially established in the east section and that the former was the one on which the (resorbed) west section had centered. The case was not that simple, however. On March 9, the sixth nomadic day (N-6), I probed into the bivouac, found a partially physogastric callow queen, and removed her. On March 12 (N-9) the colony was raiding on its back trail but still behaved as a unit. Therefore, on the following morning (N-10), I probed into the bivouac once again, finding, as expected, numbers of alate males and a nearly mature all-worker brood but, to my surprise, a third callow queen. This queen was more advanced in physogastry than the one taken on N-6. On March 5, therefore, after the resorption of the east section, the bivouac of this colony had contained three callow queens. Such polygynic daughter colonies are very unusual.

As a reasonable explanation, I postulated that two of these queens may have entered from the resorbed west section, the first as the leading queen of that section, the second later from a sealing-off cluster. The case illustrates, first, the instability of daughter sections just after a colony division and, second, the plasticity of gregarious behavior in a reunited colony. What would have happened had all three queens remained in the reunited colony, we cannot say. The two callow queens taken last from the bivouac were both in good worker boluses and physogastric; hence both clearly had fed well since their subsections coexisted for days in a colony that operated as a unit in raids and emigrations. The colony was a large one and showed no signs of instability (except possibly the back tracking on March 12) until the last of the three queens had been removed. Abnormal cases of this type merit further study.

Colonies of *Eciton* in the sexual-brood season commonly contain at least one dealate male, and it is probable that a callow queen, once she is established in a daughter colony, is soon fertilized. Also, we have commonly found such young queens entering physogastry much earlier in the nomadic phase than is usual for parent queens. It is likely, therefore, that the callow's great attraction for workers at that time increases their grooming her and force-feeding her well beyond that of parent queens, which, moreover, may be subject to inhibitory pheromones not effective for new queens. As a result, a callow queen after the first nomadic phase in her daughter colony is able to mature and deliver a large brood close to the species-typical size. This is a remarkable performance, especially since the young queen matures her first batch of eggs within only about one month after she has emerged from her cocoon. We tested this by carrying the study of colony '52 B-I to the time at which the young queens of both the north and south sections had laid their first eggs, which in number closely approached that usual for adult queens of the species.

Those males that manage to mate with callow or adult queens do so only by completing a difficult obstacle course (Schneirla, 1948). Counted as alates at the time they emerge in a sexual brood, they are numerous. A few days later, and after the division of the parent colony, daughter colonies of *Eciton* each start out with about one-half of the male brood or roughly 600 to 800 in *E. hamatum* and 1400 to 1500 in *E. burchelli*. In *E. hamatum*, when the divergent emigrations of division occur, these males usually have all emerged as alates and can run along in the column; in *E. burchelli*, however, as a rule many of the males are carried along in cocoons.

In the bivouacs of daughter colonies of *Eciton*, alate males usually hang in strands of workers and are frequently groomed. Commonly, these strands are also loaded with booty, on which the males may feed by pressing their mouthparts against morsels, imbibing the juices as they do in drinking. Efforts to observe male feeding have not succeeded, however, either with *Eciton* or other dorylines (Schneirla, 1948). The males may actually ingest no solid food but only fluids and otherwise subsist on their own tissues. Whatever their nutriment, some males of *Eciton* live at least a month after they emerge from their cocoons.

In the first emigrations of the young colonies, the alate males keep closely in line along the trail as they are still too weak to fly. They hold a strong attraction for the workers, which follow each of them in retinue, often running with their mouthparts pressed against the upper part of the male's thorax or gaster. Worker attraction to male pheromones is indicated here as well as at earlier times in development, such as during emergence when workers avidly consume pupal tissues still attached to the callow males.

In these emigrations, one often sees a submajor worker running astraddle a male, with her mandibles enclosing his folded wings or his gaster (Fig. 11.2). This and comparable behavior by smaller workers, which involves what we call "clinging reactions," becomes more frequent as the males move along in the emigration, suggesting that their attraction for workers may even increase as they run.

(a)

(b)

FIGURE 11.2.
Alate males of *Eciton hamatum* in nocturnal emigration of their colony. (*a*) Many workers of the males' entourage are carrying larvae of the new all-worker brood. (*b*) Close-up of male in emigration column. One worker touches mouthparts to male's gaster.

After the third or fourth night, however, the males, now stronger, begin to deviate from the described pattern (Schneirla, 1948). Then on coming out of the bivouac or during halts in the emigration, alates now and then break loose and make little skipping runs over the leaves, mounting vines and other sloping surfaces they may encounter. The strength and persistence of these responses is shown in laboratory tests in which alate males vigorously mount a narrow inclined runway after having

FIGURE 11.3.
Alate males of *Eciton burchelli* (and other species), started at the base of a narrow incline in laboratory tests, readily mount on the path. Note that the hairy tip of gaster is in contact with the substratum and that the wings are functional. (See Plate V, following page 138.)

been set down at its base (Fig. 11.3). Such changes in the males lead to increasing resistance to the clinging workers, which, in turn, undergo behavioral changes of their own. A typical note is:

> Alate males delay the emigration by running off the trail just outside the bivouac, the workers following them in numbers. At the base of the cluster are columns and groups of milling workers, concentrating here and there on males by clustering over them or holding to them as they run. On the main trail at 3 m from the bivouac, I see two alates running back and forth among workers at intervals making short hopping flights. On one branch of a nearby bush, a small cluster of workers holds a male; on another branch a column moves behind a climbing male. (Schneirla, 1948.)

Thus, gradually, increasing strength and diversity in male behavior seem to modify the worker's clinging reaction to one of restraining the males.[2] As the males become more vigorous, however, more and more of them break away and fly off with each emigration, the largest numbers probably leaving on nights around midphase. The result is that daughter colonies of *Eciton*, in their first nomadic phase, literally seed the area through which they pass with winged sperm carriers.

The males in their take-off flights usually do not seem very strong and

---

[2] It is mainly the submajor workers that accomplish this. At first, they straddle the alate males in running, opening their mandibles widely in a relaxed position and applying their mouthparts to the male's thorax or gaster. In further emigrations, as the males run more vigorously and variably, the worker's mandibles seem to tighten so that in effect they hold the males to the route by locking their wings in place.

tend to follow erratic though steep courses upward. Most of them may fly considerable distances, however, especially when those rising above the forest canopy are helped along by wind. Flights of *Eciton burchelli* males may often cover more than 300 m as I have observed series of alates and dealates running in the same emigration of a colony lacking a sexual brood of its own and operating at least that far away from any colony of the species known to have mature males.

Because doryline queens are wingless, for mating to occur a male must complete his flight, then somehow find and enter a colony bivouac of his species. Large numbers complete their flights, as records of light trapping show, but relatively few seem to survive as dealates in the bivouacs (Schneirla, 1948; Rettenmeyer, 1963; Forbes et al., in prep.). From observations and tests with males of *Eciton* and *Neivamyrmex* we may surmise what can happen with postflight males of the surface-living species. By running about on the ground, males may cross army ant trails which they can follow to the bivouacs. The chances of finding such a trail may be increased by pheromones spread by the male's ventral gaster brush as he runs (Fig. 11.3), attracting workers with whom he may reach their bivouac in column. Males that chance to follow trails of species other than their own evidently are killed[3]; also, males may be killed on the trails even by excited workers of their own species. Males still unattached on the first postflight day have little chance of surviving predators, drought, and other hazards. It is highly probable, therefore, that most of the large male populations flying from daughter colonies are lost.

When his flight is ended, the male soon loses his wings through dropping them or having them pulled off, evidently as one of the physiological changes resulting from the flight itself.[4] A dealate male of *Eciton* established in a foster colony of his species may remain in it for many days or even weeks (possibly sustaining himself on such reserve tissues as degenerate wing muscles and on booty juices) before he mates or dies. We turn later to the subject of mating.

*Neivamyrmex nigrescens,* a hypogaeic ant, releases daughter colonies

---

[3] Interspecies matings in the dorylines are unlikely as one frequently finds dead males of other species on the debris heaps of *Eciton*. In laboratory tests with circular columns, males of *E. burchelli* released before circular columns of *E. hamatum* are attacked and killed (and the reverse) whereas males of the same species generally are accepted.

[4] Effects of the flight include loss of the wings (subsequent breakdown of the wing muscles and, probably, their assimilation as food); also, as in other winged insects, it must increase the level of respiration and other physiological processes. Postflight males often have distended gasters from which a whitish material issues— Rettenmeyer (1963) suggests a fecal product. In preliminary tests with *Eciton hamatum* and *E. burchelli* (Fig. 11.5) alate males do not couple with females of their own species although postflight males do.

in emigrations on the surface or under surface cover only after the processes of queen exodus and overt division have occurred underground. This species differs from *Eciton* in at least two important ways: first, the daughter colonies move off divergently on different nights; and, second, from our evidence, the males all go with the section of the callow queen.

In our continuously studied case, colony '59 N-III, the parent queen's section emigrated to the southwest mainly on the night of August 11, but the callow queen's section moved to the east two nights later on the night of August 13, with the queen and most of the males in the column (Fig. 10.6). Section III-X, the daughter colony with the parent queen, raided nightly but did not emigrate again until the night of August 14. In the next emigration of this section, on the night of August 15, I captured its queen. This colony retained a thinly followed trail connecting section III-Y through the statary bivouac area, which, however, it ended on August 18. On the night of August 20 section III-X emigrated from our study zone.

Daughter colony III-Y, after its first emigration on the night of August 14, carried out nightly raids and emigrated on three of the following four nights. On the next four nights it remained in place, with raids but no emigrations; then began a busier schedule of raiding and more frequent emigrations. In its last emigration of a seventeen-day nomadic phase, on the night of August 29, the colony's all-worker brood was prepupal, its queen (then captured) was in early physogastry, and few of the alate males were still present.

It seems, therefore, that although in this doryline daughter sections centered on a callow queen start late from the statary site, they afterwards pass through a nomadic phase that is species-typical in duration. Its difference from *Eciton* is that in the first and last parts of the phase, emigrations may be missed more often. Raiding, although infrequent for the first few nights in the phase, later increases, after which the phase ends—as in regular colonies of the species—when the young brood reaches the mid-prepupal stage.

Results for colony '59 N-XVI (Schneirla, 1961) also indicate an initial lowered activity level in daughter colonies. This colony, which in all probability was found one or two nights after its separation from the parent colony, passed three nights without emigrating, then began a program of large nightly raids and frequent emigrations like that of colony '59 N-III-Y. Thus young colonies of this species, although at first somewhat underactive, as are those of *Eciton,* later become more active and complete the first nomadic phase in a manner close to normal.

As for *Neivamyrmex nigrescens,* the organization of daughter colonies evidently stabilizes rapidly from the time of division, as is indicated by worker responses to the young queen in emigrations. Also, as in *Eciton,* the workers seem to overfeed their young colony queen from a time early in the nomadic phase when she exhibits signs of physogastry never seen in adult queens at that time. The young queen of colony '59 N-III-Y was well advanced in physogastry at the end of this phase and no doubt would have delivered her first brood in the oncoming statary phase. The young queens are evidently fertilized at a time before their first nomadic phase is well under way.[5]

If, as our results indicate, all of the males in *Neivamyrmex nigrescens* adhere to the section of the young queen, these ants would resemble *Dorylus,* in which this also seems to occur, rather than *Eciton.* (When the parent queen is superseded in *N. nigrescens,* however, it is probable that the males divide between the sections of the callow queens.) Supporting this point, we found alate males in three daughter colonies of this species with callow queens but never in colonies with adult queens. The separation of the parent queen's section from the sexual-brood section, which these ants evidently carry out in the subterranean bivouac, seems therefore to be rigidly maintained until the colony has divided. By contrast, in the divisions of *Eciton* on the surface, the portioning of the males between the daughter colonies indicates that the (odor-based) separation breaks down before the divergent emigrations begin.

We are now in a position at least to suggest how and when alate males fly away from their colonies—a question left indefinite by earlier authors (e.g., Wroughton, 1892; Wheeler, 1900). Our suggestion is that males of surface-active species leave their colonies around dusk whereas the males of hypogaeic species leave in the evening or night, also that there are important differences in the reactions of these males to environmental stimuli that influence their behavior after landing. This idea, which we offer for *Neivamyrmex* specifically, may apply comparably to males of other dorylines.

For one thing, although we have posted light traps in the area of '59 N-III-Y and other daughter colonies of *Neivamyrmex nigrescens* (Forbes et al., in prep.) containing alate males (which by emigration counts decreased in numbers night after night), these traps captured no males of *N. nigrescens.* These males may fly off around dusk, return to ground early in the evening, and therefore not react to our lights.

These results were confirmed in an investigation[6] in the same area in Arizona, where for two years light traps were posted regularly from dusk

---

[5] In his subsequent histological studies of the young queens of daughter colonies '59 N-III-Y, '59 N-X, and '59 N-XVI, R. M. Whelden (pers. comm.) found the spermary receptacles of all three well filled with sperms.

[6] This project was carried out by Mont Cazier, professor of zoology at Arizona State University, who preserved and sent us for study all of the doryline males he trapped.

on during the summer nights. In support of our hypothesis, no males of the surface-active species *Neivamyrmex nigrescens* and *N. opacithorax* were captured, but, instead, large numbers of alates representing six definitely hypogaeic *Neivamyrmex* species were taken regularly.[7]

From these and other preliminary findings we conclude tentatively that the males of surface-active species of *Neivamyrmex* react differently to environmental stimuli and that after their flights reach colonies of their species by different means than males of hypogaeic species of the same genus. Significantly, males of surface-active species have relatively small, flat eyes and readily follow doryline trails in tests whereas males of hypogaeic species have larger, convex, and turreted eyes and seem to react weakly if at all to doryline trails. Therefore, males of surface-active species, on returning to the ground after their dusk flight, may be dominantly reactive to chemical stimuli and so reach colonies of their species by following trails (as I have suggested for *Eciton*). Males of hypogaeic species, by contrast, may fly later in the evening and on descending to the ground may be dominantly responsive to visual stimuli —and thus move toward rocks, logs, and other moonlight-silhouetted objects, to which their keen vision makes them highly responsive.[8] Beneath such objects, rather than on open ground, these males are likely to encounter trails of workers of their own species.

As the reader will recall, the three group A genera have similar patterns of colony division, with the notable exception that in *Aenictus* the males fly from the parent colony just before its division begins, but in the other two group A genera after the daughter colonies have separated. This difference may have a direct bearing on species reproduction. In the case of *Aenictus*, for example, it could reduce the chances of mating between males and the queens within the same parent colony (Chapter 14).

The similarity of these three genera in colony division extends into the early behavior of the daughter colonies as colony '64 Ae. g.-1 shows. Section B of this colony, the first daughter unit to sever its connection with the parent colony, moved away with more than half of that colony's

---

[7] Although the Cazier trappings yielded no males at all of *Neivamyrmex nigrescens* and *N. opacithorax,* they regularly contained large numbers of alates of six cogeneric species so definitely hypogaeic that they are described taxonomically from the males alone (Smith, 1942; Creighton, 1950; Borgmeier, 1955). These are: *N. harrisi, N. fuscipennis, N. pilosus mandibularia, N. swainsoni, N. oslari,* and *N. minor.* Except for *N. harrisis,* which has been synonymized with the male of *N. wheeleri* (Watkins, 1968), their worker affinities are unkown.

[8] In faint light, dark-adapted individuals of highly visual species commonly approach objects that contrast with the background as would moonlit rocks or tree stumps. This type of reaction, called "skoto-taxis" (Alverdes, 1930; Fraenkel and Gunn, 1961), would be expected in doryline males of hypogaeic species but not in those of surface-active species.

workers and half of its young brood. On the afternoon of May 24, the day after its separation, this new colony carried out a series of emigrations that took it 50 m from the statary bivouac and at least 70 m from the other two sections then operating in independent raids. The queen of section B, distinguished as a callow by her light coloration, then moved on at the head of a 2 m long entourage in a new emigration. Since at this time there was no question that daughter-colony B had separated completely (i.e., without any connecting file) from the rest of the parent colony and was a potentially thriving colony, we captured it with queen and brood for eventual study.

The two other sections dividing from this colony—section A centered on a callow queen and section C on the parent queen—were then still connected by a liaison column. After section B separated on the afternoon of May 23, each of these sections made a series of short emigrations from one clustering place to another while continuing their raids independently. The moves were not yet divergent; in fact, at one point in an emigration the queen of one of them, amid her entourage, passed close to the place in which the queen of the other section was then resting in an elevated cluster.

These two sections could not be distinguished from each other (i.e., which one contained the parent queen and which the callow queen) in the dim light of late afternoon. Both seemed small, as compared with section B, when, after dusk, the liaison column between them disappeared. They passed the night in well-separated clusters; then after dawn the two daughter sections again moved divergently—A to the west and C to the south—to places where each started to raid independently. At 7:00 A.M., when section A began to emigrate westward (still without any signs of resuming its connection with C), we captured its queen with her retinue (about 50 cm long) as she was heading away from the statary bivouac and about 30 m from it. This queen, light brown in color, was unmistakably a callow. I estimated that this colony (evidently section A) contained between 12,000 and 15,000 workers, or only about one-fourth as many as section B.[9]

Meanwhile, section C moved off to the south and raided southward. At 11:00 A.M. when the queen was moving in an emigration (with a worker entourage 2 m long), she was captured. Her dark coloration left no doubt that she was the parent queen. At that time her section was

---

[9] This callow queen was captured and preserved shortly after she left the parent bivouac of colony '48 H-12 with her worker entourage. She had been fertilized within the previous few days in the main bivouac following her emergence, as a subsequent histological study by R. M. Whelden (pers. comm.) disclosed "four or five uniformly large discrete ball-like masses of sperms in her spermatheca," together with "many unidentified fragments of loose tissue apparently introduced with the sperms."

about 40 m southwest of the statary bivouac and contained an estimated 15,000 to 18,000 workers—more workers than section A and perhaps more young brood also.

Impressive features of this case were the early separation of section B (centered on the leading callow queen) and its much larger population than the other two daughter colonies together. A lesser stability in the other two sections (organized about their queens) seems indicated by their having separated at a time well after section B was clear. Although both of these sections were small—the workers of section A were "just about enough to fill a teacup"—when the study ended, they were raiding and emigrating independently more than 60 m apart. If let alone, they might have built up their numbers and made their way successfully as independent colonies. On the other hand, perhaps as an adaptation typical of *Aenictus,* a continuing instability in either A or C might have led to their rejoining into a single colony over connecting trails.

Doryline mating has been a matter of conjecture as for some time the only known case was that of a dealate male and a queen of *Neivamyrmex carolinensis* found in copula in a bivouac that was being dug out.[10] Our records for *Eciton hamatum* indicate that the leading callow queens may be inseminated in or near the bivouac of the parent colony within the first few days after they emerge from cocoons or at least within the early part of the first nomadic phase in their new colonies (Schneirla and Brown, 1950, p. 305).

My first observation of doryline mating came about when, after having taken from the bivouac of colony '46 H-H the adult queen and a dealate male, I separated them for mating tests in the laboratory (Schneirla, 1949, p. 45). The event took a different course, however, as a second dealate male was present in or near the queen's cluster in the bivouac but was overlooked in my inspection of queen and cluster in the field.

At 4:30 P.M., when I took the queen's cluster from its container in the laboratory, this male was already "in copula with the queen, mounted over her with his mandibles holding tightly to one of her large petiolar horns and the tip of his gaster inserted so far into her abdomen as to deform her gaster considerably." The pair was still joined at 6:00 P.M.

---

[10] Rettenmeyer (1963) put together an adult queen of *Eciton burchelli* and a dealate male found in the emigration column from which she was taken. Coupling began within one minute and persisted, with the male mounted on the female, for a little more than an hour. The male was never inserted more than partially, and his external genitalia remained outside the queen's gaster. He made constant wide movements of his antennae and front legs, occasionally seeming to stroke the queen with his antennae. He held the queen by his middle and hind legs, grasping her petiole behind the horn with his mandibles. No histological examination is reported from this case, in which disturbance of either or both individuals by intense light used for photography seems to have inhibited mating.

FIGURE 11.4.
Mating pair (postflight, dealate male and young colony queen) of *Eciton hamatum,* preserved after having been coupled for 10 hours. When the pair was then preserved, the male (evidently dying) had withdrawn from the position of full insertion. Note distorted pattern of queen's gaster.

when next observed. The male seemed lethargic from the first (i.e., from 4:30 P.M.); "the queen, however, remained active, and at intervals even ran about the enclosure carrying the male with her." At 6:30 P.M. we found the union broken and the male very sluggish. He was then preserved. This queen, definitely an adult, was received normally by her colony when I returned her to the bivouac the next day.

A second case of mating in the same species (colony '46 H-L) involved the callow queen of a newly divided colony. I placed this queen in a container with her worker bolus from the bivouac. But, in this case also, when I removed her two hours later from the jar, she was in full copula with a dealate male that had remained undetected in her bolus. The insertion was maximal, with great distortion of the queen's gaster, and persisted for ten hours, near the end of which the male, then evidently dying, partially withdrew his gaster. The two were preserved (and photographed, Fig. 11.4) in this last position. Clearly, from a subsequent histological study (Schneirla, 1949, p. 46) the male had inseminated the queen and was dead when he was preserved.

In both of these cases, dealate males were present in or near the queen's cluster in the bivouac, but mating began only when capture had excited the ants and when handling might have brought them together. This suggests that the usual scene of mating may not be the bivouac but the situation of emigration, which presents intervals of excitement with opportunities for union when a dealate male nears the queen as she pauses at an obstacle or enters the new bivouac at the end of the run. From the two described cases we may conclude that mating in *Eciton* is a protracted event in which the male inserts maximally into the female, seeming literally to explode his sperms into her gaster, after which he dies.

From our limited evidence on *Eciton,* most of the matings occur in the season of sexual broods and involve adult or leading callow queens. Responses of males in tests (Fig. 11.5) indicate that when still alate they are incapable of mating. Hypothetically for mating I suggest that the male must be a dealate that has been prepared as follows: (1) Having completed his flight and shed his wings, he is physiologically ready; (2)

FIGURE 11.5.
Callow queen and alate (preflight) male of *Eciton burchelli* in mating test. The male has the odor of the queen's colony; hence she responds to him; he has not had his flight, so does not respond to her. (See Plate VI, following page 138.)

having lived for a period in the host colony, he carries its odor and is thus accepted by the adult queen or a leading callow queen of that colony; and (3) he is habituated to the host-colony odor and will mate only with a queen bearing that odor.

Events in the origin of new colonies are different and more variable in *Dorylus*. For these ants in about 70% of the cases, as Raignier and Van Boven (1955) reported, sexual-brood production ends with the males flying away in a four-to-six-day interval but evidently without any exodus of the colony.

Because these ants have subterranean nests, with long nest stays, Raignier (1959) had to study colony foundation indirectly by digging up as many colonies with sexual brood as possible. He found one case [colony 195 A—*Dorylus* (*Anomma*) *wilverthi*] evidently in the process at the time. In one of two nests (apparently from the same parent colony) he found an adult queen ("with eggs") and much worker brood (and probably also a large worker population, not mentioned). At a distance of 200 m there was a second nest of the same species containing a young queen (not yet laying), many advanced male pupae, (some workers?), and much worker brood. Raignier concluded that the parent queen's section had recently emigrated and that the young queen would lay eggs and found a colony of her own. (In similar cases, e.g., colony 172 A, he found a young queen "starting to lay" in a nest with many winged males and few workers and worker brood.)

Raignier (1959) also reported cases in which he found aggregations of sexual brood in nests from which the parent queen's section evidently had left. In colony 102 A-E of the same species there was much male brood (ranging from larvae to advanced pupae), eleven young queens at stages from advanced pupae to callows, also worker brood, but few workers and no parent queen. There were also pieces of four or five young queens. His conclusion was that the parent queen's section had left and that "biting to death of excess queens" had started.

Raignier (1959) thus infers from many "digs" that in a minority of cases (about 30%) the parent queen departs with most of the workers and all-worker brood, leaving the young queens with the (less mature) males and a variable number of workers and worker brood. After all but one of the young queens have been killed, the survivor—fertilized meanwhile—begins to lay eggs, thus founding a new colony. This process is seemingly not influenced by the males, which fly off as they mature.

From these results, the new colonies of *Dorylus* do not divide the worker population and young brood as in group A species but concentrate most of it in the parent queen's section which (from Raignier's report) is the one that moves away in a continuous column. A second colony may be formed from the sexual-brood section but in a very different manner from that of any known group A doryline, apparently involving only a minor portion of workers and brood from the parent colony. The workers remaining from the parent colony may be the agency reducing the young queens to just one[11]; they may also help found the new colony of the surviving queen by tending the eggs and first brood. The resulting colony, which presumably does not emigrate until most or all of these workers have died out, is evidently therefore truly a new colony.

Hence there exist marked differences among doryline genera in the way new colonies form although in all of them this process depends upon secretory and related divisive factors introduced by the development of a sexual brood. Differences in the actual process are greatest, however, between group A genera and the driver ants. Chief among these are: the relatively precise and genus-typical scheduling of the male exodus in group A as against great variability in *Dorylus*; the relatively even division of the parent-colony's worker population and young brood typical of group A as against a great disproportion of the sections in *Dorylus*; and the sequence of group processes whereby secondary queens are eliminated in group A as against the (evidently) outright cannibalism that operates in the driver ants.

The group A processes of divergent, concurrent sectional movements are characteristic of those dorylines that are essentially surface-adapted

---

[11] It is possible—since the evidence is circumstantial and not from direct observation of behavior—that callow queens as well as workers may be involved in the "slaughter" of supernumerary young queens.

and that have relatively small colonies. Significantly, these processes become highly complex and variable in the largest colonies of *Eciton burchelli*. In this way, the group A behavioral methods of forming new colonies are perhaps best characterized by an open and sequential communication within a well-integrated system, as against those of group B regarded as an interruptive communication between variably interrelated subsystems of great scope. We continue this comparison by examining in Chapter 12 the typical processes of colony organization in different dorylines, and in Chapter 13 the relevance of habitat and colony size to doryline functional patterns.

# 12

# Individual and Colony

An insect colony is a unified group that functions as a whole in maintaining itself and reproducing. It includes many types of individuals organized in complex functions that greatly exceed the limitations of their individual roles. Because, for example, doryline workers are in very different group contexts depending on whether they are in the nomadic or statary phase, their properties appear in striking contrast from one phase to the other. The main purpose of this chapter is to review some of the major problems of doryline behavior in terms of the individual's relation to the group.

Figure 14.1 presents a theoretical outline of colony organization, derived from evidence on interrelationships that arise when the different types of individuals—workers, brood, and queen—function together in the environment typical of the species. This is a smoothly working pattern, no important part of which can be produced by one type of individual apart from the others. These are interdependent individuals, therefore, whose collective accomplishments cannot be fathomed except by analyzing the types of interrelationships that unite them. Such mystical terms as Maeterlinck's "spirit of the hive" (Wheeler, 1928), although offering an eloquent means to represent the unity of a colony, suggest no investigations either of normal organization or of types of disorganization that may appear.

In such major functions as reproduction, communication, and movement to new territory, the colony as a whole is far more than the sum of its parts. As Emerson (1939a, 1939b) has pointed out in discussing Wheeler's concept (1911) of "superorganism," impressive analogies may be drawn between the functions of an insect social group regarded as an aggregation of unit-individuals and those of any many-celled animal conceived of as an aggregation of cellular units.

Analogies of this type are of course very useful for beginning a study and making introductory points. The queen in her reproductive capacity, for instance, may be likened to the organism's reproductive system; the workers may be likened in their responsiveness to environmental conditions to sensory tissues and in their processes of interindividual liaison to neural tissues; and so on. Those who examine such analogies critically as a basis for hypotheses in research, however, soon find their shortcomings. A main one is that they are illustrative but not really theoretical. They draw no attention to differences between the compared objects so that inconsistencies develop inevitably whenever attempts are made to apply them to any specific problem. As an example, in army ant colonies, sensing and responding to environmental conditions and to other individuals is accomplished by workers that are far more than sense organs or neural tissues and the queen is far more than a reproductive unit. Rather, these individuals possess many additional properties (including those of individual functional integration) not really matched by those of tissues in the body of a multicellular individual.

How analogies may become misleading when carried beyond the purely descriptive level may be illustrated by the implications of Wheeler's term (1925), "air of dejection," for workers of an army ant colony lapsing collectively into quiescence after their initial disturbance when Wheeler removed the queen. In reality, the colony is not behaving emotionally as would a person who realizes he has lost something precious. Rather, an episode of suddenly aroused mass excitement in these ants is always followed by a sudden change to quiescence, whether or not the queen has been removed. As a test, the ants react similarly when a screened container with highly excited workers of their species is passed over them in the air, first with general excitement, then with quiescence a few seconds later, the length of the first interval depending on the number of ants in the container and how excited they are. A good hypothesis for testing this behavior is that mandibular gland secretion from the aroused ants has caused the disturbance and that a physiological reversal of the arousal into inaction occurs when the specific strong stimulus has ceased. In Chapter 14 some further implications of the superorganism concept for army ants are examined.

As earlier chapters have shown, the complex organized functions of a doryline colony are based upon processes to which queen, brood, and workers contribute when all function together. The functional cycle, for example, results from many types of interrelationships among these properties of active agents. Major types of interrelationships are illustrated in Figure 14.1. The queen's normal egg-laying rhythm is timed, Chapters 7 and 8 show, not by an internal controller of her own, but by periodic, regular presentations of trophic and stimulative effects from the colony situation. But without her ability to respond maximally to these social effects, colony cyclic function could not continue.

Also, chemical products of the queen herself seem to constitute the key factor that both unifies the colony and normally keeps it apart from other colonies of the species. In many colonies of the group A species tested on this point, we have found that fusion with another colony of the species may begin within about sixteen hours after the queen has been removed. The results suggest two factors important for colony organization. First, normally the workers are reinforced regularly in their habituation to the colony odor containing the queen component and when so reinforced accept workers bearing their queen's odor but are disturbed by workers bearing a foreign queen odor. When this habituation to the queen's odor is sufficiently weakened by nonreinforcement, however, they mix readily with workers of another colony. Members of the queen's guard, in contrast, seem most strongly habituated of all to the specific queen odor as they usually do not mix but fight or run. Second, after the queen has been removed, workers of her colony lose enough of her odor to be accepted by workers of another colony. The brood of the absorbed colony evidently retains sufficient traces of its own queen's odor to be disturbing, however, as evidenced by the fact that after being carried into the host bivouac it is cannibalized. Through these processes, group behavioral properties develop on the basis of individual biological properties.

Queenless colonies show other indications of a decreasing integrity attributable to the absence of the distinctive queen odor. Within a few days after the queen has been taken out, the bivouac of the colony begins to lose its normally compact appearance, and before many days the colony may be forming separate clusters or even masses spread out flat on the ground. Clustering in species-typical patterns, therefore, is an aspect of colony unity that is aided by queen odor. A queenless army ant colony, unlike a honeybee colony, cannot survive by producing new queens. No static comparison of species-normal structures of bivouacs can adequately disclose how they arise functionally through individual properties and behavior.

To find the basis of a normal function, colony division and individual and subgroup relationships must be investigated. Division of colonies in group A species, first, is generally a two-way process because an odorous attraction introduced for workers by a developing sexual brood (Chapter 9) attracts a large subgroup of workers in competition with the odor of the regular colony queen. Either a large part of the worker population dissociates itself from the mother queen by affiliating with a young queen, or the entire worker population divides its affiliations between two young queens as the old queen is superseded (Chapter 10). A mother queen that is superseded ceases to hold her own as the main unifying agent of the colony, possibly through an insufficient odorous attraction with advancing age. In contrast, the attraction of a new queen in her daughter colony seems weak and variable at first, judging from the signs of colony instability in the first days. Soon, however, the colony becomes able to operate as a well-integrated whole in all of its major functions.

Doryline colony function is grounded in an intimate adult responsiveness to the brood, as the described correspondence (Chapters 6 and 7) between colony functional phase and brood condition indicates. Wheeler (1928) emphasized, in his concept of trophallaxis, unification of the colony through the exchange of exudates, secretions, and nutritive substances between brood and workers. On the basis of evidence for doryline function, I have extended this concept into a brood-stimulation theory whereby the brood energizes the worker population through reciprocal stimulative relationships.

In the nomadic phase, complex patterns of stimuli from thousands of squirming larvae excite the workers who, in their turn, feed, groom, and stimulate the larvae. Highly aroused through these feedback relationships, the workers engage busily in massive group actions within and outside the bivouac; they capture far more food, emigrate regularly, and are far more active than in the statary phase when brood-stimulative effects are relatively weak.

Still other integrative effects depend on queen and brood odors. Some hours after the queen has been removed from the colony (Chapter 8), members of her cluster start a process of back-tracking into the abandoned emigration trail rather than of movement into an available raiding trail. Tests suggest that these older workers are attracted specifically to the emigration trail because it presents to them attractive remnants of brood odor and, perhaps also, of queen odor. These members of the queen's guard, normally stationed near the queen but directly above the brood in the bivouac, seem to have an affinity for the brood odor as well as for that of the queen.

The role of the bivouac as a unifying center for the colony undoubtedly depends to a great extent on the pervasive attractive odors of both queen and brood within it. The potency of these intracolony attractants may, in fact, go a long way toward accounting for the relatively weak affiliation of a doryline colony with the locality in which its bivouac happens to be stationed. When its internal condition changes appropriately, the colony emigrates readily from the occupied zone, in sharp contrast to the fixity with which the colonies of most other ants remain in their stable nest sites. As an example, colonies of the tropical tree ant *Azteca* spp. are attached so strongly to their specific nests that the entire colony stays within the tree after it is felled and dies with it (Goetsch, 1939).

All aspects of colony integration are ultimately based on properties of individuals. The resources of the doryline worker (Chapter 2) emphasize worker contributions to inclusive, complex group functions in which individuals, as individuals, have little weight. Doryline workers make and follow their chemical trails in highly stereotyped ways, infinitely more restrictive in scope for individual action than the way many camponotine ants master individual foraging routes (Chapter 4). The army ants, although much more poorly equipped visually than most surface-active ants in other subfamilies, stand high in their contact and chemical sensitivity, which, with their secretory resources, serve them well in complex group behavior.

In all army ants, group communication patterns depend on individual properties of proximal orientation and are highly developed on this basis. Workers turn to the side of weak contact or of species odor and away from intense stimulation. Nestmates are thus followed closely on the trail with great precision by travelers that appear to be responding to the combined effect of odor from the trail and of tactual and odor stimuli from adjacent ants. By the communicative process of *pressure*—a group action whereby ants move away from the side of continued and heavy group stimulation—a zone of sparse booty in raiding is soon evacuated as traffic reverses its direction along an entire column. Through *drainage*—a group action whereby ants move toward continued low-intensity contact and into vacated, colony-odorized ground—the discovery of a new booty source attracts recruits by a reversal of traffic all along a trail from the area of arousal to the base. By comparable processes, a doryline colony when scattered collects its members and brood and re-forms its bivouac much faster than can most other ants. This might describe either the normal process by which a colony gathers its widespread forces and emigrates after a raid or reacts as a whole to a catastrophic disturbance of its bivouac by humans or some other large predator. Normal communication between workers or between workers and brood is an intimate rapport through highly varied processes of reciprocal stimulation and response.

Doryline colony responses are more complicated when the reactions of different individuals differ in the same types of situation. Individual differences in function within a doryline colony are pronounced when the worker population is polymorphic, as is the case in most army ants (Chapter 3). In *Eciton,* most of the general functions, e.g., in raiding, in feeding all but the smallest brood, and in clustering, are carried out by the intermediate workers. The minor workers are also involved in these tasks so far as the limitations of their size admit; also, they predominate in handling the eggs and in feeding the smallest larvae. Major workers, although important for colony defense and in bivouac clustering, can neither carry objects nor feed brood; the submajor workers, with their large bodies but shorter mandibles, are able to carry such large objects as heavy booty or large larvae. The major workers, however, carry out an important heavy defense function by gathering readily at scenes of intense disturbances from which other workers flee, and they contribute an important traffic function by the ways in which their clumsy actions incidentally block column movements in the development of a raid or an emigration.

These differentiations in worker function increase the complexity of organization in all group behavior of polymorphic dorylines in contrast to the simpler organization of behavior in monomorphic populations of *Aenictus* (Chapter 13). In *Eciton,* the development of an emigration normally requires carrying out a complex series of preliminary events during the day's raid. The communicative processes of *Aenictus,* by contrast, are so much simpler that actions initiating a raid, an emigration, or a period of quiescence sweep through the colony far more quickly and without the detailed, time-consuming preliminaries necessary in *Eciton.*

In worker populations of polymorphic dorylines, levels of excitability vary individually in ways characteristic of the species. Workers of *Eciton burchelli* all attack much more readily than do corresponding types in *E. hamatum,* those of *E. hamatum* somewhat more readily than in *E. mexicanum.* As a result, the booty haul of *E. burchelli* contains far more adult insects than does that of *E. hamatum,* that of *E. hamatum* more than *E. mexicanum.* The importance of worker excitability level for differences of this kind is shown by colonies of *Eciton, Neivamyrmex,* and *Aenictus* spp. in the nomadic phase, when the workers are able to gather larger numbers with far more facility and to fight much more fiercely than in the statary phase. The basis lies in the pervasive excitatory relationship between workers and brood.

All dorylines respond readily to the odors and actions of nestmates and brood in their colonies. The process of social-bond formation and the worker's strong dependence on the group may begin in the larval stage. Then, when tactual and chemical stimuli from workers are combined with effects of food and colony odors, compelling conditioned responses to these pervasive stimuli can develop. Early conditioning of insects, as other researchers (Uvarov, 1932; Thorpe, 1956) have suggested, can influence their behavior as adults. Such processes may influence behavioral development of the callow workers in their colony situation. Adult workers, as a result, respond to the antennal tappings of nestmates because the nestmate presents familiar, attractive colony odors.

Queen-removal test shows that workers from other doryline colonies are accepted when they bear the species odor without the distinctive, disturbing odor of a queen from a foreign colony. Studies of the movement of these ants in a circular column show that for a time individuals of different species run closely in file, responding well to bodily contacts and odors, provided they have been given a preliminary habituation to each other's odor (Fig. 12.3). But when the column slows or stops, fighting begins as the ants come together and evidently present each other with disturbing odor differences. Normally, the distinctive odor of each doryline colony serves as the basis of social control by reinforcing all types of communicative exchange among nestmates, but foreign doryline odors and booty odors are dealt with very differently.

The sensory, neural, motor, and secretory resources of the worker are those typical of her species and thus equivalent to the resources of her nestmates. The basic role of these factors in her behavior is not a matter of innate organization in the sense that her genes dictate a neural pattern that organizes the worker's function in different behavioral situations. Individual behavior patterns, like colony functional patterns, develop in the standard environment in which the structures and physiologic factors basic to them evolved. For example, the callows, on their emergence, do not attain characteristic adult functions at once (Chapter 6) but need a few days in which further development occurs in conjunction with stimulus-response combinations. By circulating within the bivouac at first, fed by adult workers while staying in place in the larval manner, they resume where they left off as larvae although now in a different setting of reciprocal stimulation and action from which new results develop. Gradually, on the basis of expanded feeding actions, the callows become capable of varied new functions. As brood odor begins to attract them, they start dropping food on larvae and licking unemerged callows, which they then carry about. The licking of a larva after feeding leads to the callow closing her mandibles gently on it, then lifting and carrying it. Even in the first nomadic emigration, callows carry their still unemerged, inert fellows. It is very probably an attraction to colony odor that affords a basis for responses to larvae and to other callows. The next step is leaving the bivouac when hungry and following bivouac odor into the raiding trail—initially a very tentative and easily interrupted action for the callows.

Within the callow's first few days, increases in strength and other aspects of maturation doubtless contribute to a gradual improvement in actions of trail following and raiding as day-by-day observations of their progress indicate. Theoretically (Chapter 6), these changes may arise through coalescing effects of organic maturation and of experience (i.e., patterns of stimulation encountered over time) through which the callow reaches the level of adult behavior. Within four or five days, callows are no longer readily distinguished from the adults either in color or in actions.

Collective actions thus are maintained by developmental processes. The act of foraging outside the bivouac, for example, may be derived from earlier feeding. Conceivably, feeding as a larva and then as a callow furnishes a basis for licking booty and handling it, for passing it to others, and finally for collecting it. As daily excursions on raiding trails increase in scope and duration, the callow becomes able to carry booty to the bivouac when she picks it up outside. Developmental effects, by feedback processes they introduce, thus promote integration of subgroups.

The motivation (i.e., impulsion) of a worker's departure from the bivouac may come from the stimulation of movements and odors of unfed larvae and of agitated nestmates, plus effects of the individual's own chronic hunger. In doryline foraging, somewhat as in the maze learning of *Formica* (Schneirla, 1943), outgoing and returning are stimulated by distinctive processes. For an army ant, the basis for returning may be laid in early short excursions after handling booty and in feeding larvae within the bivouac. Observations of callows in raiding behavior suggest that a hypothetical running set may depend on subtle cues from the distribution of booty odor on trails and the direction of booty-laden traffic, only gradually mastered by the new workers. In the day-long raids of *Eciton,* it is the callows that first have difficulties in the afternoon when trail cues evidently become mixed through traffic complications, suggesting that they need stronger and clearer direction signals than the adults. At the same stage in early afternoon, adult workers correct "wrong" turns with a quick reversal after only a centimeter or two of progress beyond the junction; only later in the day do they begin to have trouble with these turns.

Although we do not know how long doryline workers live, their life span in general may not be much longer than that of honeybee workers, which is around three weeks. The doryline worker's life is doubtless a hard one, as is that of the honeybee, though possibly eased somewhat by occasional day-long periods of rest in the bivouac wall in the nomadic phase and by much longer intervals of low activity when the colony is statary. The hazards of combat, however, may reduce the life span of the average worker to little more than two cycles, or only about two months, as the frequent large brood increments seem just barely to replace the losses. These considerations suggest that doryline colonies must be perennially engaged in processes of functional reorganization involving their entire personnel. This certainly happens each time a great new contingent of callow workers enters the colony.

These replenishing processes may be most complex for workers of the queen's guard, who, beside the queen, must be the longest-lived members of the colony. All workers are attracted to the queen, and these older workers may become members of her regular central group by virtue of a kind of elimination process. Perhaps workers among those first strongly affiliated with the brood are the ones most likely to shift to the adjacent queen's group. Odor is basic, without question, and, through normal processes of social stimulation, members of this group serve as agents in spreading the queen's odor through the bivouac (more effectively than volatilization alone would permit). The strength of their behavioral fixation on the queen is shown by the persistence with which these darker workers cluster about the parent queen when she is superseded in colony division.

In Chapter 4, I offered a set of hypotheses for the species-typical patterns of predatory raids in doryline ants. Species differences in the pattern of raiding are common and doubtless rest on stable structural-physiological factors in the workers. Results of observations and tests suggest, for example, that in *Eciton hamatum* certain factors in worker equipment are critical for the group pattern of column raiding that arises, e.g., (1) highly acute tactual and olfactory sensitivity, (2) workers less excitable than *E. burchelli*, and (3) with less diffuse and perhaps less concentrated trail chemicals than in *E. burchelli*. The combination of factors presented by *E. hamatum* would account for advance raiding groups heading relatively narrow columns and dividing while they are still small; those presented by *E. burchelli* would account for the advance raiding groups of that species becoming very large before splitting, with relatively broad rear columns. Significantly, when colonies of *E. hamatum* are at their peak of nomadic phase excitement, the end groups grow to their largest before dividing. This phase difference is much more pronounced in *Aenictus*. Such results suggest ideas for systematic investigations of species differences in behavior.

Forming clusters is as characteristic a type of behavior of doryline ants as is their pattern of locomotion. Clustering, although its main phase is one of passivity, begins with active responses and is not at all simple. Clusters arise through workers becoming quiescent in clumps when the general level of stimulation is low (e.g., in dim light), especially under gentle, repetitive tactual stimulation in the presence of brood odor. Workers also seem to become immobilized as a physiological reaction to being stretched—for example, when they form hanging chains in making a bivouac (Chapter 3).

Formation of ant roadways illustrates a type of clustering behavior nearly as important adaptively as the formation of bivouacs. In surface-adapted *Eciton* species, whose heavy emigration columns usually move at night, this type of clustering becomes increasingly prominent as the larvae grow during the nomadic phase. Near the end of the phase when the larvae are largest, clustered structures I call "flanges" and "fills" are often seen, usually where the route is roughest. These aspects of army ant road making are illustrated in Figure 12.1.

The ant roadways arise most often and in their best developed form when the burdens are large, as when a colony has a sexual brood in its larval stage (Schneirla, 1948, pp. 98–99). Night after night as the larvae grow, these events arise increasingly. Early in the emigration, an hour or two after nightfall, workers begin dragging the bulky male larvae from the bivouac. This action begins near the exit from the cluster, then in time expands to a broader front. At first each unwieldy larva is tugged about by workers of varied sizes. Soon the former raiding trail is obscured in a wide field of action around the bivouac base, where growing numbers of workers mill about with more and more larvae. Adding to the difficulties, as the burdens are yanked on and off the route, are such common physical obstacles as leaves, brush, and tangled vines. By degrees, however, within an hour or two, changes arise that resolve the transportation problem nicely.

As the many groups struggle around with their burdens, smaller workers are often rolled under. Gradually, more and more of them drop out of the action, huddling or stretching out where the swarming-under occurred. By small steps, a layer of clustered ants is thereby formed over the trail itself and in wide areas beside the trail near the bivouac. Over this growing roadway other workers continue to pull larvae about, incidentally smoothing the substratum by forcing ants in projecting parts of these clusters to shift position. Although the roadway is not uniformly thick, the running surface becomes level as most of the ants cluster at edges of leaves and other obstructions and in depressions where the greatest struggling and threshing about occurs (Fig. 12.1).

Watching particular ants caught in the rush, one notes that elements of the roadway are recruited mainly from among smaller workers because these are the ones most subject to being repeatedly bumped and overrun. Once they are rolled under, these ants stretch out with tarsal claws anchored to leaf edges and similar places or to the protruding legs of nestmates already clustered, then lie motionless in place except for slight vibrations of antennae.

As a level roadway forms, a moving procession carrying burdens gets under way from the bivouac. On this route, clustered workers may lie underfoot for hours as travelers move over them. Clearly, it is repeated tactual stimulation that keeps the pavement of bodies immobilized, for clusters begin to disappear from places where the column no longer passes. In places where side eddies of traffic have ceased to run or in remote depressions where ants and larvae have rolled together off the trail, inert ants at length begin to stir and shift about, disengage themselves, and begin to pull at nearby larvae, then are soon away.

The roadway, usually widest at its starting point, narrows progressively away from the bivouac. On further nights, as the brood grows, the roadway becomes wider and longer, a reminder that it is based on struggle in carrying bulky objects over rough ground. When the larvae near maturity, carrying them raises physical difficulties so great that a wide ribbon of

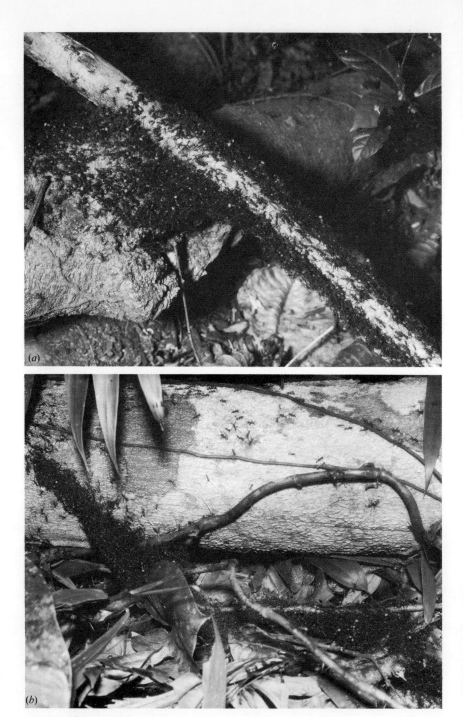

FIGURE 12.1.
Clustering behavior of *Eciton burchelli* during emigrations in the late, most excitable part of the nomadic phase of their colony. (*a*) Ramp from a rock to a vine, with flanges bordering the route on the vine. (*b*) Ramp on lower side of a route down

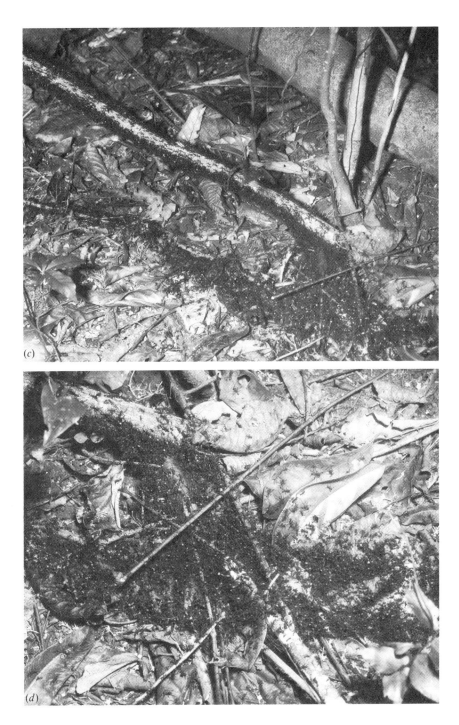

the side of a log; ant road over a rough stretch on the ground below. (c) Flanges bordering the route along a vine, with a scattered ant road formation at the turn. (d) Close-up of aggregation at the turn, illustrating how the clustering occurs.

clustered bodies may meander as far as 75 m from the old bivouac before bare spaces with only occasional ant-fills appear.

Interesting changes occur in the carrying of larvae as the roadway grows. Once the pavement has become smooth, the initial period of struggle is short as each larva is brought out; soon the smaller workers have been sloughed off, and one or two of the largest workers move off with the larva. As the carriers, usually submajors, get under way, some shifting of positions and of grips occurs until at length they can hold the burden firmly as they run along (Fig. 5.2). Usually two carriers run in tandem, one clutching the larva at its anterior end, the other near its midsection with the larva slung beneath their bodies and straddled. When the sexual larvae are mature, it is only the submajors that persist in this action as other workers are too small and their mandibles too short to permit their running well in traffic with these loads.

Once the eventual carriers are moving with their burdens, they travel steadily at a surprisingly fast rate with few stops except where traffic jams or new obstacles arise. Although the total labor cannot be called "co-operation" in the human sense that the participants are working together with a common plan and a common "purpose," it is organized in the sense that its activities are interrelated in standard ways. As these actions progress, the emigration is carried out and the brood is moved. The idea that either the road units or the porters understand the problem they are helping to solve might seem reasonable to a person seeing these impressive events for the first time but is untenable to one familiar with the detailed operations.

Most of our knowledge about insect social organization has come from studies of activity around the nest site or among small groups elsewhere, as in tending aphids. Studies of organized activities on a large scale have progressed slowly as the essentials are either lost to sight or are scattered and elusive. Alfred Emerson (1938), professor of zoology at the University of Chicago, tried to meet this problem for termites by classifying species patterns of nest building in terms of nest structures, used as "morphological indications of behavior patterns . . . [which] express the behavior of a population." On this basis he worked out aspects of nest building speculatively in relation to the evolutionary background of species behavior patterns. Although useful, these indirect methods have very limited theoretical value both as to the behavior patterns themselves and as to their emergence through natural selection. This difficulty will hold until we have better evidence on how group behavior is integrated in the actual building processes of each species.

Analytical studies of coordinated group behavior are needed. To answer this need, systematic studies can be made of the forays of doryline ants, which often reach massive proportions. This is true of the swarm

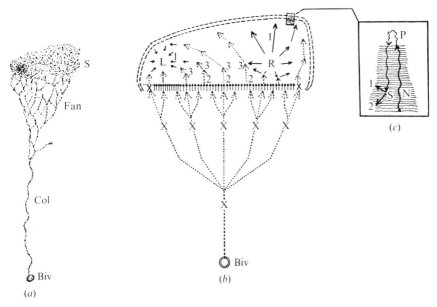

FIGURE 12.2.
Pattern of swarm raiding in *Eciton burchelli*. (a) Schema of the swarm (S), fan, and basal column (Col) during raiding. (b) Abstraction of this raid to illustrate basal-pressure effects (X), lines of communication by columns (1, 2, and 3, etc.), and concentrating (L) and expanding (R) flank actions in the swarm. (c) Small section of frontal border, to illustrate reactions of "pioneer" over scented ground (N), briefly extending the scent (P), then returning into the swarm (S).

raids carried out by *Eciton burchelli* (Schneirla, 1940), which offer the most complex instance of organized mass behavior occuring regularly outside the home site in any animal except humans. Detailed studies of consecutive raids by the same colonies emphasized: (1) the development of raids and their changes through the day, (2) contrasting general and local processes of group behavior, (3) analyses of behavior of workers in the main types of subsection, and (4) major changes of organization as in swarm splitting. All of these events were studied in relation to concurrent conditions in the bivouac. Important comparisons involved studies of raids by the same colony in successive environmental situations and at sharply differing levels of colony excitation in the nomadic and statary phases.

The typical pattern of swarm raiding in *Eciton burchelli*, sketched in Figure 12.2, involves: (1) the swarm, a massive body of workers moving broadside away from the bivouac; with which it is connected by (2) the fan, a network of columns tapering behind it; to (3) the base column leading to the bivouac. The nomadic swarm raids of *E. burchelli* begin at dawn, when the first light brings a stirring of workers on the bivouac surface. First, loose columns and groups of meandering ants expand in all directions from the bivouac. Then in an hour or two, one part of this mass becomes dominant and moves outward as an increasingly organized unit. This mass, as it grows, absorbs other sections by drainage and improves in its internal organization. Gradually, as this swarm moves outward, it enlarges, then a fan and a base column form behind it. During the morning, these three sections of the swarm raid grow more or less in step. Usually, before midday, the nomadic swarm has advanced 70 m or more from the bivouac and is 12 to 15 m wide, with perhaps between 100,000 and 500,000 ants operating in it.

To emphasize that swarm raids are carried out by individuals, let us summarize some individual properties that are involved. These include: (1) a low-grade visual sensitivity, inadequate for distance orientation but serving limited functions, e.g., arousal at dawn and withdrawal from sun-bright areas; (2) routine responses to tactual stimulation, e.g., turning toward weak contact or away from intense contact; (3) such olfactory responses as stereotyped trail following, moving toward colony odor and booty odor, or away from foreign or intense odors; (4) reactions of pouncing, biting, and stinging given to moving objects that present booty odor; and (5) secretory functions as in trail laying and in releasing mandibular gland product in amounts dependent on individual excitatory state or external stimuli.

Although individual properties are simple and stereotyped, individual behavior differs greatly according to the circumstances of collective action. Under appropriate conditions any worker may be involved in any function, from extending the odor zone to returning with booty. The odor zone is extended by workers entering new ground, in what we call the "reversal reaction" (Fig. 12.2). The pioneer's advance across colony-odorized ground ends in a shock reaction when the odor stops; then she crawls hesitantly forward for a short distance, releasing from her gaster a line of trail chemical; and finally she reverses direction quickly and returns into the swarm. Through reversal reactions, typical of ants entering new ground, the swarm zone is extended by successive pioneers in a relay fashion. Variations of these reactions occur when workers in the forward zone meet booty odor or moving booty (Chapter 4).

The problem of swarm organization is raised sharply by the following question: How is it possible for a swarm of *Eciton burchelli* to advance broadside in one direction? Results show that one factor contributing to this ability is the *frontal barrier,* i.e., the disturbing effect of new ground opposing forward progress, continuously forcing reversal reactions from workers entering it. This factor serves as a gradually yielding anvil, flattening the swarm by slowing down the advance of all workers in the forefront. The advance of many individuals against the frontal barrier is impelled by *basal pressure*—a major summation of collective tactual-odor stimulation delivered continuously by ants entering the swarm from the bivouac side—which serves as hammer.

The swarms of *Eciton burchelli* are more clearly flattened and better defined than are those of other swarm-raiding species that are subterranean nesters and whose recruits issue from underground. In *E. burchelli*, the new forces arrive in more or less continuous columns moving out from the bivouac on the surface and in a well-defined direction. These forces, spread radially in the fan, present a basal pressure across the entire rear of the swarm (X, Fig. 12.2*b*). The dotted arrows (2) in Figure 12.2*b* indicate how this effect is exerted by moving columns in the swarm in such a way as to direct most of the reversal reactions forward. Basal pressure, operating against the frontal barrier, not only compresses the mass laterally but also acts equivalently upon different sections of the swarm over a period of time; consequently, the entire great body keeps fairly well to a straight course. Swarms of *Labidus praedator*, in contrast, are fed by columns that arise so variably from underground as to exert irregular basal pressure effects; as a result these swarms (and those of driver ants also) are quite variable in their directions of progress.

The swarm raids of *Eciton burchelli* therefore have an internal organization superior to that in the raids of *Labidus praedator*. A better integrated communicative pattern prevails in *E. burchelli* because the basal pressure effect is so consistently directed forward through the mass as to steady it and press it onward in a common direction.

Another communicative effect, that of cross-liaison, is exerted by columns moving across the swarm, connecting the wings. These columns, kept going by interactions of the frontal barrier and the basal pressure factors, enable the swarm to continue operating as a unit and moving forward despite physical obstacles and frequent diversions of large groups, as in raiding up trees. Although, under these conditions, the swarm may lose headway for a time, it regains its forward momentum when its main forces are reunited and once more subject to the common basal pressure effect.

The influence of all three communicative factors is revealed in the *flanking movements* carried out by these swarms, in which the mass as a whole pivots first one way and then the other way in operations that surely increase the booty haul considerably. In effect, the net of *Eciton burchelli* swings first to one side and then to the other as lines of communication change between its two dominant subsections in the wings. In contrast, subgroups in the raids of *Labidus praedator* change position so variably that flanking movements are usually difficult to detect. In *E. burchelli* steady operation of the main communicative factors keeps the wing sections well integrated in most of their actions. The wing sections are the key parts of the mass because their positions give each of them just one fluid boundary (i.e., the central zone) across which communication effects transfer readily. The outside boundaries, on the frontal and lateral borders, are set by new ground and are relatively rigid. In each of the flank zones, energized by basal pressure, behavioral conditions that build up constantly are readily transmitted across the center to the opposite wing by cross-connecting columns, which thereby have a communicative role.

Flanking operations therefore reveal important aspects of swarm organization. In any one such movement, the critical events are behavioral changes of opposite nature, i.e., concentration *versus* expansion. While one wing (e.g., L, Fig. 12.2b) is concentrating its personnel, it serves as a pivot for a partial turn of the swarm; meanwhile, the other wing (R, Fig. 12.2b) expands while swinging mainly forward and obliquely inward. Basal pressure operates to project the expansion of the mobile wing strongly forward while the same section is pulled inward by the drainage effect of columns headed toward the concentrating wing. The fact that pressure and drainage processes coordinate in each swing toward the concentrating flank is clear from coded notes and analyses of detailed column movements in well-developed swarms. As the lateral component is given an oblique thrust forward through its interaction with basal pressure (3, Fig. 12.2b), the general outcome is a swing of the entire raiding body through about 20 degrees toward the concentrating flank. In a vigorous swarm, flanking movements reverse roughly every fifteen minutes.

In each reversal of a flanking movement, the previous pivotal wing becomes the mobile part of the mass, and vice versa. This change is normally influenced by opposite effects in the two wings, in which concentration in one wing reaches a peak and then shifts toward expansion after it reaches a maximum while expansion in the other wing shifts toward concentration after it reaches a maximum. In these changes the flanks influence each other through commissural columns.

While the communicative factors, pressure and drainage, continue in effect, subgroups are adequately integrated, and the swarm proceeds in predatory action as a unit. This condition is maintained so long as: (1) the internal condition of each flank reverses regularly; (2) each flank influences the other through central commissural columns; and (3) these processes are energized and impelled forward by a steady mass pressure from the bivouac side. When late in the morning of a nomadic raid, the second and third of these effects are weakened, both through the swarm growing oversized and through a lowering of basal pressure by booty-laden return traffic, swarm organization deteriorates. Decreased integration is shown when the flanking movements become indistinct, internal columns lose their normal patterning, and the swarm wavers in its progress. As internal communication processes break down, the wing sections first begin to operate out of phase with each other, then diverge and operate independently as separate masses connected only by columns in the rear. Through the radical modification of the initial swarm pattern, a new type of organization arises in the total swarm raid. Further divisions of the raiding bodies take place in the afternoon.

Patterns of group function observed among members of a doryline colony result from interactions among conditions both extrinsic and intrinsic to the colony (Fig. 14.1). Our analyses of roadway formation and of swarm raiding in *Eciton burchelli* have shown in different ways how the behavioral resources of the individual worker, although limited and stereotyped in themselves, in situations of group action in the forest environment contribute collectively to highly organized, adaptive total operations. "Leadership" here is a generalized property of the aggregate since individual roles in the swarm—whether in pioneering, predation, or in commissural or booty-return actions—are subject almost momentarily to change according to the circumstances of the individual in the general situation at the time.

In the described ways, individual properties that hold workers in a strictly proximal orientation to the group situation contribute to patterns of interindividual behavioral relationships. These relationships arise and change according to the collective setting, as we have shown for the ant-roadway and mass-foray situations. In the swarm situation, through the described communicative processes, the impressive phenomena of directionalized mass movement with its regular alternate flanking operations can appear. In a laboratory arena, however, the same individual properties lead to either the fractionalized operations of small, separate groups or—as the one unitary action possible in that situation—to the monotonous aggregate action of circular milling (Figs. 12.3 and 12.4). Studies of milling and similar mass activities (Schneirla, 1944c) clarify the general problem of levels of integration in social behavior by bringing to light limitations of the animal not evident in its everyday behavior.

Circling arises in army ants when many individuals yield in common to routine stimulus-response mechanisms dominating group locomotion. In this respect, it resembles emigration (Chapter 5), with the difference, however, that the pattern of emigration is adaptive whereas that of milling is likely to be maladaptive.

Analyses of milling show us that in such group functions the participants are so limited in the scope of their responsiveness to surrounding conditions as to respond in circular, reflexlike ways. The milling pattern, taken alone, may falsely suggest psychological inferiority by obscuring

FIGURE 12.3.
Excited workers from a nomadic colony of *Eciton hamatum* move in a circular column in a laboratory nest. The column began with the slow movement of a few ants around the square moisture pad, then became circular in speeding up.

resources for plastic, opportune behavior. This is true especially when patterns analogous to milling appear even on the human level.

It happens that the army ant circular column really typifies the nature of the animal (Chapter 2). Highly aroused workers in confinement readily display canalized behavior patterns. In nature, however, through the heterogeneity of forested terrain, these animals meet innumerable possibilities for varying their behavior. Environmental diversity then permits a limited repertoire of stereotyped individual resources to produce complex, adaptive behavior. In the raid, as an example, the same individual properties contribute to subgroup processes capable of progressive changes and so well integrated as to produce impressively adaptive aggregate performances.

On its face then milling and other simplified group behavior suggests only that a psychologically reduced performance can occur under certain conditions. Elementary group patterns of this type may disclose binding, life-long limitations in psychological resources, as in the case of the army ants. It is true that army ants, on the one hand, are constitutionally susceptible to circular-column behavior and are freed from it only by the circumstances of environmental variation. But humans, with a cortical basis for versatile corrective patterns (e.g., learning to counteract propa-

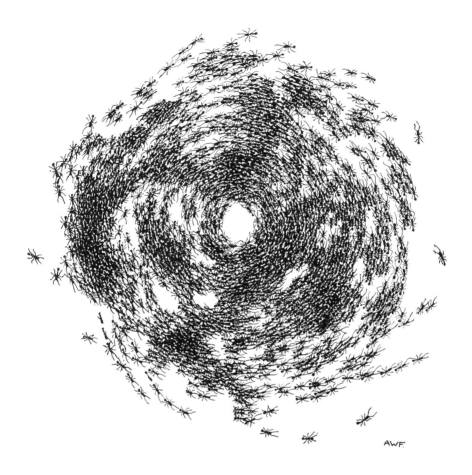

FIGURE 12.4.
Circular mill formed under natural conditions by a few thousand workers of *Labidus praedator* after they had been cut off from their raid by rain. This mill continued to rotate for more than 30 hours, stopping only when all of the ants were dead. (Facsimile drawing by Mrs. A. W. Froderstrom from a photo.)

ganda or other coercive measures) and with encouragement, should be able to reduce social behavior of the milling type to an occasional subway rush.

The rarity of milling behavior in army ants under natural conditions shows us that workers functioning in large groups in the forest can so operate as to transcend their limitations as individuals or as groups in simplified settings. The colony, when appraised objectively, rises far above individual resources, exhibiting complex collective functions attained through the coordination of elements (e.g., queen, brood, workers), which by themselves are highly limited in function and behavior. The colony therefore can be analogized with the individual only in a limited, figurative sense (Schneirla, 1946).

The systems of events described in this chapter, through their interdependence, provide a basis for impressive patterns of group organization in the doryline colony. One set of the queen's properties, her reproductive functions, is critical for the colony functional cycle but owes its rhythm to repetitive changes in stimulative and trophic effects exerted on the queen from the general colony situation. Another set, accounting for her distinctive odor, gives a basis to normal processes of communication that keep her colony functioning as a whole and apart from others of its species. Reciprocal stimulative relationships between workers and successive broods serve to energize the colony's cyclic functions and thus to account for their species-typical timing. The impulsion of group foraging and the initiation and maintenance of the nomadic and statary condition in alternating phases depend upon differences in the reciprocal-stimulative and trophic functions for which changing developmental conditions of the brood are critical. When the brood-excitation factor is high, colony operations in appropriating space and food are correspondingly high and complex. These results occur because a qualitatively superior pattern of foraging can arise in the nomadic situation, reaching a threshold at which an emigration results almost inevitably. That foraging can give rise rather directly to emigration in *Aenictus,* but only through a complex series of changes over a day-long interval in *Eciton,* is attributable especially to the simplified intrabivouac communication processes of the former in contrast to the complex, heterogeneous communication processes of the latter.

Trail following, clustering, foraging, and emigration—activities common to all dorylines—appear in very different behavioral settings in the different genera. One of the outstanding differences among the dorylines is that between the complex serial functions of group communication in polymorphic species and the relatively simple communicative functions of monomorphic species. As an example, to explain the difference between *Eciton* and *Aenictus* in the relationship of foraging and emigration, it is necessary to appeal to widely different characteristics of communication in the two genera. To explain the existence of very different patterns of raiding in closely related species of *Eciton,* however, our attention turns to the more specific, individual characteristics of sensory and excitatory thresholds to precision or diffuseness of secretory functions to differences in individual strength, and still others. Significant differences in a general communicative factor, basal pressure, have been pointed out in the organization and directionalization of swarm raiding in certain surface-adapted species as against certain subsurface-adapted species. Species in the three genera of group A differ in major ways from those of group B in the regularity of their cyclic changes as well as in the frequency with which group foraging operations lead to colony emigration. These are only a few of the group functions in which species-typical or genus-typical differences are grounded in different individual properties.

Doryline cyclic patterns are based upon many interacting structural, physiological, behavioral and environmental factors affecting group interactions. In healthy colonies operating in the optimal forest environment, these reactive systems can occur maximally. Organizations in function attained by any species through this pattern are not determined through the inherited properties of any one type of individual—queen, workers, or brood—or even through those of all individuals together. Rather, they develop in aggregations of queen, workers, and brood that are functioning under environmental conditions appropriate for working out changes in individual and interindividual behavior essential to them. The resulting functional patterns of subgroup and colony are not systems that preexist in any real sense in the participants, to be drawn out by specific environmental stimuli. Only in the sense of their development in the group situation can the adaptivity of these group processes be understood in terms of its often subtle species and generic differences.

# 13

## *Aenictus*: Army Ant on a Small Scale

*Aenictus* is remarkable in several respects. In contrast to its wide distribution (Fig. 1.3) as a genus, it is one of the smallest of known dorylines both in individual and in colony size. *Aenictus* intrigued me from the first perhaps because although little known, it seemed *Eciton*-like.

In work with *Neivamyrmex* (Schneirla, 1958) I had found it so markedly equivalent to *Eciton* in nomadic functions as to warrant grouping the two genera as functionally very similar. As this comparison of army ant genera progressed (Schneirla, 1957b), my interest in studying *Aenictus* intensified. Meanwhile, new observations by others (Chapman, 1964; Wilson, 1964) suggested that this little ant might to a large extent hold the key to the problem of understanding doryline cyclic functions in general.

There is much to be gained by comparing *Eciton,* as one of the most complex army ants, with *Aenictus,* as probably the simplest. First, *Aenictus* seems to be the Old World doryline closest to the cyclic functional patterns of *Eciton* and *Neivamyrmex;* second, it has a wide range from western Africa to New Guinea and Australia and may be the most generalized of existing dorylines in its smallness and virtual monomorphism.

Among perhaps nearly one hundred existing *Aenictus* species (Wilson, 1964), some in Asia are surface-adapted and active both in open areas and in forests; the rest, mainly hypogaeic, are virtually unknown. Workers of *Aenictus,* as Figure 2.1 emphasizes, are scarcely larger than even the minor workers of *Eciton* and *Dorylus*. The males and queens are correspondingly small. Workers of *Ae. gracilis* approximate only 3.5 mm in body length; those of *Ae. binghami,* one of the largest, are only about 5.5 mm long. Populations of even smaller workers seem to prevail among hypogaeic species.

FIGURE 13.1.
Monomorphic brood of a colony of *Aenictus laeviceps* as the colony is about to enter the statary phase. This brood, just entering the prepupal stage, has just been removed from the colony's new bivouac in a hollow log. (See Plate VII, following page 138.)

In keeping with small individual sizes in a quasimonomorphic series, *Aenictus* is probably the smallest among army ants in colony size (Table 2.1). An estimate of 100,000 for a worker population of *Aenictus laeviceps,* although perhaps a little high, is not even 1% of the estimate for driver ants and scarcely 10% that for *Eciton burchelli.* The virtually uniform adult and brood worker series, represented in Figures 2.1, 2.2, and 13.1, contrasts strongly with the diversified series of other genera. The differences in worker series between *Aenictus* and the other dorylines correlate with the relative complexity of their colony functional patterns.

My own interest in *Aenictus* dated from the time when, shortly after beginning field studies on *Eciton,* I read a paper by the eminent Italian ant taxonomist Carlo Emery (1895a). Clearly, from his comparisons of specimens, he viewed *Aenictus* as one of the existing dorylines nearest to the primitive trunk near the ponerines. Even the sparse literature on these ants (Brauns, 1901; Wheeler, 1930) hinted that relatively small size and uniform structure of workers might affect colony behavior as compared with that of the complex, polymorphic *Eciton.*

Inferences from structure to behavior require careful scrutiny especially since research on behavior is still relatively new as an approach to understanding evolutionary relationships. This is a rather different approach from that taken by taxonomists, whose classifications derive mainly from studying and comparing structures both in existing and in extinct forms. The approach through behavior is necessarily the more diversified one, for the relationship of structure to behavior is nearly always indirect and presents new difficulties in the investigation of each different type of animal. However simple the animal may seem to be, the relationship of its structure to its behavior still resists investigation. The behavior of insects (and of other invertebrates) is only relatively more closely related to, and more directly influenced by, specific structural characteristics than is that of mammals (Schneirla, 1946, 1948). From progress made in discerning the significance of behavioral data for evolutionary relationships, derivation of new methods for relating structures and functions to behavior should furnish useful clues to research and theory. Comparison of colony functions in generalized *Aenictus* with those in more specialized dorylines, the focus of this chapter, may indicate how species-typical patterns develop in existing forms and also suggest how they evolved.

Helpful in starting research on *Aenictus* was the American biologist and missionary, J. W. Chapman (1964), whose observations on this ant in the Philippines suggested a similarity to the cyclic pattern of *Eciton* as I had described it. Significant also were reports to the same effect from observations in New Guinea and elsewhere in the Far East by E. O. Wilson (1964) and in China and Australia by W. L. Brown, Jr. (in Wilson, 1964).

In 1961 I was at last able to study *Aenictus*, opening a program in the southern Philippines that was continued by the biologist, Alfredo Reyes, my collaborator, and his aides (Schneirla, 1965). We worked at first in a forested area above Dumaguete City, then by degrees expanded to a larger area, including a plateau and a deep ravine of mixed cover farther down the mountainside; also the daily period of research lengthened until it stretched—to meet the usual schedule of *Aenictus*—around the clock.

The activity schedules of army ant colonies vary greatly according to genus and species, as comparisons in Figure 14.2 emphasize. Surface-adapted *Eciton* species are day-active in raiding, but the surface-active *Neivamyrmex nigrescens* is nocturnal although both emigrate by night. *Aenictus* turned out to be very different from both of these. We found its colonies, when nomadic, capable of raiding and emigrating at any time of day or night, as a rule, with variations dependent upon colony condition in the phase. In this respect, the routine of *Aenictus* resembles the variable day-and-night activity schedules reported for driver ants (Raignier and Van Boven, 1955).

These generic differences may depend to a large extent upon sensitivity to light and associated environmental conditions. Most of the New World dorylines have small eyes; those in *Eciton* may cause arousal by reactions to increasing daylight after dawn but in *Neivamyrmex nigrescens* by decreasing light at dusk. *Aenictus,* whose workers (like those of driver ants) are eyeless, reacts to light in tests but seems far less keyed to environmental conditions in its daily schedules. As a variation from *Eciton* and *Neivamyrmex,* its receptors may mediate dawn-arousal or dusk-arousal according to current colony excitation level (Chapter 7). Hypogaeic species, distinctly light-shy, may follow patterns even less related to daily environmental changes.

Our research concentrated on *Aenictus laeviceps* and *Ae. gracilis,* the two dominant surface-adapted species in the area of study. Early in the nomadic phase, these ants usually begin their raids at dusk. Then, as the phase continues, raids start also around dawn, and soon they can begin at almost any time around the clock (Fig. 14.2). The same holds for emigrations (Schneirla and Reyes, 1969). These results suggest that a hypersensitivity to light (and other surface conditions) prevails early in the nomadic phase, curbing group activity aboveground in the daytime, but is followed in that phase by a condition of increased tolerance of surface conditions by day as well as by night. This tendency, which also appears in colony nesting, is typically reversed near the end of the nomadic phase and in the statary phase when the ants go underground.

How far these ants have progressed in surface adaptation is shown also by the frequency with which their colonies, when nomadic, bivouac in clusters on the surface itself. Their bivouacs then are simple masses of the "platter" type (Chapter 3) formed beneath surface cover or even exposed, and, as shown in Figure 13.2, they are little more than flat, disc-shaped clusters of workers on the ground, enclosing brood and queen. This type of bivouac—rare in *Eciton*—may be considered a primitive form among dorylines, and it is also common in the colonies of many ponerine legionary ants (e.g., *Leptogenys diminuta* of the same area) when active on the surface.

But in the last days of the nomadic phase, colonies of these *Aenictus* begin to occupy well-sheltered sites; then in the statary phase, they invariably enter subterranean or deeply sheltered sites (Schneirla and Reyes, 1966). A similar tendency appears in colonies of surface-adapted *Eciton* (Schneirla et al., 1954). As suggested in Chapter 3, such changes in nesting seem dependent upon altered physiological conditions in the colonies, caused by modified effects of brood. The difference is that colonies of surface-adapted *Aenictus,* in becoming statary, seem to shift much further from the behavioral and physiological condition of the nomadic phase than do those of *Eciton* and *Neivamyrmex.* This is shown particularly in the typical absence of surface raiding in their case through

FIGURE 13.2.
A platter-type bivouac of the surface-adapted *Aenictus laeviceps,* formed under dry leaves and a rock with exposed clusters on its downhill side. The ants are beginning a raid by spreading from the bivouac. (See Plate VIII, following page 138.)

most of the statary phase (Table 4.2) although such raids, however reduced in frequency and vigor, are continued by the other dorylines mentioned.

At corresponding times in the functional cycle, patterns of raiding are generally much smaller and simpler in *Aenictus* than in *Eciton* and *Neivamyrmex* (Fig. 4.6). Except when most excited late in the nomadic phase, these ants usually develop only one trail system with a single trail from the bivouac in contrast to the three typical in *E. hamatum.* Also, colonies of *Aenictus* seem to use traces of preexisting trails, remaining perhaps from previous raids by other colonies, far more often than do

*Eciton* and *Neivamyrmex*. This difference is indicated in *Aenictus* by the frequent passing of emigrating columns well beyond the zone of colony raiding. This suggests a primitive condition of trail making in *Aenictus*, with the chemical substances less specific to particular colonies than in more specialized dorylines.

*Aenictus*, we have found, varies far more than *Eciton* and *Neivamyrmex* in its program of predatory forays. The start of a raid in *Aenictus*, by day or by night, is usually marked by a local turbulence at the margin of the bivouac, followed soon by the appearance of a group of workers on the ground in a rushing column or spreading mass. Although the forays of *Aenictus* usually develop much faster, they are broadly similar to those of *E. hamatum* and *N. nigrescens* (Fig. 4.6) in that they also involve an exodus from the bivouac in which a column divides and redivides while the raiders capture prey and build up caches at trail junctions; then a reversal of traffic takes place in which booty may be carried to the bivouac. But between the stages of exodus and reversal in *Aenictus*, traffic on the base trail is often very light; also many of the raids, especially early in the nomadic phase, may end with just one exodus and a general return. This is the characteristic pattern, even though *Aenictus* is also capable of complex forays in which there are successive operations from the bivouac, each reinforcing the raid and extending its front, before a lasting reversal of traffic toward the bivouac (or an emigration) occurs. Such raids become common as the nomadic phase advances and the expeditions are featured by the increasing numbers and excitement of their participants.

Nomadic phase raids of *Eciton hamatum*—to emphasize their vast difference from those of *Aenictus* in pattern and scope—increase steadily in complexity from dawn to dusk, spreading widely and often reaching out more than 250 m from the bivouac, with long and complex series of worker caravans returning booty to the colony base as well as to caches. The much simpler raids of *Aenictus* vary far more, as they may last from less than an hour to a few hours; also they extend on the average only 25 m from the bivouac. A relatively complex raid carried out by a colony of *Ae. laeviceps* late in a nomadic phase involved a series of three overlapping operations of exodus, each of these with successive branches headed by column networks and spreading groups of foragers and with numerous caches at trail junctions. A still larger raid by the same colony on the next day involved two lengthy trail expansions of about four hours each, in which local terminal groups often grew to widths of more than 5 m and came to resemble miniature swarm raids (Fig. 4.6). The few raids carried out by colonies of *Aenictus* on the surface in the statary phase are small and weak in comparison with nomadic raids by the same colonies (Schneirla and Reyes, 1969).

As a major difference in predatory pattern, colonies of *Aenictus* usually operate—even in the nomadic phase—on single systems of branching trails with just one basal route to the bivouac. Only at the height of the nomadic phase do they approach *Eciton* and *Neivamyrmex* in their complex patterns by developing one or two additional trail systems from the bivouac. Their raids commonly involve forays of varying durations, overlapping at times, or, at other times separated by periods without any raiding. This contrasts to the ten hours of continuous predatory action usual for *Eciton* and the six to eight hours usual for *N. nigrescens*. As continuous records show, nomadic colonies of *Aenictus* nevertheless spend about the same amount of raiding time per twenty-four-hour interval as colonies of *Eciton* in their day-long raids, i.e., ten to twelve hours. The first raids of *Aenictus* in the nomadic phase are generally smaller in numbers and scope (and in booty captured) than are the later raids of this phase.

Thus the daily activity programs of colonies in these three genera differ greatly. In *Eciton*, in the nomadic phase the sequel to each long daily raid is usually a nocturnal emigration; in the statary phase it is always a general return to the bivouac. In *Neivamyrmex nigrescens*, also, a raiding exodus with its subsequent operations, including the emigration or the general return, constitutes one continuous set of actions. But with *Aenictus*, a raiding foray may be followed by a further surge of workers from the bivouac, by an emigration, or by a general return to the base and a subsequent interval in which all ants stay in the bivouac. "Quiescent" intervals in this army ant may come at almost any time and last from a few minutes to several hours. This does not mean quiescence within the cluster, for, as laboratory observations show, worker actions with brood, centered on feeding, may proceed to an extent depending upon food supply, the brood stage, and how much feeding may have gone on just before.

To sum up, the group forays of *Aenictus* often begin suddenly with a rushing column or spreading mass from the bivouac. They may follow raids or emigrations or may overlap or follow intervals of quiescence. Activities taking place at a distance from the bivouac, such as pillaging one productive area after another with the booty accumulated in caches, may be connected with the bivouac by only a slender column on the base trail. Commonly, one surge of raiding may be overlapped or followed by a further one, an emigration, or a general return and subsequent quiescent interval. In certain respects this pattern bears interesting similarities to the forays of legionary ponerines and may be considered primitive.

Turning now to a comparison of nomadic patterns, I wish to show how booty is more closely related to emigrations in *Aenictus* than in more specialized dorylines. *Aenictus* takes not only ants and their brood but

FIGURE 13.3.
Workers of *Aenictus gracilis* posed to match a field sketch of an attack on a worker of *Polyrachis bihamata*, a tree ant that is a frequent victim of raids by *Aenictus*.

also a wide range of other arthropods including wasps, roaches, and termites. For such a small ant it moves with great speed, and with potent stings, strong biting, and gang pulling, it can overcome prey much larger than itself. Among the frequent victims are workers of the tree ant *Polyrachis*—easily overcome by groups despite advantages of large size and strong armor (Fig. 13.3)—and workers and brood of *Camponotus, Formica,* and *Pheidole*. The little army ants transport a variety of booty over their trails, including many sizable pieces (Fig. 13.4) which are hauled along by crowds of workers with surprisingly good headway despite much countertugging. As the nomadic phase advances the raids grow in numbers of recruits and in duration and distance and collect larger amounts of booty in an increasing variety. Whether the booty is returned in time to the bivouac or stored in caches depends particularly upon the stage of the raid and the current situation of the colony center or bivouac. The last factor, depending chiefly upon the current condition of the brood, seems a principal determinant of whether an emigration intervenes.

*Aenictus* has the simplest colony emigrations of any doryline we have studied (Schneirla, 1965). Nothing in the literature led us to believe that the colonies of this little ant could change quarters at nearly any time of day or night and with far less preparation than that needed by *Eciton* and *Neivamyrmex*. Emigrations of *Aenictus* usually, but not always, begin over a previous raiding trail of the colony. An emigration generally may be expected soon after a local stirring about takes place on the bivouac surface. My typical field notes were: "Workers circulate in a whorl near the base of the cluster, next extend their excited actions to running back and forth short distances on the raiding trail itself. Then,

FIGURE 13.4.
Section of a raiding trail of *Aenictus gracilis* expanded over a leaf bridge on which some of the workers pause to drink at damp spots, others carry booty, and a group (*lower left*) struggles along with an especially large piece.

from the bivouac, within a very few minutes, there issues a procession of workers carrying larvae." All of this happens without any indications that booty has run out in the area. Matters may be very different within the bivouac itself.

Emigrations at times begin within ten to fifteen minutes after laden traffic has tapered off, or after a long procession of mainly unladen workers has returned, or after an interval has passed without any group action whatever outside the bivouac. These conditions, with field and laboratory observations, suggest that the status of brood feeding and the alimentary condition of the brood within the bivouac is the major factor governing the start of an exodus and particularly the promptness with which larva carrying in column begins at the bivouac.

Colonies of *Eciton* usually emigrate after each successive day-long raid in the nomadic phase, but with *Aenictus* the relationship of raiding and emigration seems far more variable. One colony of *Ae. laeviceps,* in a total of 27 nomadic days in two phases, carried out a total of 45 emigrations, of which 26 began during raids, 7 after intervals of extrabivouac quiescence ranging from ten minutes to six hours in duration, and 12 while a preceding emigration was still in progress. The latter condition of overlapping emigrations resembles somewhat the successive movements in a column-carrying brood from cache-cluster to cache-cluster that feature the emigrations of *Eciton* and *Neivamyrmex* late in a nomadic phase. As carried out by these specialized dorylines, however, such "leap-frog" advances are coordinated sections of one continuous nocturnal change of colony base whereas with *Aenictus* they are at times well separated, at times linked, and usually far less predictable than in the polymorphic dorylines.

With the entire emigration lasting between two hours (early in the phase) and five hours (late in the phase), the carrying of larvae generally begins early in the movement. In our records for many colonies in each of the two species, *Aenictus laeviceps* and *Ae. gracilis,* larva carrying nearly always began with twenty minutes after the first local commotion at the bivouac. Of seventeen successive emigrations observed in one colony of *Ae. laeviceps,* larva carrying began in eight cases within five minutes, and in all of the others within fifteen minutes, after the first agitation was seen in the cluster wall.

Once larva carrying is under way, the emigration continues, barring rain, wind, or other drastic interruptions, until the entire colony has moved. The stream of laden ants, usually very evenly spaced (Fig. 5.2) runs along at a rate near two meters per minute. At the height of the movement the column may be four or five ants across (i.e., about 2 cm), though often it may thin out ahead of traffic blocks or groups of workers struggling along with large pieces of booty.

One regular instance of column thinning in an emigration arises ahead of the queen. The queen may be caused to leave the bivouac by a stirring of workers around her, increasing as more and more of the brood is moved from below them. Her appearance usually comes after the exodus is more than one-third completed and often after most of the brood has been moved out. On emerging from the cluster, she moves off promptly, always running steadily in the column under her own power. She generally needs about a half-minute to cover one meter, with her progress slowed somewhat most of the way by workers that surge around and over her. Many of the workers, usually unladen, that are close to the queen may be members of the queen's guard, like those described in Chapter 8 for *Eciton*.

Anyone accustomed to seeing the sizable queens of *Eciton* in their emigration runs is hardly prepared for the appearance of the little queen of *Aenictus* in her column. The entourage of *Aenictus* seldom is more than one meter in length, compared with the much longer and broader masses usual for *Eciton* and *Neivamyrmex*. These queen groups seem proportional to the size of the participants, however, and the diminutive queen of *Aenictus* seems as highly attractive to her workers as the large queen of *Eciton* is to hers. The workers of both these dorylines, and *Neivamyrmex* as well, similarly form entourages about their queens in the column and cluster quickly over them when they stop.

The facile transformation from a raid or other colony situation to an emigration, distinctive of *Aenictus,* may be considered primitive. In comparison with the time-regulated and highly coordinated raiding procedures prerequisite to emigrations in *Eciton* and *Neivamyrmex,* activities necessary to emigration in *Aenictus* seem to be very simple. Extra-bivouac activities coming before the emigrations of *Aenictus* are so variable as to suggest that the really essential events occur within the bivouac. From our results, these are direct processes of communication between brood and workers.

As the larvae of *Aenictus* grow, emigrations of the colony become more and more frequent, until five or more of them may take place in one twenty-four-hour interval. Colony '63 Ae. l.-I, in the first six days of a nomadic phase, averaged only 2.9 hours per day in emigrating, but in days 7 to 12 of the same phase, when emigrations increased in number and duration, the colony averaged 11 hours per day in emigrating. Although emigrations early in the phase require little more than two hours as a rule, the time increases until these movements take more than five or six hours to run their course. The extra time is used for operations connected with the growing brood, for transporting the larger larvae over greater distances, and for longer stops with more feeding by brood at caches than before.

Although the emigrations of *Aenictus* usually press their way along relatively simple trail systems without the traffic complications common in *Eciton,* colony activity, nevertheless, builds up to an increasingly high pitch. From the shorter movements around 20 m common at first, the distance increases two- or threefold as longer movements become common. Also, an exodus then often surges on so strongly that it overruns the zone of raiding into new terrain in such a way that long treks of more than 80 m may occur (Schneirla and Reyes, 1969). Appropriate tests show that the procession is following a chemical trail; but numerous records indicate that such extension trails are those developed earlier by colonies other than the ones using them. This behavior, which incidentally insures unworked terrain beyond the new bivouac, seems to offer another indication of the generalized aspects of raiding and emigration in *Aenictus*.

As the nomadic phase advances in *Aenictus,* emigrations occur more and more often in series with raids of increasing complexity. A pattern typical of *Ae. laeviceps* is illustrated in Figure 4.5. At the same time it becomes increasingly difficult to predict emigrations on the basis of specific colony actions. Often two or more reversals of traffic may occur, each with a procession of laden workers returning to the bivouac—and each followed by a further exodus that extends the raid—before an outpouring of workers carrying larvae marks the start of an emigration. At times the exodus with brood may begin rather suddenly, either after an interval of scanty booty-laden returns to the bivouac or after a period of external inaction with the entire colony in the bivouac. Such intervals are not "quiescent" in the sense that the colony is then dormant since at these times workers within the cluster are busy with the brood and with brood feeding. Our observations also suggest that when food runs low in the bivouac, behavioral conditions leading into a raid or an emigration can arise. These conditions, however, seem to arise within the bivouac itself and—contrary to the hypothesis that territorial food exhaustion causes emigrations—have no necessary relation to what booty may remain in the environs.

We may view an emigration in *Aenictus* as a rather sudden large-scale excitation of workers by a mass of larval brood that has run short of food. It is significant that emigrations can be aroused much sooner and more simply in these dorylines, with their relatively small worker and brood populations of roughly uniform behavioral resources, than in the larger, polymorphic populations of *Eciton* and *Neivamyrmex*.

Chapters 4 and 5 showed in detail that the raids of nomadic colonies of *Eciton* are each an outcome of a complex progressive mosaic of conditions that needs many hours to reach its climax, linked specifically with conditions at dawn for its daily initiation and with conditions at dusk for its transformation into emigration. In *Neivamyrmex nigrescens*, comparably, each nomadic raid has a complex progression, in its case set off by decreasing light and associated conditions at dusk but ending in a nocturnal transition into emigration evidently based on intracolony processes alone. In the much simpler activity sphere of *Aenictus*, raids and emigrations are far less regularly scheduled without important relations to environmental events in the day-night sequence and seem much more exclusively and directly related to current stimulative conditions within the colonies.

Observations on *Aenictus* suggest that within bivouacs in the nomadic phase, a condition of food shortage can arise that activates the larval brood widely and soon leads to reactions of larva grabbing by workers across the cluster; or under other conditions the prevalence of satiation in the brood quiets workers and inhibits any exodus. The behavioral outcome may be quiescence, brood feeding, or emigration, depending upon whether the brood is, respectively, well fed, plied with food, or starved without food at hand. The more prompt and more direct rise of emigrations in *Aenictus* thus seems to depend upon simpler, more uniform, and more direct communicative processes between brood and workers than prevail in the heterogeneous bivouac situations and complex raiding progressions of *Eciton* and *Neivamyrmex nigrescens*.

Although this theory of emigration in *Aenictus* is preliminary, it is based upon brood-stimulative theory which accounts reasonably for the differences among group A genera. A central idea is that the essentially monomorphic brood and worker populations of *Aenictus* are both composed of similar individuals that react in much the same manner under the same conditions. The similarity of behavioral resources in *Aenictus,* effective in much smaller colony centers of far simpler pattern, promote impromptu-type changes in colony behavior but not the day-long sequences preceding emigrations in *Eciton*.

In the heterogeneous, much larger populations of *Eciton,* such changes as the return of a long procession of booty carriers or a local condition of food shortage in the cluster are small episodes in the complex events of the day. Commonly there results a new wave of exodus from the bivouac, but *Eciton* soon absorbs these group reactions in the expansion of its great raiding systems. In these bivouacs, during raids, local brood groups inevitably differ in alimentary condition according to their size types, and the workers attending them differ correspondingly in their reactions to brood. The result is that a complex pattern of intrabivouac communication prevails during the day, in which each section (e.g., central versus marginal) is engaged in distinctive behavioral operations and so responds very slowly to cues from other sections. The colony thereby advances only gradually to the internal state of uniform stimulation and response essential to emigration (Chapter 5). Significantly, in *Neivamyrmex nigrescens* these intracolony effects themselves lead to emigration, but in *Eciton* their summation with dusk-associated environmental stimuli seems necessary for a collective exodus with brood.

*Eciton, Neivamyrmex,* and *Aenictus* thus are fundamentally alike in that in all three emigrations are initiated through intracolony arousal effects centered on the brood. But in the polymorphic dorylines, the role of these processes in the specific arousal of emigrations is obscured by the long, complex behavioral routines that regularly precede colony movements. Only in *Aenictus* can we glimpse the simpler, and probably ancestral, pattern of food-deficient larvae directly stimulating workers to collective actions of raiding and of changing bivouac.

Notwithstanding these generic differences in colony communicative patterns, *Aenictus* has a nomadic phase similar to that of *Eciton* and *Neivamyrmex,* in which the persistence of colonies in vigorous raids and periodic emigrations depends equivalently upon the sweep of a developmental interval because active brood stimulates the workers intensively. When this effect falls sharply at larval maturation (in *Eciton*) or at a stage shortly thereafter (as in *N. nigrescens* and *Aenictus*), colonies lapse similarly into the statary condition in all three genera.

From our results, *Eciton, Neivamyrmex,* and *Aenictus* have comparable functional cycles of equivalent functional phases that change regularly depending upon brood condition. An outstanding similarity between *Aenictus* and the distinctly more specialized *Eciton* and *Neivamyrmex* is the length of the nomadic phase, which in the species of all three is generally about two weeks (Table 7.3). This similarity in nomadic durations is all the more remarkable because it depends upon similar processes controlled by larval stages of much the same length in all three genera.

By contrast, a striking difference appears in durations of the statary phase. In *Aenictus* this phase lasts about twenty-eight days in the two species studied whereas in *Eciton* it lasts only around three weeks and in *Neivamyrmex nigrescens* a day or two less. In this respect also phase duration rests on the timing of a major stage in brood development—this time the pupal stage.

There are interesting generic differences in both the beginning and the ending of the statary phase. *Eciton* and *Neivamyrmex* are similar in its termination, which in both of these army ants depends upon the emergence of the advanced brood as callow workers (Fig. 7.1). But in *Aenictus* the statary phase is not so ended since the pupal brood emerges after about the twenty-third statary day, a few days short of the phase end.

Comparisons of functional events in the statary phases of the three genera should tell us how the brood-stimulative theory advanced in Chapter 7 for dorylines in general may apply to *Aenictus* specifically. (For simplicity, we are talking here only of colonies with all-worker broods.) In *Eciton* this phase begins with the maturation and enclosure in cocoons of a larval brood. In *Neivamyrmex,* however, the all-worker larvae, on maturing, spin no cocoons, and the statary phase begins when this brood reaches its *mid-prepupal development.* This change occurs in *Aenictus* as well in the last-described way, with the slight difference that the statary phase begins as a rule when the advanced brood has entered its *early prepupal stage of development.* In *Eciton, Neivamyrmex,* and *Aenictus,* similarly, the pupal stage of the brood is normally about three weeks long although its relationship to detailed colony functions in the statary phase involves some subtle differences.

Examining these differences is a useful way of studying doryline cyclic function. In all investigated species of *Eciton, Neivamyrmex,* and *Aenictus* the queen enters physogastry in the last few days of the nomadic phase, then after about one week in the statary phase becomes fully gravid and begins to lay a batch of eggs (Fig. 13.5). In all three genera, with limited species and colony differences, this laying then advances continuously and

FIGURE 13.5.
Queens of *Aenictus laeviceps*. (*a*) Contracted condition typical of the nomadic phase (live specimen). Scale: Length of queen 8.1 mm. (*b*) Gravid or physogastric condition of egg production early in the statary phase (preserved specimen).

reaches its end around or shortly after the fifteenth statary day. Conditions indicate that in maturing this batch of eggs the colony queen in all three genera responds equivalently to stimulative and trophic conditions introduced to the colony through changes in brood condition. In *Aenictus,* also, a colony-feedback theory is therefore a more convincing explanation of the timing of the queen's reproductive processes than the idea of an "innate rhythm" (Chapter 8).

*Aenictus* arrests the young brood after egg laying with the result that a highly synchronized population is produced in which all members are nearly the same. This ant has a relatively uniform, monomorphic series in each generation of young in contrast to the polymorphic brood populations of other known dorylines. Polymorphism in these ants, still to be investigated in its detailed processes, seems attributable to certain advantages held in development over those next in the series, these over the next ones, and so on. For *Eciton* and *Neivamyrmex* the advantages may begin before laying, perhaps through a differential drain on oöcytes—or some other inhibitory process—in the queen; later, doubtless, this occurs through differences in worker behavior affecting the feeding of larvae and social stimulation at all stages of brood development.

The external timing of brood delivery is much the same in *Aenictus* as in *Eciton* and *Neivamyrmex*. In all three of these dorylines the queen, entering physogastry late in the nomadic phase, becomes fully gravid after about one week in the statary phase and delivers a large batch of eggs on a schedule much like that of the other group A army ants. The contrast between the contracted and physogastric conditions in the small queen of *Aenictus* is particularly impressive (Fig. 13.5). Notwithstanding the similarities, *Aenictus* differs sufficiently from *Eciton* and *Neivamyrmex* in developmental conditions finally to produce monomorphic, rather than polymorphic, broods.

We can only guess about the basis of this striking difference. First, the diminutive size of the queen in *Aenictus* may be associated with processes through which all of the oöcytes are reduced to nearly the same low nutritive level before they are laid as eggs. Second, the leveling processes seem to continue after laying, when we find the young brood in a slowly changing embryonic condition from a time shortly after it appears as eggs until around the twenty-fourth statary day. It may be held back through inhibitory pheromones, possibly from the queen, as well as by the inaction of the workers and unresponsiveness to the brood in this interval. Only within the last four or five days of the statary phase do the embryos transform into microlarvae.

It may be a combination of these deterrent conditions that operates in *Aenictus* virtually to wipe out the developmental advantage of priority in egg laying that underlies polymorphic broods in *Eciton* and *Neivamyrmex*. To sum up, our hypotheses for monomorphic broods in *Aenictus* include: (1) low physiological resources in the queen, which may reduce all of the oöcytes to about the same low level; (2) possible queen-produced inhibitory pheromones; (3) a low level of brood attraction; and (4) inactivity on the part of the workers and their relative unresponsiveness to the young brood. Through these and perhaps other factors, the all-worker broods of *Aenictus* are highly synchronized and develop as virtually monomorphic populations.

Therefore, notwithstanding its similarities to the other group A genera in the functional cycle, *Aenictus* presents important differences. Although the advanced (i.e., pupal) brood matures similarly after about three weeks in the statary phases of all three group A genera, its emergence as callows does not contribute in *Aenictus* to the strong arousal effect that stirs colonies of *Eciton* and *Neivamyrmex* into a new nomadic phase. Instead, in its case, the colonies are only partially activated and remain so for some days after the callows appear.

Although colonies of all three group A genera are highly activated in the nomadic phase and all much reduced in activity level in the statary phase, those of *Aenictus* seem the most reduced. This last difference is emphasized by our results on statary raiding, summarized in Table 4.2. Statary raids, usually small, are absent on some of the days in *Eciton* and more frequently in *Neivamyrmex* but are consistently absent through the long intermediate part of the phase in *Aenictus*. Any underground raids that might occur in this interval would indicate a lower excitatory level than with surface raiding.

In this sense, it is significant that surface raiding resumes in colonies of *Aenictus laeviceps* around the twenty-third statary day when the pupal brood is completing its maturation. This is a distinct sign of increasing brood excitation which, although weaker than is indicated by colonies of *Eciton* and *Neivamyrmex* in their last statary days, seems to rise steadily to the end of the phase. Colonies of *Ae. gracilis* by contrast, generally carry out no surface raids in this interval and are inactive aboveground except for entrance clusters. Further research may reveal a rising level in subsurface raids, paralleling the surface increase in *Ae. laeviceps* thus revealing a stronger inhibition by surface conditions to the end of the phase.

All of these differences, including especially the long statary phase, point to the prevalence of an especially low metabolic level in colonies of *Aenictus* during the entire statary phase. Such a condition would explain why the emergence of callows does not arouse nomadism as in the other group A genera. At the most, as conditions in colonies of *Aenictus* dug up between the eighteenth and twenty-sixth statary days indicate, maturation of the pupal brood may be a factor in stirring the young brood from the embryonic to the microlarval condition. This may happen directly (e.g., through hormonal effects) or indirectly (e.g., by enlivening the workers stimulatively and trophically). When the nomadic phase does begin in these *Aenictus* species, the older brood is already so advanced as mature callows that its members can hardly be distinguished from adult workers.

The callow-arousal effect, therefore, is clearly weaker in *Aenictus* than in the other group A genera, in which it is the major factor initiating the nomadic phase. At best it has a secondary, facilitative role in *Aenictus,* in which the young brood, on attaining the microlarval condition (Fig. 7.1), energizes the colony into a new nomadic phase.

It is interesting to note that *Aenictus,* at the much lower physiological levels to which its colonies evidently sink in the statary phase, carries out a rather distinctive pattern of brood coordination through which further monomorphic brood populations are produced with great regularity. In normal colonies, this is a remarkably precise pattern, through which cyclic colony functions are maintained as efficiently in this smallest member of group A as in the larger, polymorphic genera.

How effectively the functional pattern of *Aenictus* promotes nomadic colony behavior may be suggested by the itinerary of colony '63 Ae. 1.-I (Schneirla and Reyes, 1969). In one nomadic phase of eighteen days, this colony carried out twenty-six emigrations, in which it covered a total ant-trail distance of more than 1000 m. The booty return was plentiful as in most of this course the colony passed through areas in which ant nests were especially numerous.

In the behavioral system of *Aenictus,* similarities appear to the legionary ponerines, on the one hand, and to the highly specialized dorylines, on the other. Likeness to the former is indicated by the directness with which a summated excitatory effect from underfed larvae seems to arouse forays and emigrations in *Aenictus.* From our studies of this ant, I suggest that food supply at the colony center may have been ancestrally the major factor underlying nomadism and that its efficiency depended upon the celerity with which masses of workers responded to summated stimulation from active brood.

A similarity to the specialized dorylines, those of large, polymorphic colony populations, is clearly marked out by the presence of equivalent functional cycles in *Aenictus* and the other group A genera. This remarkable little ant, if it were extinct and unknown, might still have been inferred as a possible transitional form, from comparisons of legionary ponerines with the dorylines and of group A with group B genera among the dorylines.

# 14

# The Doryline Colony as an Adaptive System

Colonies of all army ants carry out large-scale predatory forays related to patterns of nomadic behavior representative of the species. These characteristics appear through the functioning of many adaptive resources possessed by these ants. Adaptive resources are structures and functions that promote the life of the colony in its environment and aid species survival. The dorylines owe the efficiency of their group behavior to their strong repertoire of adaptive resources contributed to colony function by all types of individuals (Chapter 2). The evolution of mechanisms promoting large-scale group actions in capturing prey and forming temporary nests clearly involved queens and males as well as workers—all contributing directly or indirectly to the operations of their complex colony adaptive systems. The tarsal claws of workers, for example, aid both group attacks on prey and the clustering reactions essential to nest formation. In the actions of raiding, these structures serve closely in combination with those of biting, stinging, and responding to odors; in operations of nesting they combine with mechanisms including tonic immobility through stretching.

Army ants are successful because they are able to carry out raids and emigrations in species-typical ways. They have evolved a variety of functional systems in different genera and species, ranging from those adapted mainly to surface life to those adapted to subterranean life. Individual interdependence in the colony is the most outstanding of these functional systems and is expressed in many ways, from the organization of subgroups in raiding and emigration to whole-colony processes essential to the functional cycle.

The sense in which each doryline colony may be viewed as an adaptive system is schematized for *Eciton* in Figure 14.1. Here the adjustments of a colony to its environment are represented by the individual properties and subgroup interrelationships. This schema applies in appropriate ways to all dorylines, which exhibit these principal adaptive characteristics: (1)

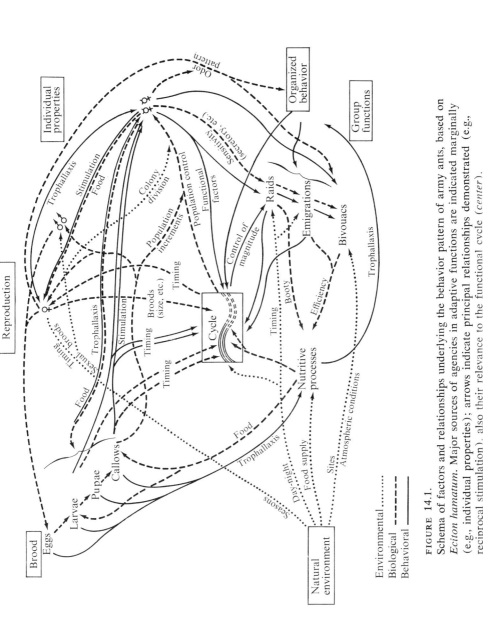

FIGURE 14.1.
Schema of factors and relationships underlying the behavior pattern of army ants, based on *Eciton hamatum*. Major sources of agencies in adaptive functions are indicated marginally (e.g., individual properties); arrows indicate principal relationships demonstrated (e.g., reciprocal stimulation), also their relevance to the functional cycle (*center*).

large, well-integrated colony populations; (2) trail development in predatory forays of species-typical patterns; (3) temporary nests as shelters, operating centers, and brood incubators; and (4) mechanisms of brood-initiated excitation basic to initiating and carrying out raids, emigrations, and other collective functions.

Among existing dorylines, two major types of colony-adaptive systems are discernible: that of the predominantly surface-adapted genera *Eciton, Neivamyrmex,* and *Aenictus,* or group A, and that of the essentially subsurface-adapted *Dorylus* and others (e.g., probably *Labidus* also) or group B. In each army ant genus, the species present a range of functional systems differing in their degree of surface adjustment. Based on nesting, raids, and emigrations of their most surface-adjusted species, we classify the genera *Eciton* and *Aenictus* as surface-adapted, *Neivamyrmex* as surface-related, and *Dorylus* as subsurface-adapted. Later in this chapter we will discuss the reasons for believing that the functional patterns of the most surface-related species in each genus indicate the probable origins of the genus.

The colony populations of all dorylines are larger than those of most other ants (Chapter 2). In these ants, attaining efficiency in group operations means that the colony must maintain a magnitude of population that meets the level of its species. Important differences in colony functional patterns and in habitat are correlated with species population ranges, hence with minimal colony levels essential for survival. This is shown by the smaller colony populations in *Eciton* and other group A genera—in which surface adaptation and regular nomadic function are prominent—in contrast to *Dorylus* and other group B genera—in which subsurface adaptation and less regular nomadic function predominate.

Except for *Aenictus,* polymorphic worker populations prevail among the dorylines, which have smoothly graded series of individuals differing both quantitatively and (usually also) qualitatively in structure. They are all, therefore, relatively specialized in their adjustments to environmental conditions as structural differentiation is basic to functional diversification in many operations from nesting to diurnal and seasonal adjustments (Fig. 14.1). The degree of specialization in colony operations is correlated appreciably with worker polymorphism and associated resources as is shown by the complexity and depth of integration in the colony functions of *Eciton* as compared with their relative simplicity in the monomorphic *Aenictus* (Chapter 13). As a comparison of *Eciton* with *Dorylus* shows however, degree of worker polymorphism and magnitude of colony population cannot in themselves determine functional patterns characteristic of genera and species. The colony adaptive pattern is determined by a mosaic of factors contributed by all types of individuals in the colony.

How it is possible for very different colony functional patterns to arise on this basis in the various dorylines is indicated by evidence reported on these patterns in earlier chapters. How the similarities arise that mark dorylines as a group is also suggested and can be reviewed in terms of the schema in Figure 14.1.

The roles of adaptive resources giving rise to the unique doryline temporary nest (Chapter 3) are emphasized by generic and species differences in the extent of surface or subsurface adaptation. In *Eciton* the patterns and situations of bivouacs differ according to modifications in the condition of colony populations through the functional cycle. Colonies of surface-adapted *Eciton* form exposed surface clusters when the workers are at their highest metabolic level and most tolerant of surface conditions but form sheltered bivouacs when they are at their lowest metabolic level. The relationship of bivouac properties to the current (brood-determined) colony condition seems much more precise in *Eciton* and in other group A genera than in *Dorylus* or other group B genera studied. Our results show that bivouac-making and bivouac-changing operations are more precise as the degree of surface relevance in species adaptive patterns increases. Thus, comparable types of colony environmental control seem much more highly developed in the surface-nesting *E. hamatum* and *E. burchelli* than in the subterranean-nesting *D. (Anomma) wilverthi*. In the latter, however, Raignier and Van Boven (1955) found indications that these functions are better developed than in the more subterranean *D. (A.) nigricans*.

Although all dorylines form temporary nests, the properties of these nests differ from genus to genus (Chapter 3). Also, each genus presents a series ranging from the most surface-related to the most subsurface-related species. For colonies in corresponding (physiological) conditions, the bivouacs of *Eciton burchelli,* for example, are always far more surface-exposed than those of *E. vagans;* hence statary colonies of the former often occupy virtually the same sites and ecological situations taken by nomadic colonies of the latter. Although in all army ants the ecological properties of nesting change according to current colony condition, the range of phasic conditions in bivouacking differs greatly according to genus and species. The bivouacs of any colony differ less in these respects in the epigaeic *Eciton,* for example, than in the epigaeic *Aenictus* species, whose colonies when nomadic form open surface clusters but when statary occupy well-sheltered and subterranean sites.

From considerations in Chapter 3, the difference between the adaptive systems of group A and of group B army ants seem to center on nesting. Adaptations of group A species to surface-related niches are indicated in the distinctive properties of their colonies for controlling the bivouac microclimate. Through worker responses to extrinsic stimulative effects depending upon the day-night succession or seasonal conditions, colonies modify the form and (to limited extents) the position of their bivouacs and so readjust their degree of exposure. Such responses, although relatively continuous, normally involve colony displacements that are slight and secondary compared with those effected through emigrations in the nomadic phase. Colonies of group B species probably modify their stable subsurface conditions less on the whole than group A colonies under more variable ecological conditions. These contrasts we may attribute to differences in sensory thresholds and related physiological properties of workers controlling the species optima for bivouacking.

All army ant bivouacs serve four functions for the colony: a base of operations, a shelter, a brood incubator, and a population reservoir. The first of these functions is expedited—in surface-active species, especially—through conditions of internal organization dependent upon distribution of the brood and workers and placement of the queen in the cluster. Through these properties, which are most specialized in surface-adapted species, the bivouac contributes vitally to the efficiency of raiding and other colony operations.

The dorylines, in their colony raids, present a wide array of species-typical differences (Chapter 4). The raids differ according to species in size and pattern, in modes of individual and group attack, in their timing and typical booty, and in the extent to which they invade surface or subsurface territories.

The forays of *Eciton* and of *Dorylus,* for example, differ strikingly in all of these respects (Schneirla, 1957b). The terminal attack groups of *Eciton* generally involve smaller numbers of workers, operating one-layer deep and often surface-anchored as they bite and sting prey; those of *Dorylus* frequently operate in layers-deep masses, emphasizing the biting and shearing jaw movements and the counterpulling actions of great numbers against booty. In *Eciton,* a greater specialization of predatory operations than in any other army ant genus is emphasized by radical differences between the swarm raids of *E. burchelli* and the column raids of *E. hamatum,* the former featuring mass attacks against a wide range of prey, the latter many small-group attacks against a limited range of prey.

The species-typical raids of *Eciton* display higher degrees of internal organization—featuring more precise types of subgroup liaison—than the relatively amorphous raids of *Dorylus.* Among other differences, the rear columns of *Dorylus* are typically much more crowded and uneven than those of any *Eciton;* in surface stretches they are often walled or covered over with clustered workers or structures of pellets and usually soon disappear underground behind the swarm as it advances. These features, also typical of *Labidus praedator,* are absent from the raids of group A species studied. They are attributable to reactions involving lower degrees of tolerance to surface conditions in group B than in group A dorylines.

The army ants differ widely in the territory covered and the booty haul of their raids. Although the forays of *Eciton hamatum* generally extend over a wider expanse of terrain and reach farther from the bivouac than those of other dorylines including even the driver ants, the swarm raids of *Anomma* doubtless return a far greater mass of booty than in *Eciton*. The raids of group A dorylines range from the largest in *Eciton* to the smallest in *Aenictus,* with those of *Neivamyrmex* intermediate. These differences correspond roughly, as in most dorylines, to differences in individual and colony size, raiding pattern, and colony condition.

The raids of group A species are best developed and largest in the nomadic phase and least developed in the statary phase. The contrast is greatest in surface-adapted *Aenictus,* which often have several surface raids daily in the nomadic phase but generally carry out none at all through the long intermediate part of the statary phase (Table 4.2). Comparably, the raids of *Dorylus* are heavy in the days just preceding and just following an emigration but are lighter in the intervening period (Cohic, 1948; Raignier and Van Boven, 1955) when the colonies are presumably least excited.

Colonies of surface-active *Eciton* and *Neivamyrmex* in the nomadic phase are regularly keyed to surface conditions in their raiding schedules, as those of the former begin at dawn and stop at dusk whereas those of the latter begin at dusk and end in the night. The forays of *Aenictus,* by contrast, exhibit only a limited day-night relationship, as they begin both in the day and the night except at times near the beginning and the end of the phase; then they begin most often around sundown. In this respect, *Aenictus* seems to resemble *Dorylus (Anomma),* whose workers also lack eyes.

Our evidence on doryline raiding (Chapter 4) reveals interesting differences in the extent to which the forays are timed according to day-night changes. Regular dawn arousal occurs in the predominantly surface-adapted *Eciton* species. Dusk arousal occurs in species of *Neivamyrmex,* which are not usually surface-active by day. Generalized day-night action with limited dusk arousal occurs in the surface-adapted but eyeless *Aenictus* and in *Anomma* which is also eyeless but less surface-active. In each doryline genus, it is probable that colonies weaken in the degree of day-night synchronization of their raids according to the subterranean tendency of the species.

Species differ often in the type and range of booty. *Dorylus (Anomma) nigricans, Labidus praedator, Eciton burchelli,* and others that raid in swarms take in a wide variety of prey; *E. hamatum, Neivamyrmex pilosus,* and other column raiders take a narrower range of booty and mainly soft-bodied forms. One important factor underlying these differences (Chapter 4) is a higher excitation level in the former than in the latter. Type of booty is evidently also influenced by size, strength, chemosensory thresholds, and secretory equipment, as well as by other factors affecting the pattern of raiding and the habitat of a species.

The pattern of raiding crosses generic lines, with the swarm type indicated in all genera, including *Eciton,* in which most species are column raiders. Importance of habitat is emphasized in the capturing of termites by species of *Dorylus, Neivamyrmex,* and *Aenictus,* which have strong subsurface affiliations whereas *Eciton,* a dominantly surface-related genus, seems even to shun termites. By contrast, even surface-adapted species of *Aenictus* include termites in their booty; all species in this genus, however, nest underground in the statary phase.

Illustrating excitability level as a factor, colonies of group A species in the nomadic phase take booty of larger size including many adults but in the statary phase take mainly soft-bodied forms. To illustrate size, species of small workers generally take smaller booty than those of larger workers. As one exception, *Aenictus,* a relatively small but quick, strong, and excitable ant, takes a fairly large range of prey although small objects predominate. All of these matters are of course important for competitive relationships on all levels from the intraspecific to the intergeneric.

Although all army ants emigrate, the relationship of nest-changing operations to ecological conditions and to conditions within the colony differs greatly among the genera. Both environmental and intracolony conditions contribute significantly to influence emigrations in the group A genera *Eciton* and *Neivamyrmex.* For *Aenictus,* however, extrinsic effects play only a minor and variable role.

In nomadic colonies of surface-adapted *Eciton,* relationships between the long dawn-initiated raid and the nocturnal emigration are intricate, as successive stages of traffic developing through the raid determine the emigration route and the location of the new bivouac. Regularly, changes in environmental stimuli near dusk aid the transition from raiding to emigration by combining with intracolony stimuli resulting from the long raid. In surface-active *Neivamyrmex,* emigrations are also conditioned by day-night changes but follow the dusk-initiated raid only by virtue of intracolony conditions developed in the foray. Notwithstanding their differences, both of these army ants regularly establish the colony basis for emigration through a long raid and in some relationship to day-night conditions. In both of them, particularly in *Neivamyrmex,* species of increasing subterranean tendency may deviate relatively (but probably not absolutely) from the described pattern.

Surface-adapted *Aenictus* differs strongly from other group A genera in its schedules of raiding and emigration as both of these activities occur by day and by night and, as the nomadic phase advances, more frequently within each twenty-four-hour interval. Representative schedules of no-

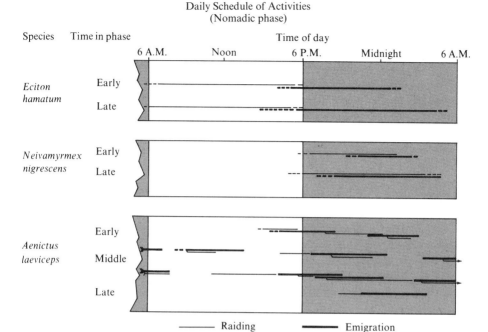

FIGURE 14.2.
Twenty-four-hour activity schedules typical of nomadic colonies in species representing two genera of New World army ants (*above*), for comparison with the schedule typical of *Aenictus laeviceps* of southeastern Asia. Actual colony records are represented for times early, middle, and late in the nomadic phase.
*Thin lines:* continuous raids; *broken lines:* transitional activities; *heavy lines:* emigrations.

madic colonies in these three genera differ prominently, as Figure 14.2 shows. Furthermore, in *Aenictus,* emigrations are usually initiated rather quickly, without the complex preliminaries in raiding typical of *Eciton* and *Neivamyrmex.* Our findings indicate for *Aenictus*—as against the other two group A genera—a loose and variable relationship between raiding and emigration, marked often by colony movements not definitely correlated with raiding, either in time succession or in route.

Even so, our results point to a functional equivalence between the nomadic behavior of *Eciton, Neivamyrmex,* and *Aenictus,* featured by the carrying out of vigorous raids and frequent emigrations through species-typical sequences of days constituting a nomadic phase. This group A pattern contrasts sharply with that of *Dorylus* and others of group B, in which the colonies carry out single emigrations of long duration at well-separated and variable intervals. As a significant behavioral difference, group B army ants emigrate largely through underground channels, linked by shorter stretches aboveground in which columns run in steadily deepening trenches (Fig. 5.9) walled or roofed by pellet masonry or clustered ants.

Our analysis supports two generalizations: first, that a *necessary* precondition for emigration in all dorylines is a potent intracolony stimulative condition initiated by the brood; second, that normally the colony leaves as a unit without the attachments to the particular site effective in most social insects. These generalizations are supported by results discussed in Chapters 6, 7, and 8.

The nomadic behavior of all army ants has its necessary basis in the production of successive large all-worker broods (and in group A of sexual broods also) by the colonies (Chapter 6). Doryline broods are distinctively large populations, capable of strongly exciting their colonies through the massive stimulative effects they deliver at certain developmental stages. Characteristics of the broods are essential for the normal operation of all doryline colony adaptive systems and offer the main key to understanding similarities and differences in schedules of nomadic behavior within the subfamily.

Let us begin with a comparison of cyclic functions in the group A genera. In *Eciton, Neivamyrmex,* and *Aenictus,* alike, large all-worker broods are produced at regular intervals, and each of these broods is a unitary, well-synchronized generation of individuals. Also, in all three genera, these successive broods are *coordinated,* i.e., they overlap in time and are functionally interrelated. The stimulative effects of these broods on colony behavior, consequently, are broadly equivalent in the three genera.

The effects of the brood on colony behavior differ secondarily, however, according to whether the all-worker broods are polymorphic as in *Eciton* and *Neivamyrmex,* or monomorphic, as in *Aenictus.* These differences (Chapter 6) offer a difficult problem in caste determination. In all three of these genera the eggs producing each all-worker brood evidently are all fertilized and genetically equivalent. In *Eciton* and *Neivamyrmex,* however, important developmental differences are introduced through variations in the queen's condition in the course of egg laying and through differences in later treatment by workers, corresponding to differences in the order of egg laying. The polymorphic all-worker broods of most army ants may comparably result from stimulative, trophic, and other differences dependent upon the order of egg laying. For *Aenictus,* however, I have suggested in Chapters 6 and 13 that such conditions are virtually negated by agencies peculiar to that genus so that a virtually monomorphic population results with important consequences for colony function.

Because of these conditions, centering on the laying of the eggs in a consecutive series by the queen and on being treated as a unitary population by the workers, the degree of synchronization within each brood is high in all group A dorylines and is highest in *Aenictus*. Through differential growth the brood range in all genera is narrowed appreciably during the larval stage. The smaller larvae, although starting latest, advance more rapidly in their relative growth (Fig. 6.4) and thus catch up appreciably on the larger members of the brood. As a result, there prevails a condition that I call "developmental convergence," whereby each all-worker brood contributes a stronger, more unitary brood-stimulative effect on the colony than would hold if all of its members developed at the same rate. Within group A developmental convergence is largest in *Eciton* and smallest in *Aenictus* and may be greatest of all army ants in *Dorylus*.

Investigations on *Eciton* (Schneirla, 1938, 1945) and other dorylines (Schneirla, 1957b) have indicated two main types of brood-stimulative effect: the callow-arousal factor and the larval-stimulative factor. These factors—through mechanical, chemostimulative, and pheromonal agencies introduced by the brood and their colony feedback effects—initiate changes in colony function and maintain them through the functional phase. Although any brood influences the colony in various ways throughout its development, the two factors named above represent its maximal effects.

Brood-stimulative theory provides a broad basis for studying the colony adaptive systems of army ants. In group A, for example, although cyclic colony patterns are broadly similar for all genera, differences appear both in the scheduling of raids and emigrations and in functional cycles. These differences point to variations in the roles of the brood-stimulative factors in the three genera.

Although in all three group A genera the colony activity level remains high through brood excitation during the nomadic phase, there are differences in the relationship of intracolony to extrinsic conditions. In Chapters 4 and 5 we have indicated the facile way in which intracolony stimulation arouses raids and emigrations in *Aenictus,* i.e., in general both by day and by night, as compared with its devious functions in the more complex colony schedules of the other genera. The role of food can thus be studied fairly directly in *Aenictus* although not in the others. Members of a larval brood of *Aenictus,* monomorphic and closely similar physiologically, tend to be disturbed at nearly the same time when food runs low in the bivouac and then begin a mass agitation which spreads readily through the cluster. The workers, also monomorphic and capable of similar responses under equivalent conditions, respond to the wave of brood stimulation with excited larva-grasping responses throughout the cluster so that the colony soon launches an exodus with brood (Chapter 13). Presumably, with food shortages of lesser degree, less intense communicative processes of the same type can initiate a new wave of raiding.

In the bivouac situations of *Eciton* and *Neivamyrmex,* by contrast, diversified alimentary conditions among the polymorphic larvae and diversified response conditions among the polymorphic workers promote far more complex and diversified communicative situations. Thus, any excitatory effects that may arise through local food shortages are quickly absorbed either within the bivouac in varied feeding operations with brood or outside in minor additions to the raid. In these army ants, the arousal of an emigration requires a multistaged summation of intra-bivouac excitations which can occur only through the effects of a long raid—plus, in *Eciton,* day's-end environmental changes as well.

All dorylines are characterized by cyclic changes in colony function and behavior (Chapter 7). These are best marked and most regular in the group A genera, in which nomadic and statary phases alternate with marked regularity. Emigrations and related operations in group B genera, as represented by *Dorylus,* differ from those in group A both in form and in degree of variability.

In group A, the nomadic phase has comparable durations in the investigated species of *Eciton, Neivamyrmex,* and *Aenictus:* roughly of about 2–2½ weeks (Table 7.3). Durations of the statary phase, by contrast, show less colony variation within any species and are usually close to three weeks in *Eciton* and *Neivamyrmex* and close to four weeks in *Aenictus*. Very different controlling conditions are indicated for the driver ants, representing group B, in which colonies of *Dorylus (Anomma) wilverthi* (Raignier and Van Boven, 1955) carry out single emigrations of long duration separated by lengthy nest stays of from six days to around two months. Similar long emigrations at variable and usually much longer intervals arise in colonies of the deeper nesting *D. (A.) nigricans*.

Notwithstanding these differences in species schedules, brood-stimulative theory accounts for nomadic behavior in both group A and group B dorylines. Through the callow-arousal factor, nomadic phases are regularly initiated in *Eciton* and *Neivamyrmex,* separate emigrations are variably initiated in *Dorylus,* and nomadism is secondarily facilitated in *Aenictus*. I consider the nest stays of *Dorylus* between emigrations functionally equivalent to the statary phases of group A colonies, in the sense that in both cases the colony threshold of excitation is too low—i.e., through insufficient brood-excitatory effects—to initiate and maintain nomadic behavior in the colony. The failure of the callow-arousal factor to excite nomadism in *Aenictus* may be a specialization attributable to inhibitory effects associated with small queens and monomorphic populations. This point, possibly very significant for doryline origins, is referred to later.

Functional differences underlying nomadism in the various army ant genera may not be as great as they seem. Thus, although the callow-arousal factor sustains only well-separated colony movements in *Dorylus,* its effect lasts each time through an interval of from two to four days, or for about the same time as in *Eciton* and *Neivamyrmex.* In group B colonies, however, the excitatory effect is not sustained at a high level as in these ants the oncoming young brood may be too poorly synchronized to deliver a strong stimulative impact. But when a larval brood happens to be sufficiently synchronized in the drivers, its stimulative effect can arouse an emigration.

In *Eciton* and *Neivamyrmex,* by contrast, a comparatively precise pattern of brood coordination prevails, through which an increasing stimulative effect from the young brood overlaps the waning callow effect and continues the nomadic phase smoothly. In *Aenictus,* through a pattern of brood coordination more precise than those of other group A genera, entrance of the larval-stimulative effect initiates each nomadic phase and then maintains this phase.

The nomadic phase ends after much the same time in the three group A genera through equivalent abrupt falls in excitation from the brood. In *Eciton,* the nomadic phase closes concurrently with larval maturation; in *Neivamyrmex* and *Aenictus,* however, it ends only after this brood has entered its prepupal stage. In the last two genera, the naked brood may continue to excite workers through pheromonal effects less involved or not at all involved in *Eciton* after the larvae spin their cocoons.

The statary interval, a time of low brood stimulation (Schneirla, 1933, 1965), is recognizable in all dorylines studied. It is most regular in its duration and in its alternation with colony nomadic behavior in surface-adapted species of group A and is least regular in species of group B. By and large, cyclic nomadism seems most regular in the colonies of surface-adapted dorylines evidently because in their case successive broods are more closely coordinated throughout development than in hypogaeic species of the same genus.

The pacemaker of doryline cycles at first sight might be either the queen or the brood, or both of these. I have emphasized the brood in this role because available evidence points that way. But let us consider the queen's functions in the colony.

One of the army ant queen's important contributions to colony life is her distinctive odor pattern, essential for the colony to operate as a self-consistent unit apart from others of its species (Chapter 8). This unique factor thereby opens the way for processes of emigration and nest resettlement essential to the colony adaptive system. Queenless colonies of group A dorylines begin to bivouac abnormally, show signs of decreasing organization, and are lost unless they fuse with other colonies of their species. Although a colony that loses its queen must end its separate existence, it can at least aid species survival by joining another colony. The queens of *Dorylus* may have comparable properties (Raignier, 1959).

Essential to the colony adaptive systems of group A army ants is the concurrence in the statary phase of a highly stable bivouac environment, low colony feeding, and the queen's interval of physogastry and egg laying. The queen is thus well buffered when she is gravid, and the reproductive resources of the colony are thereby protected. But when the colony is nomadic and brood feeding is at a high level, the queen is contracted and evidently feeds little until the end of the phase. My observations on *Eciton* suggest that the capsule of guard workers which then encloses her may restrict her feeding by screening her off from the rest of the colony.

Chapter 8 leads us to conclude that the periodic gravid condition of the doryline queen is timed not by processes within her but by effects of the colony situation and particularly by variations in social stimulation and in feeding through the cycle. Secretory agents from workers and brood may be involved, exerting inhibitory or stimulative effects upon her according to time in the cycle. A colony-situation control of the queen's reproductive functions is indicated by a variety of evidence (Chapters 6 through 9) concerning: (1) relationships of phase durations to the rate and duration of stages in brood development, to colony booty intake, and to ecological conditions; (2) the normal functional relationships of the queen with her colony (Fig. 8.6); (3) a drastic shortening of the nomadic phase when a larval sexual brood is present; and (4) aspects of the initiation of function in young colony queens. Such conditions suggest (Schneirla, 1957a) that the timing of phases in the functional cycle depends upon trophic and stimulative feedback effects from the colony situation, varying according to processes of brood coordination in particular. Both the subtle timing of phases in group A genera and the relatively crude processes of timing evident in schedules of emigration in *Dorylus* may be equivalently dependent upon colony-situation control.

Light may be thrown on the queen's normal function by the circumstances of a seasonal adaptation in certain group A species: the resurgence of cyclic function after winter dormancy (Schneirla, 1963). Colonies of *Neivamyrmex nigrescens* in the southern United States, quiescent during the winter, begin to raid nocturnally as soon as environmental temperatures begin to rise in the spring, then gradually increase the scope of their raids (Fig. 14.3) and their nightly hauls of booty. There must be a crescendo of social stimulation and feeding of the queen by workers in this interval to which the queen responds by maturing and laying her

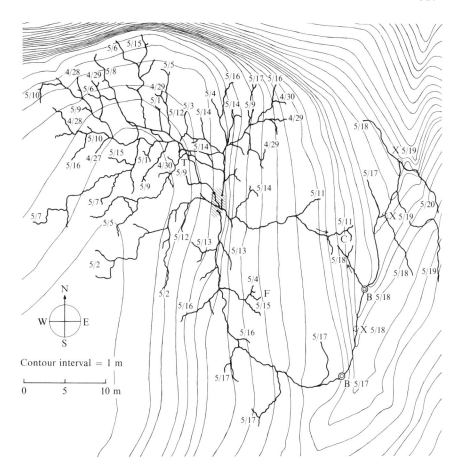

FIGURE 14.3.
Area bordering a canyon in southeastern Arizona in which a colony of *Neivamyrmex nigrescens* overwintered; sketched to show the main routes of nocturnal raids during the spring resurgence of this colony until May 13 when it became nomadic. *Shaded area:* site of subterranean bivouac; B 5/17 and B 5/18: bivouacs of first and second nomadic days; F: area where nests of *Formica* were raided; C: area of raids on *Campanotus;* X 5/18 and X 5/19: probable bivouac sites; T: an area of frequent raiding.

first large batch of fertilized eggs in that season. Entrance of this brood into the larval stage starts the colony in a nomadic phase, and cyclic functions follow through the summer. When nocturnal temperatures begin to fall in the autumn, the process is apparently reversed: Raiding is progressively reduced, ovulation stops in the queen, cyclic function ends, and the colony enters winter dormancy (Schneirla, 1963).

In the colonies of *Dorylus,* whose queen generally delivers an immense number of eggs after each emigration is ended (Raignier and Van Boven, 1955), equivalent processes may occur. As the colony becomes excited through the maturation of a pupal brood, its trophic level rises to the point at which the queen is stimulated to become fully gravid and deliver her maximal series of eggs. Significantly, the general term of development of an all-worker brood in *D. (Anomma) wilverthi* is roughly twenty-five days, which coincides with the most frequent interval of emigration in that species, and the queen seems most often to receive her maximal social stimulation and feeding and thus to lay her peak quantity of eggs around that time.

Thus I view the nomadic cycle of doryline colonies, whatever the species pattern may be, as the composite result of complex trophic and stimulative processes initiated by the brood, next involving the workers reciprocally, and soon involving the queen. This theory holds that effects from the brood set off the first stage of each major change in the colony functional cycle but that other subgroups of the colony soon become reciprocally involved and so contribute in essential ways to the outcome. The essence of the doryline adaptive system, as the schema in Figure 14.1 indicates, is its rise in the standard environment through the intricate functional interrelationship of all types of individuals.

Existing army ants, as this review shows, all have fundamentally similar predatory—nomadic patterns that distinguish them from other ants. How did these patterns come to be? Wheeler (1936) believed that dorylines arose from ponerinelike forms that carried out group raids and perhaps also emigrated. Wilson (1958a), reviewing the evidence for existing group predatory ponerines, outlined three hypothetical stages by which ancestral legionary ants gave rise to modern dorylines: (1) by enlarging their colonies, (2) by increasing the efficiency of their group predatory behavior, and (3) by becoming nomadic. His schema is reasonable but leaves open two questions: how any one of these stages may have led to the next, and how related operations of raiding and emigration may have evolved.

Comparisons of existing legionary ponerines with *Aenictus,* the simplest doryline, suggest that they may have shared remote ancestors capable of collective attacks on varied prey or of emigrations under arousal through excitation both from within the colony (e.g., brood) and from the environment (e.g., light). Basic to these adaptations may have been group actions of forming and following trails of the workers' own fecal products, a resource stemming from feeding relationships of brood and workers. Worker responses to brood may have been central to the causation of both raiding forays and emigrations, which may thereby have evolved together. Once the basic legionary ant stock was launched in this pattern, its increase and selective variation would have been assured in tropical and semitropical habitats bearing an insect life already thriving as a rich source of food.

On this basis, adaptive radiations may have occurred throughout the world in such habitats, producing many stocks varying in their degrees of success. Existing ponerine legionaires, in their relatively simple raiding patterns (Wilson, 1958a) perhaps combined with sporadic nomadism, and with their small populations of similar workers and their ergatoid (e.g., *Leptogenys*) or small dichthadiigyne female reproductives (e.g., *Simopelta*), may give us clues as to their own ancestry as well as to that of the dorylines. But although they may indicate early trends, existing legionaires of varied patterns perhaps represent the termini of divergent branches rather than the stages whereby the main ancestral trunk produced the dorylines. Consider as an example Brown's skepticism (1954) about so regarding *Cerapachys,* which Emery (1901) proposed for this role.

Comparisons of existing legionary ants with the dorylines and of *Aenictus* as a simpler doryline with *Eciton* as highly specialized (Chapter 13) give rise to the hypothetical schema of doryline origins offered in Figure 14.4. The pre-doryline ancestors, like those of the ponerine legionaires, probably were surface-adapted, lived in small colonies of monomorphic workers with compound eyes, and had ergatoid female reproductives. *Aenictus,* although well specialized in some respects, possesses many behavioral characteristics that may have been present in species representing the main ancestral trunks of the dorylines. These characteristics, which are considered primitive, are: (1) relatively small colonies of monomorphic workers, (2) simple patterns of raiding and emigration in surface-adapted representatives, (3) a single relatively small (cf. *Eciton*) colony dichthadiigyne, and (4) relatively simple brood-worker communicative processes underlying raids and emigrations.

The main scene of early doryline evolution may have been the Old World; the time: the latter part of the Cretaceous period (perhaps 80 million years ago) when, under selective influences of plentiful booty and adequate shelter, an accelerated modification of stocks could have involved all types of individuals. The modifications presumably centered on reproduction and colony size, group attack power against booty and predators, and colony mobility. As one focal resource, *sexual-type broods* may have emerged as the basis of both female and male specialization. Ergatoids, developing under these exceptional trophic conditions, could have specialized rapidly in their size, reproductive capacity, secretory attractiveness to workers, and their resources for becoming gravid under increases in stimulative and trophic conditions in the colony. With the differentiation of sexual-type broods, young colonies could reduce their

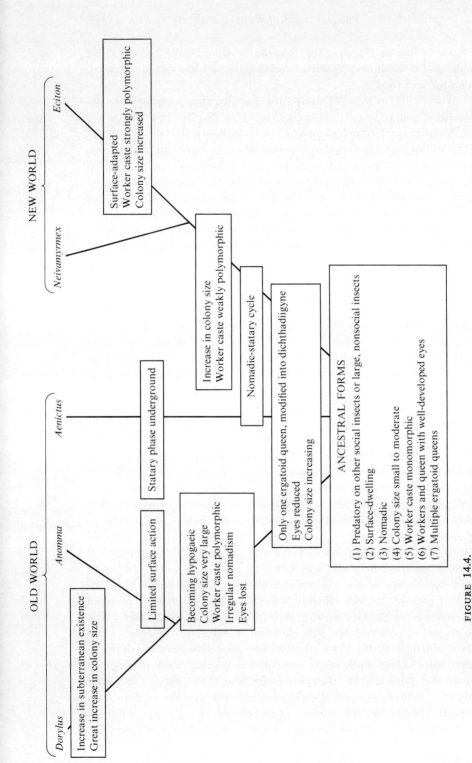

**FIGURE 14.4.**
Schema of modifications and specializations postulated for the rise of existing genera of army ants described for groups A and B.

ergatoid series in competitive processes of briefer and briefer durations approaching those of callow-queen elimination in existing dorylines (Chapters 10 and 11).

A second center of modifications, the emergence of *specialized all-worker broods,* may have advanced colony size and efficiency in many ways. Wilson (1958a), from his survey of the legionary ponerines, thought that specializations in group predation may have been primary and those of nomadism secondary. But if sexual broods and all-worker broods began to differentiate and specialize early in doryline evolution, mechanisms underlying raids and emigrations could have been modified selectively together as colonies increased in their sensitivity to stimulation from food-depleted larvae. These resources in combination may have provided the strategic basis for large colonies with adaptive systems equal to surviving peak environmental hardships through life spans of increasing duration.

It is reasonable to think that these key types of specialization could have arisen best in one immense area in which were differentiated two main genetic stocks: (A) an *Aenictus*-like ant living in colonies of moderate size capable of mobility and efficient action mainly on the *surface,* and (B) a *Dorylus*-like ant with larger queens living in larger colonies mainly adjusted to *subterranean* conditions. The differentiation and dissemination of such stocks could have occurred in the late Cretaceous period when numerous termites, important items of food, were expanding both through the Old World tropics and (from Asia) into the Americas (Emerson, 1955, 1967).

The case for a common origin of the main doryline genera, although still necessarily circumstantial, derives support from Seevers' studies (1959, 1965) of the dorylophiles and seems to explain best with fewest assumptions the significant functional relationships I have described for group A genera. Although a polyphyletic origin is conceivable for the dorylines (W. L. Brown, 1954, pers. comm.) and may be supported by further evidence, the concept of a monophyletic origin best fits available functional and behavioral evidence, seems consistent with taxonomic considerations, and is therefore used here.

Of various characteristics that may have been crucial to further doryline evolution, let us discuss two very different ones: (1) presence or absence of eyes, and (2) type of functional cycle. Significantly, the workers and queens of all existing Old World dorylines lack eyes although their primary ancestors surely had eyes. Thus, the forerunners of both *Aenictus* and *Dorylus* most often have been subjected to harsh surface conditions that forced them underground for long intervals, during which time eyes and visual apparatus would sustain a progressive degeneration. What was lost in this respect could never be regained.

On the other hand, retention of only partially reduced eyes by workers and females of nearly all New World dorylines could have followed an early separation of *Aenictus*-like stock from Asia and its dissemination in the Americas under more favorable surface conditions. The existence of *Eciton* as the one predominantly surface-adapted genus of army ants points to the prevalence of forest cover and equable surface conditions in the New World during the rise of this genus there. *Neivamyrmex,* probably derived earlier from the basic A stock, may have been exposed so long to variable and often austere surface conditions as to greatly reduce its surface affiliations in comparison with *Eciton. Labidus,* perhaps derived even earlier from the A stock, may have adapted so strongly to subsurface conditions as to approximate convergently the group B functional pattern of *Dorylus* (Chapter 7). A similar hypothesis may apply to *Nomamyrmex.*

To deal with the problem of cyclic functions, let us return to early doryline origins and the divergence of our hypothetical stocks A and B. A great dissemination of both these stocks westward toward Africa and eastward across the Bering land connection could have occurred late in the Cretaceous period when the Northern route was still tropical. Basic features differentiating stocks A and B could have arisen early through the specialization of A to surface habitats and of B to underground habitats. The general course of modifications, sketched in Figure 14.4, is inferred on grounds of ecological and behavioral considerations related to the potentialities for further differentiation of ants living in smaller colonies and of those living in larger colonies.

In the early stages, both stocks may have thrived on the surface, with the B type gradually increasing colony size and subterranean adaptations more than the A type. This differentiation could have accelerated with the coming of variations toward periodically severe conditions aboveground (e.g., drought) that forced colonies down for lengthy intervals. The continuation of such conditions over long geologic time could have tended the A stock (of smaller queens and smaller colonies) toward adjustments to intervals of high-level action and mobility on the surface and to intervening stays of low-level action underground during which time the queens initiated new broods. Studies of *Aenictus* suggest that the doryline statary phase represents an archaic adaptation retained by all A type descendants but modified according to what selective conditions dominated their later habitats.

When representatives of the A stock began disseminating in the New World, colony cyclic adaptations presumably were already well under way with eyes only partially degenerated. Adjustments to surface conditions superior to those effected by the pre-*Aenictus* stock that remained are reflected by *Eciton* in regular day-night schedules of raids and emigrations and open microclimatically controlled bivouacs in the nomadic phase.

These specializations, attained through epigaeic adjustments aided by worker polymorphism, admit alternate statary intervals of brood initiation by the queen and of low-level action and well-sheltered bivouacs by the workers. Existing *Aenictus,* with its longer statary phases tending strongly toward colony dormancy, reflects a much longer background of adjustment to more austere surface conditions.

Stronger adaptations of both *Eciton* and *Neivamyrmex* to the surface are revealed in their nomadic phase activity schedules correlated with day and night. These adjustments show a much stricter day-by-day schedule of nomadic behavior than those of *Aenictus,* which evidently held to the pattern of regular functional cycles only through retaining small monomorphic worker populations of simpler behavioral properties.

Comparisons of the smaller *Neivamyrmex,* more subterranean in habitat and strongly night-active, with the larger *Eciton,* epigaeic and day-active on the surface, suggest an early specialization of the former under more rugged and variable surface conditions—perhaps combined with daytime competitive difficulties—then held for the latter. That the successful surface adaptation of *Eciton* occurred when mammalian and other predators were on the rise is indicated by the prevalence in this genus of the strongly stinging polymorphic worker series, including fishhook-jawed majors (Chapter 4). Close connections between worker polymorphism and colony function are indicated in *Eciton* by plastic brood-tending processes adequate to restore colony populations to their species-typical patterns after natural catastrophes or disease have distorted them.

Common doryline origins under surface conditions are indicated by the characteristics of sexual broods and of colony division processes based on these broods (Chapters 9, 10, and 11). First, the sexual broods of all three group A genera are seasonally conditioned, even in species that are (probably secondarily) hypogaeic. Second, they present in all three similar patterns of relationships between the young queens and the males in development and between the sexual broods and the colonies. Of great interest are the patterns by which the males of these genera are released from the colony, significantly different as isolating mechanisms and evidently archaic, as with any adjustments involving the males (Forbes et al., in prep.).

Comparisons of the group A genera with *Dorylus* are rewarding. Evolutionary divergences of these ants toward surface and subsurface adjustments, respectively, are indicated throughout their colony adaptive systems. Colonies of surface-adapted *Aenictus,* for example, emigrate far more regularly and around twenty times as often as those of the most surface-related driver ant, *D. (Anomma) wilverthi.* We find these differences grounded in distinctively different adaptive resources, from those of the brood to those of the queen and workers (Chapters 7 and 8). Even so, similarities between group A genera and *Dorylus* in the brood-stimulative basis of nomadic behavior—featuring the role of callow arousal—and of brood and worker relationships underlying colony division seem difficult to explain save in terms of a common ancestry in remote times.

Although considerations of phyletic origins are not essential for the comparative investigation of function and behavior in living organisms, I have found this approach valuable, as the foregoing discussion may show. If a reasonable theory of origins can be worked out from the same evidence that gives rise to one's theory of adaptive systems in living representatives, without inconsistencies between these approaches or between either of them and sound taxonomic theory, then progress has been made.

Progress cannot be expected when we depend upon analogies or upon direct inferences from structure alone. Anthropomorphism, for example, is not a useful tool in an investigation of this kind because its presumptions cannot be tested and lead us astray. Ants are not little humans, any more than humans are giant ants—except in our moments of unscientific thought-play. As one alternative, the word "communication" was used in Chapter 13 as a means of comparing group processes in two different types of army ant, but not because of any implied similarity to human communication. As another, we have used the concept of "social bond" in comparing the group behavior of army ants and cats (Schneirla and Rosenblatt, 1961) as a means of relating the broad similarities of these two types of organism to the extremely important differences found in their social life. In the development of their behavior patterns, army ants are *biosocial* in the sense that structural and physiological mechanisms are dominant whereas mammals are *psychosocial,* in that learning and related abilities *can* dominate (Chapter 12).

This book offers evidence on the behavior of known army ant species and derives from this evidence a theory to account for similarities and differences between genera and species. This is really a theory of their instinctive behavior because it is designed to show how their species-typical behavior patterns come about. This theory does not imply the instinct concept, however, because it is not based upon concepts of their "inherited urges" or "genetically determined behavior." Rather, I consider the complex behavior patterns I have described as resulting from species-typical developmental processes, to which all types of individuals in the colony contribute, and I do not find any reason for resorting to the use of either of these classical thought patterns. Instead, to denote the major causal factor dominating colony behavior, my theory employs the concept of brood-stimulative effects, then applies this concept to know army ant species as a means of understanding similarities and differences in their *adaptive systems*. Progress to date illustrates abundantly the strong potentialities of this theory as a breeding ground for hypotheses-stimulating research.

Among the many reasons for studying army ants, one is to work out a self-consistent system of explanations for their behavior, i.e., a valid theory, based on investigations of what these insects actually do. As one of its chief uses, such a theory provides a basis for comparing this type of behavior with that of other animals—of humans, for example (Chapter 12). Among others, a theory of army ant adaptive systems should lead to useful comparisons with types of adaptive system that are convergent, i.e., alike but only very remotely related in evolution. Honeybees and army ants, for example, are strikingly similar and at the same time strikingly different in their modes of forming new colonies (which are convergent). As another example, comparisons of the seasonal movements of birds and the nomadic-statary cycles of army ants may be useful in the general study of animal migration (Chapter 7), despite the gulf that separates these animals in their evolution. Students of cyclic behavior also will find in the adaptive systems of army ants an interesting array of internally controlled cycles, unique in the animal kingdom.

Another important problem for investigation, and one with a more obvious practical side, concerns the relationships of army ants to other animals in the give-and-take of everyday living. The massive raids, carried out day after day by army ants in tropical and subtropical wilds around the world, surely make these insects a major factor in the natural balance of every area in which they live. Clearly, the reproductive conditions of a wide variety of insects and other arthropods that serve as perennial victims of the dorylines must be heavily adapted to army ants in major adjustments of species population. The world's insect population, however, is not really threatened by the dorylines as these animals have evolved along with the army ants and are naturally adjusted to their inroads in a variety of ways. The same cannot be said of their relationship to the ways of humans.

A chief aim of this book, represented by its subtitle, has been to examine the army ants as a problem in social behavior. One of the main doryline characteristics, in this regard, is the ability to create—in various ways, according to genus and species, but always similarly at basis—series of highly efficient group operations by compounding the resources of many low-grade individuals. The individuals are not low grade because they are simple; instead, in their internal make-up, they are beautifully complex. The point is rather that army ant adaptive systems have evolved through the reduction of individuals and of their functions to just the ones that contribute in one way or another to efficiency in collective operations. In this book we have studied a variety of collective operations that illustrate how complex the aggregate behavior of such low-grade individuals can be.

# Glossary

*Adelphogamy:* Mating between male and female progeny of the same queen mother in a social insect.

*Allometric growth:* Differential growth. In doryline ants, the ratios of general development to the growth of local parts vary regularly through an all-worker population.

*Antbirds:* Tropical birds of the family Formicariidae and others, attending the mass forays of swarm-raiding army ants, feed on the prey stirred up by the ants.

*Bivouac:* A doryline colony living in its temporary nesting situation.

*Brood:* In social insects, all of the young present in a colony at a given time. In dorylines, a population of synchronously developing young.

*Brood coordination:* In doryline ants, successive broods usually overlap in time, progress in parallel, and are integrated (i.e., physiologically interactive).

*Caches:* In the raids of army ants, temporary storage heaps of booty not directly transported to the bivouac accumulate, usually at trail junctions.

*Callow:* In ants, the condition of the adult shortly after emergence from the pupal stage of development, when it is not entirely hardened or fully of the mature color.

*Chemotactic sensitivity:* Sensing tactual and chemical stimuli together.

*Claustral:* The type of colony founding common among ants in which a fertilized female, living in seclusion, lays the eggs and then feeds the first members of her colony.

*Colony:* A group of social insects composed of interdependent individuals, capable of reproductive functions and existing together as a unified whole.

*Column raid:* A doryline predatory foray carried out in branching columns, each terminal column headed by a relatively small group laying chemical trails and capturing prey.

*Commensal:* A relationship of species in which one feeds on the nutritive supplies of another without harming the host.

*Communication:* Behavior and its by-products (e.g., secretions) operating to influence the behavior of the same and other individuals.

*Coordinated (e.g., broods):* Overlapping in time and, to an extent, interrelated.

*Cycle:* Periodic events in which reciprocal phases of differing (or opposite) nature are coordinated and recur alternately. (In group A dorylines, nomadic and statary phases recur regularly in alternation.)

*Dealate:* A doryline male that has lost his wings after his flight. Doryline queens and workers never have wings.

*Developmental convergence:* In all-worker broods of doryline ants, particularly in the larval stage, differences in relative growth (e.g., the smallest members grow at the fastest rate) reduce the brood time range at pupal maturity materially below the range at egg laying.

*Dichthadiigyne:* A special type of gynaecoid female (queen) present in the doryline ants and some related forms; wingless, without ocelli (some lack eyes), and capable of periodic physogastry and a high reproductive output.

*Division of labor:* Functional differentiation among workers in a colony—also between workers, queens, and males—based mainly upon polymorphic differences.

*Dorylophiles:* Insect guests, ranging from symbionts to parasites, living regularly with the colonies of doryline ants.

*Dys-synchronous:* Weakly or variably synchronous. Applicable to all-worker broods in *Dorylus* and other groups B dorylines with a wide growth range in early stages and with variable degrees of interbrood coordination.

*Dzierzon's rule:* Fertilized eggs develop into females (workers or queens) and unfertilized eggs develop into males. This rule applies in general to most social Hymenoptera.

*Emigration:* In the doryline ants, an exodus in which the entire colony changes its base, transporting the brood, in response to intracolony excitation. (Distinguished from a *shift*.)

*Epigaeic:* A species adapted to living on the surface of the earth or arboreally.

*Ergatoid:* A workerlike female, wingless and usually capable of laying eggs.

*Fission:* In bees, new colonies form by swarming; in doryline ants usually (in group A dorylines) through a two-way division of the parent colony.

*Fusion:* In army ants, a queenless colony merges with the first colony of its species it may encounter.

*Gaster:* The abdomen or hindmost body section of an ant; the part posterior to the petiole.

*Group A dorylines:* The genera *Eciton, Neivamyrmex,* and *Aenictus,* in which all colonies of investigated species exhibit well-marked functional cycles of regularly alternating nomadic and statary phases.

*Group B dorylines:* The genus *Dorylus* and possibly others (e.g., *Labidus*) in which emigrations occur as single events separated by non-nomadic intervals.

*Gynaecoid:* A workerlike female present in a colony of social insects, physiologically capable of laying eggs.

*Habituation:* Ceasing to respond to an initially disturbing stimulus or getting used to an attractive stimulus so that it promotes adaptive responses.

*Holometabolous:* An insect that passes through a complete metamorphosis (i.e., change of developmental state) from the egg and embryonic through larval and pupal stages, preceding the adult form.

*Homologous:* Structures that have a common evolutionary origin and (usually) are similar in their general patterns.

*Hypogaeic:* Adapted to an underground mode of existence—in the dorylines to subterranean (or highly sheltered) nests with or without surface activity (other than by males).

*Intermediate workers:* In polymorphic army ants, members of the graduated all-worker series between the submajors and minors.

*Legionary:* In ants, predatism combined with nomadism as in colonies of certain ponerine and all doryline species.

*Major worker:* In doryline ants, the largest member of the series of polymorphic workers in a colony, often called a "soldier."

*Malaxation:* The action of biting, squeezing, and otherwise working a morsel of food in the jaws, forming a pellet.

*Microclimate:* The local environmental conditions of an animal or group, e.g., the specific temperature, humidity, and other conditions prevalent in and near an army ant bivouac.

*Migration:* Movements of individual animals or groups from one ecologically distinct habitat to a different one, with a periodic reversal and return to the alternate type of habitat, based on cyclic, reversible reproductive processes. (Cf. *Emigration.*)

*Minor worker:* In polymorphic army ants, the smallest members of the graduated all-worker population, below the intermediate size groups.

*Monogyny:* Having only one queen in a colony. In the dorylines, the exception is the brief period in which a sexual brood is produced.

*Nest:* The place in which a colony is established and (in social insects) where the brood is generally reared; its living situation, stable in most social insects, is temporary in doryline ants. (See *Bivouac.*)

*Neuter:* A loose term for a worker female in an ant, bee, or wasp; in army ants, incapable of being fertilized or (in most cases) of laying eggs.

*Nomadic:* In doryline ants, a condition through which nesting sites are changed in emigrations of the entire colony.

*Ocellus, lateral:* In doryline ants, a degenerate compound eye characterizing workers and queens of most of the New World species.

*Pedicel:* In doryline ants, the one (e.g., *Dorylus*) or two (e.g., *Eciton*) usually nodiform segments joining the thorax to the abdomen.

*Pheromone:* A substance secreted to the outside of the body in a social insect, influencing interindividual communication and the behavior of other individuals.

*Photodermatic sense:* Sensitivity to light through the exoskeleton; present in workers and queens of all army ants, including those of *Dorylus, Aenictus,* and others without eyes.

*Physogastric:* An enlarged condition of the gaster, associated in doryline queens with a recrudescence of fatty tissues and a large-scale maturation of eggs.

*Polymorphism:* In doryline ants, the condition of graduated quantitative and, usually, qualitative differences within populations of adult workers and all-worker broods (See *Division of labor*.)

*Proprioception:* An animal's sensitivity to its own movement and postural changes. In doryline ants and other insects, as the insect moves, a buckling of exoskeleton stimulates sensory endings beneath.

*Proventriculus:* In the dorylines, a rather simple, straight tube with muscular walls, joining the crop and the ventriculus of the digestive tract in the gaster.

*Pseudogyne:* An apterous workerlike ant close to a worker in its general size and the size of the abdomen, but with a thorax resembling that of a queen.

*Pygidium:* The upper plate of exoskeleton (tergum) of the last abdominal segment, located above the anus.

*Queen's guard:* Members of a group of workers closely affiliated with the queen in feeding and in the emigration and clustered about her in the bivouac.

*Raid:* An exodus of doryline workers in numbers from the colony base in which booty is captured.

*Requeening:* In certain social bees, the process of developing in the colony one or more new queens, one of which may replace a dead or superannuated queen. In army ants this can occur only through production of a sexual brood.

*Sealing off:* A group operation in which army ant workers hem in a queen that disturbs them and inhibit her movements—in contrast to clustering about one that clearly attracts them.

*Sedentary:* Settled to remaining in one place, e.g., in most social insects, in an established permanent nesting situation. Not equivalent to the *statary* condition of army ants.

*Shift:* An exceptional change of bivouac by a doryline colony, usually over a short distance, and aroused by disturbances (i.e., distinguished from emigrations) from extracolony sources.

*Siesta effect:* A midday depression or lull in mass activities, commonly observed in the raids of surface-active species of army ants.

*Statary:* A phasic condition in which doryline colonies are lowered in their physiological level, are reduced in raiding, do not emigrate, and occupy a sheltered bivouac.

*Surface-active:* Suited for limited functions above the surface although not for nesting there. (As distinguished from *Surface-adapted*.)

*Surface-adapted:* Epigaeic. Suited through evolution for nesting and related activities on or above the surface of the ground.

*Swarm raiding:* Doryline predatory forays advancing in one or more large masses of workers, behind each of which a network of columns and a basal column connects with the bivouac.

*Sympatric:* Species with coextensive or overlapping ranges.

*Synchronous:* The many individuals in each all-worker or sexual brood of doryline ants develop more or less in step with one another.

*Termitophagous:* Preying on termites and using them extensively as food, general among hypogaeic army ants.

*Trophallaxis:* Wheeler's term (1928) for exchange of food and other substances between members of a colony of social insects. Here considered one aspect of reciprocal stimulation.

*Trophogenic:* Related to feeding or colony nutritional processes. For example, polymorphic differences among workers in the same colony are mainly determined trophogenically, i.e., through differential feeding as larvae.

*Trophorhynium:* Roughened surfaces or opposable plates (often striated) of the mouthparts useful in rubbing or mallaxating food.

*Trophosphere:* The area in which food is available to a colony and in which nutritive processes may be carried out. In army ants, this area changes in relation to operations of cyclic colony nomadism.

# References

Akre, R. D., and C. W. Rettenmeyer. 1966. Behavior of Staphylinidae associated with army ants (Formicidae: Ecitonini). *J. Kansas Entomol. Soc.* 39: 745–782.
Allee, W. C., A. E. Emerson, K. P. Schmidt, O. Park, and T. Park. 1949. *Principles of animal ecology.* Philadelphia: W. B. Saunders.
Alverdes, F. 1930. Tierpsychologische Analyse der intracentralen Vorgänge, welche bei decapoden Krebsen die locomotorischen Reaktionen auf Helligkeit und Dunkelheit bestimmen. *Z. Wiss. Zool.* 137: 403–475.
André, E. 1885. Species des Hyménoptères d'Europe et d'Algérie. *Les Fourmis* 2:838–840.
Arnold, G. 1914. Nest-changing migrations of two species of ants. *Proc. Rhodesia Sci. Assoc.* 13: 25–32.
Arnold, G. 1915. A monograph of the Formicidae of South Africa. *Ann. S. African Mus.* 14:1–756.
Autuori, A. 1956. La fondation des sociétés chez les fourmis champignonnistes du genre *"Atta"* (Hym. Formicidae). In P.-P. Grassé, ed., *L'Instinct dans le comportement des animaux et de l'homme.* Paris: Masson. Pp. 77–104.
Bates, H. W. 1863. *Naturalist on the river amazons,* vol. 2. London: J. Murray.
Beebe, W. 1919. The home town of the army ants. *Atlantic Monthly* 124: 454–464.
Belt, T. 1874. *The naturalist in Nicaragua.* London: J. Murray. (*Eciton,* pp. 17–29.)
Bequaert, J. 1922. The predaceous enemies of ants. *Bull. Am. Mus. Nat. Hist.* 45: 271–331.
Bingham, C. T. 1903. *The fauna of British India, including Ceylon and Burma. Hymenoptera,* Vol. II, *Ants and cuckoo-wasps.* London: Taylor & Francis.
Blum, M. S., and C. A. Portocarrero. 1964. Chemical releasers of social behavior. IV. The hindgut as the source of the odor trail pheromone in the neotropical army ant genus *Eciton. Ann. Entomol. Soc. Am.* 57: 793–794.
Blum, M. S., and E. O. Wilson. 1964. The anatomical source of trail substances in formicine ants. *Psyche* 71: 28–31.
Bodenheimer, F. S. 1937. Population problems of social insects. *Biol. Rev.* 12: 393–430.
Borgmeier, T. 1950. A fêmea dichthadiiforme e os estádios evolutivos de *Simopelta pergandei* (Forel) e a descrição de *S. bicolor* n. sp. *Rev. Entomol.* 21: 369–380.

Borgmeier, T. 1955. Die Wanderameisen der neotropischen Region (Hym. Formicidae). *Studia Entomol.*, no. 3., Petrópolis, R. J., Brasil: Ed. Vozes Ltda.

Borgmeier, T. 1957. Die Maxillar und Labialtaster der Neotropischen Dorylinen (Hym., Formicidae). *Rev. Brasil. Biol.* 17: 387–394.

Brauns, J. 1901. Ueber die Lebensweise von *Dorylus* und *Aenictus* (Hym.). *Z. Syst. Hymenopt. Dipterol.* 1: 14–17.

Brian, M. V. 1956. Studies of caste differentiation in *Myrmica rubra* L. 4. Controlled larval nutrition. *Insectes Sociaux* 3: 369–394.

Brian, M. V. 1957a. Caste differentiation in social insects. *Ann. Rev. Entomol.* 2: 107–120.

Brian, M. V. 1957b. Food distribution and larval size in cultures of the ant *Myrmica rubra* L. *Physiol. Comp. Oecol.* 4: 329–345.

Brian, M. V. 1957c. Serial organization of brood in *Myrmica*. *Insectes Sociaux* 4: 191–210.

Brian, M. V. 1965. *Social insect populations*. New York: Academic Press.

Brian, M. V., and J. Hibble. 1963. Larval size and the influence of the queen on growth in *Myrmica*. *Insectes Sociaux* 10: 71–82.

Brian, M. V., and J. Hibble. 1964. Studies on caste differentiation in *Myrmica rubra* L. 7. Caste bias, queen age, and influence. *Insectes Sociaux* 11: 223–238.

Brown, W. L., Jr. 1954. Remarks on the internal phylogeny and subfamily classification of the family Formicidae. *Insectes Sociaux* 1:21–31.

Brown, W. L., Jr. 1960. The release of alarm and attack behavior in some New World army ants. *Psyche* 66: 25–27.

Bruch, C. 1923. Estudios mirmecologicos. *Rev. Mus. de la Plata* 27: 172–220.

Bruch, C. 1934. Las formas femeninas de *Eciton*. Descripción y redescripción de algunas especies de la Argentina. *Ann. Soc. Ciencia Arg.* 108: 113–135.

Brues, C. T. 1930. The food of insects viewed from the biological and human standpoint. *Psyche* 37: 1–14.

Brun, R. 1914. *Die Raumorientierung der Ameisen und das Orientierungsproblem im allgemeinen*. Jena: G. Fischer.

Brun, R. 1959. Le cerveau des fourmis et des insectes en général comme instrument de formation des réflexes conditionnés. *Animal Psychology Seminars, International Union of Biological Sciences*. New York: Pergamon Press.

Bursell, E. 1964. 7. Environmental aspects: temperature. 8. Environmental aspects: humidity. In M. Rockstein, ed., *The physiology of insecta*, vol. 1. New York: Academic Press. Pp. 284–321, 323–361.

Butler, C. G. 1962. *The world of the honeybee*. London: Collins (1st ed., 1954).

Butler, C. G., and E. N. Fairey. 1963. The role of the queen in preventing oogenesis in worker honeybees. *J. Apicult. Res.* 2(9): 14–18.

Carpenter, F. M. 1930. The fossil ants of North America. *Bull. Mus. Comp. Zool.* (Harvard) 70: 3–66.

Carr, C. A. 1962. Further studies on the influence of the queen in ants of the genus *Myrmica*. *Insectes Sociaux* 9: 197–211.

Chapman, F. M. 1929. *My tropical air castle*. London: D. Appleton.

Chapman, J. W. 1964. Studies on the ecology of the army ants of the Philippines—genus *Aenictus* Shuckard (Hymenoptera: Formicidae). *Philippine J. Sci.* 93:551–595.

Chapman, R. N. 1931. *Animal ecology with especial reference to insects.* New York: McGraw-Hill.

Cloudsley-Thompson, J. I. 1961. *Rhythmic activity in animal physiology and behavior. Theoretical and experimental biology*, vol. 1. New York: Academic Press.

Cohic, F. 1948. Observations morphologiques et écologiques sur *Dorylus (Anomma) nigricans* Ill. (Hym. Dorylinae). *Rev. Franc. Entomol.* 14: 229–276.

Cole, A. C. 1953. Studies of New Mexico ants. III. The ponerines and dorylines (Hymenoptera: Formicidae). *J. Tenn. Acad. Sci.* 28: 84–85.

Creighton, W. S. 1927. The slave raids of *Harpagoxenus americanus*. *Psyche* 34: 11–29.

Creighton, W. S. 1950. The ants of North America. *Bull. Mus. Comp. Zool. Harvard Coll.* 104: 1–585.

Crozier, W. J., and T. J. B. Stier. 1928. Geotropic orientation in arthropods. I. *Malacosoma* larvae. *J. Gen. Physiol.* 11: 803–821.

Crozier, W. J., and T. J. B. Stier. 1929. Geotropic orientation in arthropods. II. *Tetraopes*. *J. Gen. Phsiol.* 12: 675–693.

Curran, C. H. 1934. Review of the tachinid genus *Calodexia* van der Wulp (Diptera). *Am. Mus. Novitates* no. 685: 1–21.

Dobrzanski, J. 1961. Sur l'éthologie guerrière de *Formica sanguinea* Latr. (Hymenoptera, Formicidae). *Acta Biol. Exptl.* 21: 53–73.

Dunham, W. E. 1931. Hive temperature for each hour of a day. *Ohio J. Sci.* 31: 181–188.

Eidmann, H. 1936. Okologisch-faunistische Studien en südbrasilianischen Ameisen. *Arb. Physiol. Angew. Entomol.* 3: 26–48.

Eisner, T., and W. L. Brown, Jr. 1956. The evolution and social significance of the ant proventriculus. *Proc. 10th Intern. Congr. Entomol. Montreal* 2: 503–508.

Emerson, A. E. 1938. Termite nests—a study of the phyolgeny of behavior. *Ecol. Monogr.* 8: 247–284.

Emerson, A. E. 1939a. Populations of social insects. *Ecol. Monogr.* 9: 287–300.

Emerson, A. E. 1939b. Social organization and the superorganism. *Am. Midland Naturalist* 21: 182–209.

Emerson, A. E. 1955. Geographical origins and dispersions of termite genera. *Fieldiana, Zool.* 37: 465–521.

Emerson, A. E. 1967. Cretaceous insects from Labrador. 3. A new genus and species of termite (Isoptera: Hodotermitidae). *Psyche* 74: 276–289.

Emery, C. 1887. Le tre forme sessuali del *Dorylus helvolus* L. e degli altri Dorilidi. *Bull. Soc. Entomol. Ital.* 19: 344–351.

Emery, C. 1895a. Die gattung *Dorylus* Fabr., und die systematische Eintheilung der Formiciden. *Zool. Jahrb. Abt. Syst.* 8: 685–778.
Emery, C. 1895b. Studi sulle Formiche delle fauna neotropica. *Bull. Soc. Entomol. Ital.* 28: 33–105.
Emery, C. 1895c. Le polymorphisme des fourmis et la castration alimentaire. *Compte Rendu III Congr. Intern. Zool.* 3: 395–410.
Emery, C. 1901. Notes sur les sous-familes des Dorylines et Ponérines (famille des Formicides). *Ann. Soc. Entomol. Belg.* 45: 32–54.
Emery, C. 1920. La distribuzione geografica attuale delle Formiche. *Real. Acad. Lincei* 13: 3–98.
Engelmann, F., and M. Lüscher. 1956. Zur Frage der Auslösung der Metamorphose bei Insekten. *Naturwissenschaften* 43: 43–44.
Ezikov, J. 1922. Uber den Character der Variabilität der Ameisen-Ovarien (Russian-German summary). *Rev. Zool. Russ.* 3: 333–357.
Fielde, A. M. 1904. The power of recognition among ants. *Biol. Bull.* 7: 227–250.
Flanders, S. E. 1946. Control of sex and sex-limited polymorphism in the Hymenoptera. *Quart. Rev. Biol.* 21: 135–143.
Flanders, S. E. 1962. Physiological prerequisites of social reproduction in the Hymenoptera. *Insectes Sociaux* 9: 375–388.
Flanders, S. E. 1965. On the sexuality and sex ratios of hymenopterous populations. *Am. Naturalist* 99: 489–494.
Forbes, J., M. Cazier, and T. C. Schneirla. In prep. The dispersal of male army ants of the genus *Neivamyrmex,* in relation to species ecology.
Forbes, J., and D. Do-Van-Quy. 1965. The anatomy and histology of the male reproductive system of the legionary ant, *Neivamyrmex harrisi* (Haldeman) (Hymenoptera: Formicidae). *J. N.Y. Entomol. Soc.* 73: 95–111.
Forel, A. 1891. Uber die ameisen subfamilie der Dorylinen. *Verhandl. Deut. Naturforsch.* 63: 162–164.
Forel, A. 1893. Sur la classification de la famille des Formicides, avec remarques synonymiques. *Ann. Soc. Entomol. Belg.* 37: 161–167.
Forel, A. 1899. Excursion myrmécologique dans l'Amérique du Nord. *Ann. Soc. Entomol. Belg.* 43: 438–447.
Fraenkel, G. 1932. Die Wanderungen der Insekten. *Ergeb. Biol.* 9: 1–238.
Fraenkel. G., and D. L. Gunn. 1961. *The orientation of animals.* New York: Dover.
Free, J. B. 1955. The behavior of egg-laying workers of bumblebee colonies. *Brit. J. Animal Behav.* 3: 147–153.
Freeland, J. 1958. Biological and social patterns in the Australian bulldog ants of the genus *Myrmecia. Australian J. Zool.* 6: 1–18.
Frisch, K. von. 1950. *Bees: their vision, chemical senses, and language.* Ithaca: Cornell University Press.
Gallardo, A. 1915. Observaciones sobre algunal hormigas de la Républica Argentina. *Ann. Mus. Nacl. Buenos Aires* 27: 1–35.
Goetsch, W. 1939. Die Staaten argentinischer Blattschneiden-Ameisen. *Zoologica* (Stuttgart) 96: 1–105.

Gotwald, W. H., Jr., and W. L. Brown, Jr. 1966. The ant genus *Simopelta* (Hymenoptera: Formicidae). *Psyche* 73: 261–277.
Green, E. 1903. Note on *Dorylus orientalis* West. *India Mus. Notes* 5: 39.
Gregg, R. E. 1942. The origin of castes in ants with special reference to *Pheidole morrisi* Forel. *Ecology* 23: 295–308.
Grout, R. A., ed. 1949. *The hive and the honey bee*. Hamilton, Ill.: Dadant & Sons.
Hagan, H. R. 1954. The reproductive system of the army ant queen (*Eciton*). Part 1: General anatomy, *Am. Mus. Novitates* no. 1663: 1–12; Part 2: Histology, no. 1664: 1–17; Part 3: The oöcyte cycle, no. 1665: 1–20.
Haldeman, S. S. 1849. On the identity of *Anomma* with *Dorylus*, suggested by specimens which Dr. Savage found together, and transmitted to illustrate his paper on the driver ants. *Proc. Acad. Nat. Sci. Phila.* 4: 200–202.
Harker, J. E. 1958. Diurnal rhythms in the animal kingdom. *Biol. Rev.* 33: 1–52.
Haskins, C. P., and E. F. Haskins. 1950. Notes on the biology and social behavior of the archaic ponerine ants of the genera *Myrmecia* and *Promyrmecia*. *Ann. Entomol. Soc. Am.* 43: 461–491.
Haydak, M. 1943. Larval food and development of castes in the honeybee. *J. Econ. Entomol.* 36: 778–792.
Heape, W. 1931. *Emigration, migration, and nomadism*. London: Heffer & Sons.
Herter, K. 1924. Untersuchungen über den Temperatursinn einiger Insekten. *Z. Vergleich. Physiol.* 1: 221–288.
Heyde, K. 1924. Die Entwicklung der psychischen Fähigkeiten bei Ameisen und ihr Verhalten bei abgeänderten biologischen bedingungen. *Biol. Zentr.* 44: 623–654.
Holliday, M. 1904. A study of some ergatogynic ants. *Zool. Jahrb. Abt. Syst.* 19: 293–328.
Hollingsworth, M. J. 1960. Studies on the polymorphic workers of the army ant *Dorylus* (*Anomma*) *nigrescens* Illiger. *Insectes Sociaux* 7: 17–37.
Huxley, J. S. 1927. Further work on heterogonic growth. *Biol. Zentra.* 47: 151–163.
Huxley, J. S. 1932. *Problems of relative growth*. London: Methuen & Co.
Ihering, H. von. 1894. Die Ameisen von Río Grande do Sul. *Berlin Entomol. Z.* 39: 321–446.
Ihering, H. von. 1912. Biologie und Verbreitung der brasilianischen. Arten von *Eciton. Entomol. Mitt.* 1: 226–235.
Jackson, W. B. 1957. Microclimate patterns in the army ant bivouac. *Ecology* 38: 276–285.
Jander, R. 1957. Die optische Richtungsorientierung des Roten Waldameise (*Formica rufa* L.). *Z Vergleich. Physiol.* 40: 162–238.
Johnson, R. A. 1954. The behavior of birds attending army ant raids on Barro Colorado Island, Panama Canal Zone. *Proc. Linnean Soc. N.Y.* nos. 63–65: 41–70.
Kempf, W. W. 1961. *Labidus coecus* as a cave ant. *Studia Entomol.* 4: 551–552.

Kennedy, J. S. 1951. The migration of the desert locust (*Schistocerca gregaria* Forsk.). *Phil. Trans. Roy. Soc. London, Ser. B* 235: 163–290.

Kenoyer, L. A. 1929. General and successional ecology of the lower tropical rain forest at Barro Colorado Island, Panama. *Ecology* 10: 201–222.

Kistner, D. 1958. The evolution of the *Pygostenini* (Coleoptera, Staphylinidae). *Ann. Mus. Roy. Congo Belge, Ser. 8, Zoology* 68: 5–198.

Lappano, E. 1958. A morphological study of larval development in polymorphic all-worker broods of the army ant *Eciton burchelli*. *Insectes Sociaux* 5: 31–66.

Le Masne, G. 1953. Observations sur les relations entre le couvain et les adultes chez les fourmis. *Ann. Sci. Nat. Zool.* 15: 1–56.

Lindauer, M. 1961. *Communication among social bees*. Cambridge: Harvard University Press.

Lorenz, K. 1965. *Evolution and modification of behavior*. Chicago: University of Chicago Press.

Loveridge, A. 1922. Account of an invasion of "Siafu" or red driver ants—*Dorylus (Anomma) nigrescens* Illig. *Trans. Entomol. Soc. London*. Pp. XXXIII–XLVI.

Ludwig, D. 1945. The effects of atmospheric humidity on animal life. *Physiol. Zool.* 18: 103–135.

Luederwaldt, H. 1918. Notas myrmecologicas. *Rev. Mus. Paulista* 10: 29–64.

Luederwaldt, H. 1920. Formigas nocivas Brasileiras. *Almanak Agricol. Brasil.* Pp. 277–278.

Luederwaldt, H. 1926. Observacões biologicas sobre Formigas Brasileiras. *Rev. Mus. Paulista* 14: 185–304.

Lutz, F. E. 1929. Observations on leaf-cutting ants. *Am. Mus. Novitates* no. 388: 1–21.

Mann, W. M. 1916. The ants of Brazil. *Bull. Mus. Comp. Zool.* 60: 399–490.

Mayr, G. 1886. Ueber *Eciton-Labidus*. *Wien Entomol. Z.* 5: 33–36, 115–122.

Mergelsberg, O. 1934. Uber den Begriff der Physogastrie. *Zool. Anz.* 106: 97–105.

Müller, W. 1886. Beobachtungen an Wanderameisen (*Eciton hamatum* Fabr.). *Kosmos* 1: 81–93.

Norton, E. 1868. Remarks on Mexican Formicidae (*Eciton*). *Trans. Am. Entomol. Soc.* 2: 44–46.

Pardi, L. 1948. Dominance order in *Polistes* wasps. *Physiol. Zool.* 21: 1–13.

Pardi, L. 1951. Ricerche sui Polistini. 12. Studio della attività e della divisione di lavoro in una società di *Polistes gallicus* (L.) dopo la comparsa delle operaie. *Arch. Ital. Zool.* 36: 363–431.

Patrizi, S. 1948. Contribuzioni alla conoscenza delle Formiche e dei mirmecofili dell' Africa Orientale. IV. Descrizione di un nuovo genere e di una nuova specie di stafilinide dorilofilo dello scioa e relative note etologiche (Coleoptera Staphylinidae). *Boll. Ist. Entomol. Univ. Bologna* 17: 158–167.

Pickles, W. 1937. Populations, territories, and biomasses of ants at Thornhill, Yorkshire, in 1936. *J. Animal Ecol.* 6: 54–61.

Plateaux-Quénu, C. 1961. Les sexués de remplacement chez les insectes sociaux. *Année Biol.* 65: 177–216.
Pullen, B. E. 1963. Termitophagy, myrmecophagy, and the evolution of the Dorylinae (Hymenoptera, Formicidae). *Studia Entomol.* 6: 405–414.
Raignier, A. 1959. Het ontstaan van kolonies en koninginnen bij de Afrikaanse trekmieren. *Mededel. Koninkl. Vlaam. Acad. Wetenschap. Belg.* 21: 3–24.
Raignier, A., and J. Van Boven. 1955. Etude taxonomique, biologique et biométrique des *Dorylus* du sous-genre *Anomma* (Hymenoptera, Formicidae). *Ann. Mus. Roy. Congo Belge, Tervuren, New Ser., Sci. Zool.* 2: 1–359.
Reichensperger, A. 1934. Beitrag zur Kenntnis von *Eciton lucanoides* Em. *Zool. Anz.* 106: 240–245.
Rettenmeyer, C. W. 1963. Behavioral studies of army ants. *Univ. Kansas Sci. Bull.* 44: 281–465.
Rösch, G. A. 1925. Untersuchungen über die Arbeitsteilung im Bienenstaat. I. Teil: Die Tätigkeiten im normalen Bienenstaate und ihre Beziehungen zum Alter der Arbeitsbienen. *Z. Vergleich. Physiol.* 2: 571–631.
Roth, L. M., and T. Eisner. 1962. Chemical defenses of arthropods. *Ann. Rev. Entomol.* 7: 107–136.
Rothenbuhler, W. C. 1967. Genetic and evolutionary considerations of social behavior of honeybees and some related insects. In J. Hirsch, ed. *Behavior-genetic analysis.* New York: McGraw-Hill.
Roubaud, E. 1910. Recherches sur la biologie des Synagris. Evolution de l'instinct chez les guêpes solitaires. *Ann. Soc. Entomol. Franc.* 79: 1–21.
Savage, T. S. 1847. On the habits of the "drivers" or visiting ants of West Africa. *Trans. Roy. Entomol. Soc. London* 5: 1–15.
Savage, T. S. 1849. The driver ants of West Africa. *Proc. Acad. Nat. Sci. Phila.* 4: 195–200.
Scharrer, E., and B. Scharrer. 1963. *Neuroendocrinology.* New York: Columbia University Press.
Schips, M. 1920. Über Wanderameisen. *Naturw. Wochschr., N.F.* 19: 618–619.
Schneider, F. 1962. Dispersal and migration. *Ann. Rev. Entomol.* 7: 223–242.
Schneirla, T. C. 1929. Learning and orientation in ants studied by means of the maze method. *Comp. Psychol. Monogr.* 6: 1–143.
Schneirla, T. C. 1933. Studies on army ants in Panama. *J. Comp. Psychol.* 15: 267–299.
Schneirla, T. C. 1934. Raiding and other outstanding phenomena in the behavior of army ants. *Proc. Nat. Acad. Sci.* 20: 316–321.
Schneirla, T. C. 1938. A theory of army-ant behavior based upon the analysis of activities in a representative species. *J. Comp. Psychol.* 25: 51–90.
Schneirla, T. C. 1940. Further studies on the army-ant behavior pattern. Mass organization in the swarm raiders. *J. Comp. Psychol.* 29: 401–460.
Schneirla, T. C. 1943. The nature of ant learning. II. The intermediate stage of segmental maze adjustment. *J. Comp. Psychol.* 35: 149–176.

Schneirla, T. C. 1944a. The reproductive function of the army ant queen as pace-makers of the group behavior pattern. *J. N.Y. Entomol. Soc.* 52: 153–192.

Schneirla, T. C. 1944b. Studies on the army ant behavior pattern—nomadism in the swarm-raider *Eciton burchelli. Proc. Am. Phil. Soc.* 87: 438–457.

Schneirla, T. C. 1944c. A unique case of circular milling in ants, considered in relation to trail following and the general problem of orientation. *Am. Mus. Novitates* no. 1253: 1–26.

Schneirla, T. C. 1945. The army-ant behavior pattern: nomad-statary relations in the swarmers and the problem of migration. *Biol. Bull.* 88: 166–193.

Schneirla. T. C. 1946. Problems in the biopsychology of social organizations. *J. Abnorm. Soc. Psychol.* 41: 385–402.

Schneirla, T. C. 1947. A study of the army-ant life and behavior under dry-season conditions with special reference to reproductive functions. 1. Southern Mexico. *Am. Mus. Novitates* no. 1336: 1–20.

Schneirla, T. C. 1948. Army-ant life and behavior under dry-season conditions with special reference to reproductive functions. 2. The appearance and fate of the males. *Zoologica* 33: 89–112.

Schneirla, T. C. 1949. Army-ant life and behavior under dry-season conditions. 3. The course of reproduction and colony behavior. *Bull. Am. Mus. Nat. Hist.* 94: 7–81.

Schneirla, T. C. 1950. The relationship between observation and experimentation in the field study of behavior. *Ann. N.Y. Acad. Sci.* 51: 1022–1044.

Schneirla, T. C. 1953. 1. Basic problems in the nature of insect behavior. 2. Insect behavior in relation to its setting. 3. Modifiability in insect behavior. 4. Collective activities and social patterns among insects. In K. D. Roeder, ed., *Insect physiology*. New York: J. Wiley. Pp. 656–779.

Schneirla, T. C. 1956a. A preliminary survey of colony division and related processes in two species of terrestrial army ants. *Insectes Sociaux* 3: 49–69.

Schneirla, T. C. 1956b. Interrelationships of the "innate" and the "acquired" in instinctive behavior. In P.-P. Grassé, ed., *L'Instinct dans le comportement des animaux et de l'homme*. Paris: Masson. Pp. 387–452.

Schneirla, T. C. 1957a. Theoretical consideration of cyclic processes in Doryline ants. *Proc. Am. Phil. Soc.* 101: 106–133.

Schneirla, T. C. 1957b. A comparison of species and genera in the ant subfamily Dorylinae with respect to functional pattern. *Insectes Sociaux* 4: 259–298.

Schneirla, T. C. 1957c. The concept of development in comparative psychology. In D. B. Harris, ed., *The concept of development*. Minneapolis: University of Minnesota Press. Pp. 78–108.

Schneirla, T. C. 1958. The behavior and biology of certain nearctic army ants. Last part of the functional season, southeastern Arizona. *Insectes Sociaux* 5: 215–255.

Schneirla, T. C. 1961. The behavior and biology of certain nearctic doryline ants: sexual broods and colony division in *Neivamyrmex nigrescens. Z. Tierpsychol.* 18: 1–32.

Schneirla, T. C. 1963. The behavior and biology of certain nearctic army ants: springtime resurgence of cyclic function, southeastern Arizona. *Animal Behav.* 11: 583–595.

Schneirla, T. C. 1965. Dorylines: raiding and in bivouac. Part II. *Nat. Hist. Mag.* 74: 44–51.

Schneirla, T. C. 1966. Behavioral development and comparative psychology. *Quart. Rev. Biol.* 41: 283–302.

Schneirla, T. C., and R. Z. Brown. 1950. Army-ant life and behavior under dry-season conditions. 4. Further investigations of cyclic processes in behavioral and reproductive functions. *Bull. Am. Mus. Nat. Hist.* 95: 265–353.

Schneirla, T. C., and R. Z. Brown. 1952. Sexual broods and the production of young queens in two species of army ants. *Zoologica* 37: 5–32.

Schneirla, T. C., R. Z. Brown, and F. C. Brown. 1954. The bivouac or temporary nest as an adaptive factor in certain terrestrial species of army ants. *Ecol. Monogr.* 24: 269–296.

Schneirla, T. C., R. Buchsbaum, and J. Walker. 1966. Army ants: a study in in social behavior. Film No. 2437. 19 minutes. Chicago: Encyclopedia Britannica Education Corp.

Schneirla, T. C., R. R. Gianutsos, and B. S. Pasternack. 1968. Comparative allometry of larval broods in three army-ant genera in relation to colony behavior. *Am. Naturalist* 102: 533–554.

Schneirla, T. C., and A. Y. Reyes. 1966. Raiding and related behavior in two surface-adapted species of the Old World doryline ant, *Aenictus*. *Animal Behav.* 14: 132–148.

Schneirla, T. C., and A. Y. Reyes. 1969. Emigrations and related behavior in two surface-adapted species of the Old World doryline ant, *Aenictus*. *Animal Behav.* 17: 87–103.

Schneirla, T. C., and A. Y. Reyes. In ms. Colony division and related behavior in two surface-adapted species of the Old World doryline ant, *Aenictus*.

Schneirla, T. C., and J. S. Rosenblatt. 1961. Behavioral organization and genesis of the social bond in insects and mammals. *Am. J. Orthopsychiat.* 31: 223–253.

Seevers, C. H. 1959. North American Staphylinidae associated with army ants. *Coleopterists Bull.* 13: 65–79.

Seevers, C. H. 1965. The systematics, evolution, and zoogeography of Staphylinid beetles associated with army ants (Coleoptera, Staphylinidae). *Fieldiana, Zool.* 47: 139–351.

Simpson, J. 1958. The factors which cause colonies of *Apis mellifera* to swarm. *Insectes Sociaux* 5: 77–95.

Smith, M. R. 1942. The legionary ants of the United States belonging to *Eciton* subgenus *Neivamyrmex* Borgmeier. *Am. Midland Naturalist* 27: 537–590.

Sudd, J. H. 1967. *An introduction to the behavior of ants*. New York: St. Martin's Press.

Sumichrast, F. 1868. Notes on the habits of certain Mexican Hymenoptera presented to the American Entomological Society. *Trans. Am. Entomol. Soc.* 2: 39–44.

Tafuri, J. F. 1955. Growth and polymorphism in the larva of the army ant (*Eciton hamatum* Fabricius). *J. N.Y. Entomol. Soc.* 63: 21–41.

Talbot, M. 1948. A comparison of two ants of the genus *Formica*. *Ecology* 29: 316–325.

Thorpe, W. H. 1956. *Learning and instinct in animals*. London: Methuen.

Topoff, H. R. 1969. A unique predatory association between carabid beetles of the genus *Helluomorphoides* and colonies of the army ant *Neivamyrmex nigrescens*. *Psyche* 76: 375–381.

Topoff, H. R. In press. Population characteristics of three genera of army ants in relation to cyclic colony function. *Am. Naturalist*.

Topog, H. R. In preparation. Reversible physiological conditions and behavior in the army ant *Neivamyrmex nigrescens*.

Topoff, H. R., W. Trakimas, and M. Boshes. In preparation. Trail-following behavior and its development in the army ant, *Neivamyrmex nigrescens*.

Uvarov, B. P. 1931. Insects and climate. *Trans. Entomol. Soc. London* 79: 1–247.

Uvarov, B. P. 1932. Conditioned reflexes in insect behavior. *5th Intern. Congr. Entomol. Paris* 1: 353–360.

Vosseler, J. 1905. Die Ostafrikanische Treiberameise. *Der Pflanzer* 1: 289–302.

Vowles, D. M. 1958. The perceptual world of ants. *Animal Behav.* 6: 115–116.

Vowles, D. M. 1961. Neural mechanisms in insect behavior. In W. H. Thorpe and O. L. Zangwill, eds., *Current problems in animal behavior*. New York: Cambridge University Press. Pp. 5–29.

Wallis, D. I. 1960. Spinning movements in the larvae of the ant *Formica fusca*. *Insectes Sociaux* 7: 187–199.

Watkins II, J. F. 1964. Laboratory experiments on the trail following of army ants of the genus *Neivamyrmex* (Formicidae: Dorylinae). *J. Kansas Entomol. Soc.* 37: 22–28.

Watkins II, J. F. 1968. The rearing of the army ant male, *Neivamyrmex harrisi* (Haldeman) from larvae collected from a nest of *N. wheeleri* (Emery). *Am. Midland Naturalist* 80: 273–275.

Watkins II, J. F., and T. W. Cole. 1966. The attraction of army ant workers to secretions of their queens. *Tex. J. Sci.* 18: 254–265.

Weaver, N. 1966. Physiology of caste determination. *Ann. Rev. Entomol.* 11: 79–102.

Weber, N. A. 1941. The rediscovery of the queen of *Eciton* (*Labidus*) *coecum* Latr. *Am. Midland Naturalist* 26: 325–329.

Weber, N. A. 1943. The ants of the Imatong Mountains, Anglo-Egyptian Sudan. *Bull. Mus. Comp. Zool.* 93: 265–389.

Weir, J. S. 1959. Egg masses and early larval growth in *Myrmica*. *Insectes Sociaux* 6: 187–201.

Werringloer, A. 1932. Die Sehorgane und Sehzentren der Dorylinen nebst Untersuchungen über die Facettenaugen der Formiciden. *Z. Wiss. Zool.* 141: 432–524.

Wesson, L. G. 1940. An experimental study of caste determination in ants. *Psyche* 47: 105–111.

Wheeler, G. C. 1943. The larvae of the army ants. *Ann. Entomol. Soc. Am.* 36: 319–332.

Wheeler, W. M. 1900. The female of *Eciton sumichrasti* Norton, with some notes on the habits of Texan Ecitons. *Am. Naturalist* 34: 563–574.

Wheeler, W. M. 1910. *Ants—their structure, development, and behavior.* New York: Columbia University Press.

Wheeler, W. M. 1911. The ant colony as an organism. *J. Morphol.* 22: 307–325.

Wheeler, W. M. 1921. Observations on army ants in British Guiana. *Proc. Am. Acad. Arts Sci.* 56: 291–328.

Wheeler, W. M. 1922. Ants of the American Museum Congo Expedition. A contribution to the myrmecology of Africa. 7. Keys to the genera and subgenera of ants. *Bull. Am. Mus. Nat. Hist.* 45: 631–710.

Wheeler, W. M. 1923. *Social life among the insects.* New York: Harcourt, Brace.

Wheeler, W. M. 1925. The finding of the queen of the army ant *Eciton hamatum* Fabr. *Biol. Bull.* 49: 139–149.

Wheeler, W. M. 1928. *The social insects, their origin and evolution.* New York: Harcourt, Brace.

Wheeler, W. M. 1930. Philippine ants of the genus *Aenictus* with descriptions of the females of two species. *J. N.Y. Entomol. Soc.* 38: 193–212.

Wheeler, W. M. 1933. *Colony founding among ants.* Cambridge: Harvard University Press.

Wheeler, W. M. 1936. Ecological relations of Ponerine and other ants to termites. *Proc. Am. Acad. Arts Sci.* 71: 159–243.

Wheeler, W. M., and I. W. Bailey. 1925. The feeding habits of Pseudomyrmine and other ants. *Trans. Am. Phil. Soc.* 22: 235–279.

Whelden, R. M. 1963. Anatomy of adult queen and workers of army ants *Eciton burchelli* Westw. and *E. hamatum* Fabr. (Hymenoptera: Formicidae). *J. N.Y. Entomol. Soc.* 76: 14–30, 90–115, 158–171, 246–261.

Whiting, P. W. 1945. The evolution of male haploidy. *Quart. Rev. Biol.* 20: 231–260.

Wigglesworth, V. B. 1965. *The principles of insect physiology,* 6th ed. London: Methuen.

de Wilde, J., and D. Stegwee. 1958. Two major effects of the corpus allatum in the adult Colorado beetle (*Leptinotarsa decemlineata* Say). *Arch. Neerl. Zool.* 13: 277–289.

Williams, C. B. 1958. *Insect migration.* New York: Macmillan.

Willis, E. 1960. A study of the foraging behavior of two species of ant tanagers. *Auk* 77: 150–170.

Willis, E. 1967. The behavior of bicolored antbirds. *Univ. Calif. Publ. Zool.* 79: 1–132.

Wilson, E. O. 1953. The origin and evolution of polymorphism in ants. *Quart. Rev. Biol.* 28: 136–156.

Wilson, E. O. 1958a. The beginnings of nomadic and group-predatory behavior in the ponerine ants. *Evolution* 12: 24–31.

Wilson, E. O. 1958b. Observations on the behavior of the cerapachyine ants. *Insectes Sociaux* 5: 129–140.

Wilson, E. O. 1958c. Patchy distributions of ant species in New Guinea rain forests. *Psyche* 65: 26–38.

Wilson, E. O. 1964. The true army ants of the Indo-Australian Area (Hymenoptera: Formicidae: Dorylinae). *Pacific Insects* 6: 427–483.

Wilson, E. O., F. M. Carpenter, and W. L. Brown, Jr. 1967. The first mesozoic ants, with the description of a new subfamily. *Psyche* 74: 1–19.

Wilson, E. O., and R. W. Taylor. 1964. A fossil ant colony: new evidence of social antiquity. *Psyche* 71: 93–103.

Wroughton, R. C. 1892. Our ants. Part II. *J. Bombay Nat. Hist. Soc.* 7: 175–202.

# Index

Akre, R. D., 34, 87, 108n6
Allee, W. C., 23, 54n10, 62, 100
Allometry. *See* Population characteristics
Alverdes, F., 257n3
Anatomy
 male, 34
 queen, 171, 187–188
 worker, 28–32
André, E., 169n
Ant bridge, 272–276
Antbird, 86–87. *See also* Dorylophiles
Arnold, G., 50, 69
*Atta*, 51, 99n, 159, 212
Autuori, A., 212
*Azteca*, 267

Bates, H. W., 2, 5, 47, 89, 92, 101, 120
Beebe, W., 55, 143, 201n
Beetles. *See* Dorylophiles
Behavioral cycle. *See* Cyclic behavior
Belt, T., 5, 6, 45, 83, 92, 101
Bequaert, J., 86n, 87
Bingham, C. T., 69
Biosocial, 324
Bivouac, 6, 44–68, 289, 307–308
 construction, 6, 54–58, 60
 definition, 44
 differences during cycle, 47–48, 58
 in environmental regulation, 62–66
 functions, 61
 as incubator, 62–64, 148
 location, 6, 47–51
 types, 47–52
Blum, M. S., 71, 76n4, 147
Bodenheimer, F. S., 42, 44
Booty, 292–293, 309–310
 differences during cycle, 97–98
 types, 3, 73, 77n5, 85, 97–99, 100
Borgmeier, T., 3, 5, 7, 29, 47, 50, 52, 52n4, 59, 69, 76, 89, 97, 99, 101n1, 112, 117, 120, 169n, 172n, 181, 198, 202n1, 257n1

Brauns, J., 102, 198, 213, 287
Brian, M. V., 23, 25, 42, 44, 62, 125n2, 127n, 131, 142, 198, 211n1
Brood, 123–148
 callows, 147
 cocoon, 144
 as colony excitation, 35, 37, 124, 140–143, 145, 157, 158, 160
 in colony integration, 123–124, 265–267
 coordination, 127, 156, 166, 186, 302
 duration of development, 130
 eating by adults, 18n, 145
 eclosion, 139
 incubation, 62–64, 148
 and nomadism, 124
 number of individuals, 125–127
 synchronization, 127
 *See also* Population characteristics
Brown, R. Z., 33, 43n, 47, 77, 95n, 103, 107, 108n7, 117, 118, 121, 145n, 151, 154, 157, 159, 161, 162, 163, 169, 174, 175, 180, 182, 183, 189, 198, 199, 200, 200n, 201, 201n, 203, 204, 208, 209, 210, 211, 213, 218, 219, 220, 223n3, 229, 230, 233, 234, 242, 245, 246, 248, 259
Brown, W. L., Jr., 69, 141, 156n6, 167, 168, 288, 319, 321
Bruch, C., 49, 76, 169n, 182, 198
Brues, C. T., 97, 98
Brun, R., 28, 76
Bursell, E., 54n9, 62
Butler, C. G., 125, 146, 176, 194, 198, 212, 212n

Camponotus, 77n5, 100
Carabidae. *See* Dorylophiles
Carpenter, F. M., 11
Carr, C. A., 194
Caste determination, 131–134, 209–213. *See also* Population characteristics
Cazier, M., 256n6, 257n7

*Cerapachys*, 167
Chapman, R. N., 54n9, 156n, 202, 286, 288
Circular mill, 281–284
Classification. *See* Taxonomy
Cloudsley-Thompson, J. I., 157, 189
Cohic, F., 9, 54, 70n, 76, 80, 83, 97, 99, 119n, 309
Cole, A. C., 10, 178n
Colony division, 217, 218–244, 245–263
  distribution of adults, 243, 247
  *See also* Sexual brood
Colony founding, 218. *See also* Queen
Colony odor, 22–23
Colony size. *See* Population characteristics
Communication, 37, 264–285
  with queen, 180–181
  in raiding, 79
  *See also* Orientation, Reciprocal stimulation
Conditioning, 94, 146
Creighton, W. S., 3, 7, 70n, 257n7
*Crematogaster*, 100
Crozier, W. J., 74
*Cryptocerus*, 100
Cyclic behavior, 149–168, 305
  cause of, 97, 151, 157–158, 162, 165, 312, 318
  duration of phases, 151–156, 162–163, 299
  functional season, 316–317
  role of brood, 124, 157, 160, 162

Dauerneste, 45
Daughter colonies, 245–263. *See also* Colony division
Developmental convergence, 138, 166
Distribution. *See* Zoogeography
Division of labor, 39–41. *See also* Population characteristics
Dobrzanski, J., 70n
Dorylophiles, 22, 34–35, 86–87, 108, 321
Do-Van-Quy, D., 34
Dunham, W. E., 63
Dzierzon rule, 125, 209

Eclosion. *See* Brood
Eidmann, H., 50, 124
Eisner, T., 100, 141
Emerson, K. P., 23, 174n, 264, 276, 321
Emery, C., 5, 7, 8, 11, 38, 134, 173, 287, 319
Emigration, 101–122, 293–298
  causation, 13, 101–103, 312
  development, 103–112

distance covered, 6, 114–116
  schedule, 117–118, 311
Engelmann, F., 188
Ergatoid, 319
Evolution, 11, 12, 37–38, 69–70, 173, 319–324
Ezikov, J., 125n2, 131, 210

Fairey, E. N., 194
Fielde, A. M., 146
Flanders, S. E., 209, 211n14, 212
Food. *See* Booty
Forbes, J., 34, 202, 254
Forel, A., 7, 27, 70n
*Formica*, 18, 22, 28, 29, 70n, 76, 176, 293
Formicariidae. *See* Antbird
Fraenkel, G., 103, 149, 257n8
Free, J. B., 216n
Freeland, J., 140
Frisch, K. von, 147n

Gallardo, A., 50, 72n, 198
Gianutsos, R., 134n
Goetsch, W., 267
Gotwald, W. H., 69, 167, 168
Gregg, R. E., 125n2, 131
Grout, R. A., 63, 212n
Gunn, D. L., 257n8

Hagan, H. R., 175, 178n, 182, 183, 186, 187, 190, 191
Haldeman, S. S., 5
Haplodiploidy, 125
Harker, J. E., 157, 189
*Harpagoxenus*, 70n
Haskins, C. P., 12
Haydak, M., 212, 212n
Heape, W., 13, 103, 121, 149
Herter, K., 54n9
Heyde, K., 146, 147n
Hibble, J., 194
Histeridae. *See* Dorylophiles
Holliday, M., 172n3, 187
Hollingsworth, M. J., 25, 134
Huxley, J. S., 25, 134

Ihering, H. von, 11, 45, 52n, 102, 120
Instinctive behavior, 18–19, 146, 270
*Iridomyrmex*, 101

Jackson, W. B., 64
Jander, R., 76
Johnson, R. A., 86n, 87

Kaiser, J., 206
Kempf, W. W., 52
Kennedy, J. S., 122

Kenoyer, L. A., 64
Kistner, D., 34

Lappano, E., 134n, 136, 143
Le Masne, G., 140, 142, 212
Leakey, L. S., 11
*Leptogenys,* 37–38, 69, 167
Lindauer, M., 147n
Loveridge, A., 3, 86
Ludwig, D., 54n9, 210
Luederwaldt, H., 99, 138, 166
Lüscher, M., 188
Lutz, F. E., 39

Maeterlinck, G., 264
Male, 33–34, 198–217
  anatomy, 34
  in emigration, 251–253
  flight, 253–254, 256–257
  mating, 251, 254, 259–261
  size, 205
  trail following, 251
  *See also* Sexual brood
Mann, W. M., 98
Maturation and experience, 146
Mayr, G., 7, 198
*Megaponera,* 69
Mergelsberg, O., 188
Migration, 121–122. *See also* Emigration
Mill. *See* Circular mill
*Monomorium,* 101
Müller, W., 6, 45, 47, 103, 124, 149
*Myopone,* 167
*Myrmecia,* 12
*Myrmica,* 125, 128

Nest. *See* Bivouac
Nomadism. *See* Cyclic behavior, Emigration
Norton, E., 45

Ocellus. *See* Orientation
*Odontomachus,* 100
*Oecophylla,* 11
Olfaction. *See* Orientation
*Onychomyrmex,* 69, 167
*Opthalmapone,* 69
Orientation
  olfaction, 28, 71–72, 79
  proprioception, 74
  trail following, 27, 28, 30–31, 37, 71–72, 77n4, 79, 94, 112, 147, 251, 268
  vision, 28–29, 76, 257
  *See also* Raiding

Pardi, L., 216, 216n
Park, O., 89n

Pasternack, B., 134n
Patrizi, S., 36
Phase duration. *See* Cyclic behavior
*Pheidole,* 100, 118, 293
Pheromone, 37. *See also* Brood, Orientation, Queen
*Phyracaces,* 69, 70
Physogastry. *See* Queen
Pickles, W., 25
Plateau-Quénu, C., 198
*Pogonomyrmex,* 98, 159
*Polistes,* 216n
*Polyergus,* 70n
Polymorphism. *See* Population characteristics
*Polyrachis,* 100, 293
Population characteristics
  allometry and polymorphism, 24–27, 38–42, 125, 130–138, 268–269, 301, 313–314
  density of colonies, 43n
  number of individuals, 23, 25, 26–27, 126, 287
  size of individuals, 24–27, 38–39, 286
Portocarrero, C. A., 71, 76n4, 147
*Pseudomyrmex,* 100
Psychosocial, 325
Pullen, B. E., 29

Queen, 32, 169–197
  age, 174–175, 189–190
  anatomy, 171–172, 187–188
  colony founding, 42
  colony integration, 23, 32, 35, 176–177, 178, 265–267, 315
  effect of removal, 170, 173–174, 178–179, 188, 191–192
  in emigration, 108–109, 120, 177–178, 189–190, 295
  mating, 251, 254, 259–261
  number, 22
  odor, 23, 32, 145, 180–181, 227, 271–272, 315
  physogastry, 183–185, 188, 190, 210
  position in bivouac, 176, 219–221, 223
  retinue, 37, 108–109, 176, 296
  size, 170, 182–185, 205
  *See also* Sexual brood

Raiding, 69–100, 277–281, 291–292, 308–311
  column, 73, 77, 80–81, 98
  development of, 89–93, 277–280
  differences during cycle, 87–88, 95–96, 159
  distance covered, 87, 277
  effect of temperature, 32

Raiding (*cont.*)
  frequency, 161, 311
  running speed, 74n
  swarm, 5, 73, 77, 82–87, 89–93, 98, 276–280
Raignier, H., 9, 23, 25, 52n6, 58, 59, 66, 70n, 83, 89, 116, 118, 119n, 126, 127, 130, 138, 164, 165, 166, 169n, 196, 202, 202n8, 206, 213, 213n, 216, 218, 242, 261, 262, 288, 307, 309, 314, 316, 318
Rain ant, 52n4, 102
Reciprocal stimulation, 35, 37, 123–124, 139
  in raiding, 79, 159, 270, 305
Reichensperger, A., 108, 174
Research methods, 15–16, 18
Rettenmeyer, C. W., 23, 34, 47, 59, 68, 70n, 76, 85, 87, 101n1, 107, 108n6, 113, 117, 120, 126, 127, 154, 175, 178n, 201, 248, 254, 254n5, 259n
Reyes, A. Y., 13, 23, 50, 54, 59, 66, 70n, 80, 88, 89, 97, 98, 99, 111, 113, 114, 118, 121, 126, 142, 156, 157, 162, 165, 202n7, 203, 203n9, 206, 213, 288, 289, 291, 296, 303
Rösch, G. A., 147n
Rosenblatt, J. S., 35, 146, 148, 324
Roth, L. M., 100
Rothenbuhler, W. C., 125, 209, 211, 211n14
Roubaud, E., 123, 218

Savage, T. S., 3, 4n, 54, 80n, 86, 99, 101
Scharrer, B., 188
Scharrer, E., 188
Schips, M., 54n10, 80n
*Schistocerca*, 122
Schneider, F., 122
Sealing off, 221
Seevers, C. H., 35, 321
Sex determination, 125, 209. *See also* Population characteristics
Sexual brood, 198–216
  cause of, 199–202, 210–213
  colony excitation, 209, 214–215
  coordination, 208
  development, 206, 209
  effect on phase duration, 207–208, 216
  number of individuals, 203, 212–213, 235
  position in bivouac, 219–221, 223
  size of individuals, 204, 207
Shift, 117, 161
*Simopelta*, 69, 167
Simpson, J., 198

Size of individuals. *See* Population characteristics
Smith, M. R., 7, 10, 50, 198, 198n
Social bond, 264–285. *See also* Communication, Reciprocal stimulation
Staphylinidae. *See* Dorylophiles
Stegwee, D., 188
Stier, T. J. B., 74
Sudd, J. H., 157, 189
Sumichrast, F., 6, 45, 101n, 102, 120
Superorganism, 264–265
Supersedure, 233–234, 247
Synonymy, 3, 5, 169n

Tafuri, J. F., 134n, 135
Talbot, M., 25
Taxonomy, 7, 9, 19–21
Taylor, R., 11, 167, 168
Termites, 23
*Termitopone*, 69, 70
Thorpe, W. H., 269
Topoff, H. R., 26, 40, 54, 108n6, 132, 146, 158, 160, 188n
*Trachymyrmex*, 239n
Trail following. *See* Orientation
Trophallaxis. *See* Reciprocal stimulation

Uvarov, B. P., 62, 146, 210, 269

Van Boven, J., 9, 23, 25, 52, 52n6, 59, 66, 70n, 83, 89, 101n1, 116, 118, 119n, 126, 127, 130, 138, 164, 165, 166, 169n, 196, 261, 288, 307, 309, 314, 318
Vision. *See* Orientation
Vosseler, J., 45, 52, 83, 101, 149
Vowles, D. M., 28

Wallis, D. I., 143
Wanderneste, 45
Watkins, J. F., 71, 76n4, 147, 178n, 202n6, 257n7
Weaver, N., 132, 134, 194, 209
Weber, N. A., 52, 52n5, 138, 157n, 166, 169n
Weir, J. S., 140
Werringloer, A., 29, 53
Wesson, L. G., 125n2, 131
Wheeler, W. M., 3, 5, 7, 11, 12, 13, 34, 37, 42, 47, 50, 52, 69, 72n, 97, 99, 102, 123, 124, 134, 140, 148, 167, 169n, 172n, 173, 178, 181, 182, 188, 198, 201, 218, 256, 264, 265, 267, 287, 318

Whelden, R. M., 172n3, 188, 202n8, 203n10, 211n12, 229n, 247, 256n5, 258n
Whiting, P. W., 209
Wigglesworth, V. B., 54n9, 62, 128, 146, 188, 189, 210, 211, 211n14
de Wilde, J., 188
Williams, C. B., 122

Willis, E., 86n, 87, 154n, 163
Wilson, E. O., 7, 9, 11, 12, 25, 63, 69, 70, 97, 134, 147, 156n, 167, 168, 240n15, 286, 288, 318, 319, 321
Wound suturing, 3
Wroughton, R. C., 73, 256

Zoogeography, 7–11, 322